*The Best American Science
and Nature Writing 2010*

GUEST EDITORS OF
THE BEST AMERICAN SCIENCE
AND NATURE WRITING

The Best American Science and Nature Writing™ 2010

Edited and with an Introduction
by Freeman Dyson

Tim Folger, Series Editor

A Mariner Original
HOUGHTON MIFFLIN HARCOURT
BOSTON · NEW YORK 2010

www.hmhbooks.com

ISSN 1530-1508

ISBN 978-0-547-32784-6

Printed in the United States of America

DOC 10 9 8 7 6 5 4 3 2 1

Contents

Part Three: Natural Beauty

Part Four: The Environment: Gloom and Doom

Part Five: The Environment: Small Blessings

Part Six: The Environment: Big Blessings

Foreword

REMEMBER THE WAY the future was supposed to be? The path to tomorrow once seemed so clear, its trajectory limned for the entire world to see in billowing plumes of rocket exhaust in the blue sky over Cape Canaveral. We would become, President Kennedy declared in 1962, a spacefaring nation, destined—even obligated—to explore what he called "this new ocean." I had barely entered grade school at the time, but I vividly remember watching the launches of the Mercury and Gemini missions and the equally dramatic splashdowns—in those days all manned American spacecraft landed in the ocean, their descent slowed by enormous orange-and-white-striped parachutes. By the late 1960s, astronauts—and cosmonauts—had walked in space; the first moon landing was at hand. We had come so far so quickly: less than sixty years separated the Wright brothers' first flight and John Glenn's solo orbit of Earth in *Friendship 7*, his closet-size Mercury capsule. No doubt we'd make even greater leaps in the next sixty years. By the year 2000? A spinning, spoked space station staffed by hundreds was a given; travel to the moon routine; footprints on the red sand of Mars—of course. To my second-grade mind it all seemed closer and more imaginable than my own adulthood.

The space odyssey that once seemed so inevitable never came to pass. Today, more than forty years after Neil Armstrong stepped onto the moon, we're unable to follow him there. The thirty-six-story-tall Saturn rockets that made such trips possible no longer exist, and nothing of comparable power has replaced them. For now — perhaps quite a long now — we're confined to orbiting Earth. Mars, the moons of Jupiter and Saturn, and destinations beyond will have to wait for their first human visitors. My eight-year-old self would have been sorely disappointed. He certainly didn't get the future he expected. On the other hand, he would have been astonished to learn that he would one day own a computer far more powerful than the ones carried aboard the Apollo spacecraft.

Science has a seemingly bottomless capacity to astonish, a quality unmatched, I think, by any other human endeavor. It thrives on unanticipated results and anomalous data. Some months ago, before I knew that Freeman Dyson would be the guest editor for this anthology, I had the pleasure of interviewing him while working on an assignment for *Discover.* Toward the end of our conversation I asked if any single discovery had most surprised him during his long career. (He will turn eighty-seven in December 2010.)

"Everything has been surprising," he said. "Science is just organized unpredictability. If it were predictable it wouldn't be science. Everywhere you look . . . I didn't expect personal computers. Like you, I thought by now we would be tramping around on Mars with heavy boots. I did not foresee that we'd be sending unmanned instruments with huge bandwidth into space. It's all been a surprise in a way. It's even more true in biology. I had no conception of the fact that we would actually read genomes the way we're doing it now. I remember when it took a year to sequence one protein. Now it's done in a few seconds. I would say there's almost nothing in science that I've predicted correctly. I hope it will continue that way; I think it's very likely it will. Really important things will happen in the next fifty years that nobody has imagined."

Tom Wolfe, one of the contributors to this year's anthology, would argue that we should resume our pursuit of the future we imagined forty years ago. In "One Giant Leap to Nowhere," he decries the premature end of the "greatest, grandest . . . quest in the history of the world": America's manned space program. Even if you don't agree with Wolfe that humanity must travel to the stars, it's impossible after reading his spirited and witty story not to wonder what the world would have been like today if the Apollo missions to the moon had marked the beginning rather than the end of a dream.

While a large part of me yearns to embrace Wolfe's vision, my sensible side yields to the arguments Steven Weinberg makes in "The Missions of Astronomy." Weinberg is an eminent physicist; he won a Nobel Prize for his work in describing some of the fundamental forces of the universe. He is also a passionate and eloquent essayist. In these pages he writes that manned missions to other worlds would hinder rather than advance science. But his remarkable article covers a great deal more than the merits of manned space exploration, ranging gracefully from Socrates to sextants to the Standard Model of physics.

We'll probably never reach the stars—our fastest existing space probes would need tens of thousands of years to get to the nearest one—but we may well be close to discovering whether life exists elsewhere in the universe. To date astronomers have found more than 370 planets orbiting other stars. None of those exoplanets resemble Earth—most are gassy giants, far bigger than Jupiter. But most astronomers believe that Earthlike planets must be fairly common in our galaxy. In "Seeking New Earths," Timothy Ferris writes about the search for planets like our own and how a new generation of telescopes may be able to find signs of life on some of them.

The remaining stories in this collection are all about one planet, which is facing challenges that no one conceived of forty years ago. Elizabeth Kolbert, who edited this anthology last year, describes a dangerous organism that threatens all life on Earth: us. Several articles collected here show how we might yet reverse some of the worst aspects of the catastrophes we've set in motion. Freeman Dyson has much more to say about that in the next few pages, but I can't help wondering which of these stories will, decades from now, turn out to have accurately glimpsed our future and which will be relegated to the what-might-have-beens.

I hope that readers, writers, and editors will nominate their favorite articles for next year's anthology at http://timfolger.net/forums. The criteria for submissions and deadlines, and the address to which entries should be sent, can be found in the "news and announcements" forum on my website. Once again this year I'm offering an incentive to readers to scour the nation in search of good science and nature writing: send me an article that I haven't found, and if the article makes it into the anthology, I'll mail you a free copy of next year's edition, signed by the guest editor. I'll even sign it as well, which will augment its value immeasurably. (A true statement, by the way, there being no measurable difference between copies signed by me and those unsigned.) I also encourage readers to use the forums to leave feedback about the collection and to discuss all things scientific. The best way for publications to guarantee that their articles are considered for inclusion in the anthology is to place me on their subscription list, using the address posted in the news and announcements section of the forums.

Years ago, at about the same time that I was watching the adventures of the first astronauts, I read a fascinating article about "Dyson Spheres." (Look them up; you won't be disappointed.) I never thought I would have a chance to work, however briefly, with the legendary physicist who came up with the idea. A manned Mars landing seemed far more likely. It's a bit of unpredictability that I'm extremely grateful for. Once again this year I'm indebted to Amanda Cook and Meagan Stacey at Houghton Mifflin. And I hope to remain indebted for many years to come to my beauteous wife, Anne Nolan.

TIM FOLGER

Introduction

THE JOB OF EDITING this collection of papers was made easy for me by Tim Folger, who did the hard work of scanning the entire scientific periodical literature for the year 2009 to select 122 articles that he found interesting. My job was only to read his 122 articles, make the final choice of 28 to put in the book, and write the introduction to explain my choices. I am grateful to Tim for doing the lion's share of the work. Unfortunately, the fraction of American magazines that publish science writing is small. The science writing that is published mostly consists of brief news items rather than thoughtful essays. Not many years ago, John McPhee used to publish in *The New Yorker* wonderful pieces, twenty or thirty pages long, giving readers a deep understanding of geological science. Such pieces no longer appear, in *The New Yorker* or anywhere else. Science writing has become briefer, sparser, and more superficial. The title of this volume gives equal weight to science and nature. In fact, it is one third science and two thirds nature. Nature is now fashionable among readers and publishers of magazines. Science is unfashionable.

I have divided the book into six parts, each with a common theme. The first two parts are concerned with science, the last four with nature. The two sciences that receive serious attention are astronomy and neurology. Both are rightly valued by the public as having some important connection with human destiny. Part One deals with astronomy, and its central theme is proclaimed in Steven

Weinberg's article, "The Missions of Astronomy." Weinberg paints in four pages a glowing picture of the history of astronomy, the science that for 2,500 years led mankind to a true understanding of the way the universe works. From the beginning, instruments were the key to understanding. The first instrument was the gnomon, a simple vertical post whose shadow allowed the Babylonians and the Greeks to measure time and angle with some precision. The legacy of Babylonian mathematics still survives in the sixty-fold ratios of our units of time, hours, minutes, and seconds.

After the gnomon came the sundial, the telescope, the chronometer, the computer, and the spacecraft. Now we are living in a golden age of astronomy, when for the first time our instruments give us a clear view of the entire universe, out in space to the remotest galaxies, back in time all the way to the beginning. Our instruments, telescopes on tops of mountains and on spacecraft in orbit, are increasing their capabilities by leaps and bounds as our data-handling skills improve. It takes us only about ten years to build a new generation of instruments that give us radically sharper and deeper views of everything in the sky. Weinberg ends his article by contrasting this ongoing triumph of scientific instruments with the abject failure of the American program of manned missions in space. Our unmanned missions to explore the planets and stars and galaxies have made us truly at home in the universe, while our manned missions after the Apollo program have been scientifically fruitless. Forty years after Apollo, the manned program is still stuck aimlessly in low orbit around Earth while politicians debate what it should try to do next.

The remaining articles in Part One discuss manned and unmanned activities separately. Andrew Corsello sees a bright future for private manned ventures in space, while Tom Wolfe explains how our public manned ventures failed. Timothy Ferris's articles describe two vivid scenes from the world of modern astronomy, one using instruments on the ground and the other using unmanned instruments in space. All three authors confirm Weinberg's judgment. If you want humans in space, let them go up there to enjoy a human adventure, preferably at their own expense, and do not pretend that they are doing science. If you want to do serious science, keep the humans on the ground and send instruments to do the exploring, a job they can do tirelessly, efficiently, and much more cheaply.

The view of space activities in Part One is a purely American one. The whole book suffers from the same limitation. By selecting only American writing, we have narrowed the focus of the collection, ignoring more than half of the world's thinking and dreaming. We have missed a great opportunity to broaden our contacts with the rest of the world. If half of the articles in this book had been translated from French or Russian or Arabic or Chinese, its value for our understanding of the world would have been far greater. For practical and economic reasons, it might be difficult to prepare timely translations for an annual publication. But we could at least have included articles from the many countries around the world that publish magazines in English.

The Americans writing in this book about space all tell us that unmanned exploration is a success and manned exploration is a failure. I was lucky to be exposed to a different view when I was invited to Baikonur in Kazakhstan to observe a Russian space launch. In March 2009, Charles Simonyi took off for his second trip in a Soyuz launcher to spend two weeks on the International Space Station. To qualify as a crew member on the ISS, he had spent three months at the Russian cosmonaut training center near Moscow. My daughter Esther went through the same training and was at Baikonur as his backup, ready to fly in case he came down with swine flu or broke a leg. Charles did not get the flu or break a leg, and Esther did not fly, but my wife and I were there for the launch and got a glimpse of the Russian space culture, which is very different from ours.

American space culture is dominated by the tradition of Apollo. President Kennedy proclaimed the mission as "Get a man to the moon and back within ten years," and so it was done. After that, there were five more missions, but the decision to terminate the program had already been made. The program was unsustainable for longer than ten years. It was affordable as a ten-year effort but not as a permanent commitment. After Apollo, various other missions, manned and unmanned, were undertaken, always with a time scale of one or two decades. American space culture thinks in decades. Every commitment is for a couple of decades at most. A job that cannot be done in a couple of decades is not considered practical.

Russian space culture thinks in centuries. Baikonur, the original home of the Soviet space program, now belongs to Kazakhstan, but

Russia rents it from Kazakhstan on a hundred-year lease, as Britain in the old days rented Hong Kong from China. The lease still has eighty years to run, and Baikonur feels like a Russian town. Historical relics of Russian space activities are carefully preserved and displayed in museums. The three patron saints are the schoolteacher Konstantin Tsiolkovsky, who worked out the mathematics of interplanetary rocketry in the nineteenth century; the engineer Sergei Korolev, who built the first orbiting spacecraft; and the cosmonaut Yuri Gagarin, who first orbited Earth. Korolev and Gagarin lived side by side in Baikonur in simple homes, which are open to the public. In a public square is a full-scale model of the Soyuz launcher that Korolev designed. It is a simple, rugged design and has changed very little since he designed it. It has the best safety record of all existing launchers for human passengers. The Russian space culture says, "If it works, why change it?"

The day of Charles Simonyi's launch was rainy and windy. If the launch had been in Florida in such foul weather, it would certainly have been postponed. At Baikonur, it went up within a second of the planned time. The launch was a public ceremony in which the whole town participated. The cosmonauts paraded through the town at the head of a procession of dignitaries including an Orthodox priest, with townspeople carrying umbrellas on either side. In the main square, the mayor was waiting with other dignitaries. The cosmonauts stood facing the mayor and formally announced that they were ready to fly. Then, after a couple of speeches, they proceeded to the launch site. The whole performance had the ambience of a religious sacrament rather than a scientific mission. In Russia you do not go into space to do science. You go into space because it is a part of human destiny. To be a cosmonaut is a vocation rather than a profession. Tsiolkovsky said that Earth is our cradle, and we will not always stay in the cradle. It may take us a few centuries to get to the planets, but we are on our way. We will keep going, no matter how long it takes.

The Russian view of the International Space Station is also different from the American view. The biggest museum in Baikonur contains a full-scale model of the ISS and also a full-scale model of the Mir space station, which the Russians had built twenty years earlier. The Mir was the first space station built for long-duration human occupation. When you look at the two space stations, you can see that the ISS is an enlarged version of the Mir. The Russians are proud that they built the essential parts of the ISS as well as the Mir. The ISS is a part of their culture. They welcome American passengers, who help to pay for it, but they still feel that they own it. American scientists and space experts mostly consider the ISS to be an embarrassment, a costly enterprise with little scientific or commercial value. They regret our involvement with the ISS and look forward to extricating ourselves as soon as our international commitments to it are fulfilled. To an American visitor, it comes as a surprise to see the ISS enshrined at Baikonur together with the Mir, two emblems of national pride.

I learned at Baikonur that the American space culture as it is portrayed in this book is only half of the truth. The Russian space culture is the other half. If you think as Americans do, on a time scale of decades, then unmanned missions succeed magnificently and manned missions fail miserably. Even the grandest unmanned missions, such as the Cassini mission to Saturn, take only one decade to build and another decade to fly. The grandest manned mission, the Apollo moon landing, ends after a decade and leaves the astronauts no way ahead. The decade time scale is fundamentally right for unmanned missions and wrong for manned missions. If you think as Russians do, on a time scale of centuries, then the situation is reversed. Russian space-science activities have failed to achieve much because they did not concentrate their attention on immediate scientific objectives. Russian manned-mission activities, driven not by science but by a belief in human destiny, keep moving quietly forward. There is room for both cultures in our future. Space is big enough for both.

Part Two contains three articles about neurology, the science of human brains. For the last fifty years, most popular writing about biology was concerned with molecular biology, the study of the chemical constituents of life. This tradition began soon after the discovery of the double helix by Francis Crick and James Watson in 1953 and rose to a brilliant climax with the publication of Watson's book *The Double Helix* in 1968. For fifty years, popular writings described how biologists explore genes and genomes and how geneticists identify the molecular machinery that guides the development of an egg into a chicken. For fifty years, the progress of molecular biology was driven by the invention of marvelous new tools, allowing the explorers to handle and dissect individual molecules with ever-increasing precision. But in recent years the tools have become too complicated and the ideas too specialized to be easily explained. Molecular biology has become a mature science with many subdivisions, each with its own jargon. The readers and writers of popular science are moving from molecular biology to neurology.

Neurology is now entering its golden age, with new tools answering simple questions that ordinary readers can understand. The three articles in Part Two describe three basic questions that neurologists are on their way to answering. How do our brains give us rational control over our actions? How do our brains give us rational control over our memories? How do our brains give us rational control over our sensations of physical pain? The tools of neurology are beginning to come to grips with the working of the brain as an organ of rational control. Each of the questions is not only important scientifically but also directly illuminates our personal experiences of thinking and deciding. Within the next fifty years, the tools of neurology will probably bring us a deep insight into our own thought processes, with all the good and evil consequences that such insight may bring. The three stories, about real people with real problems, give us a foretaste of the effects of deeper insight on our lives. The stories are told with a minimum of scientific jargon and a maximum of human sympathy.

The longest section is Part Three, with seven articles describing wonders of nature. Here the quality of the writing is as important as the subject matter. The pieces are written for nature lovers, not science lovers. There are many other nature articles of equal quality in the thick pile that I discarded. In making my choices, I tried to choose pieces that were as different as possible from one another. I chose some that are outstanding in style and some that are outstanding in subject matter. But I have to confess that for me, "The Flight of the Kuaka" is in a class by itself. It is a celebration of nature's glory, going beyond science and beyond poetry.

Parts Four, Five, and Six deal with the environment, the most fashionable subject of popular writing in recent years. Environmentalism has now replaced Marxism as the leading secular religion of our age. Environmentalism as a religious movement, with a mystical reverence for nature and a code of ethics based on responsible human stewardship of the planet, is already strong and is likely to grow stronger. That is the main reason why I am optimistic about the future. Environmentalism doesn't have much to do with science. Scientists and nonscientists can fight for the environment with equal passion and equal effectiveness. I am proud to stand with my nonscientist colleagues as a friend of the environment, even when we disagree about the details. The fact that we all share the ethics of environmentalism, striving to step lightly on the Earth and preserve living space for our fellow creatures, is one of the most hopeful features of our present situation. Each of the writers in this collection shares those ethics in one way or another.

I divided the articles about the environment into three parts: gloom and doom, small blessings, and big blessings, to emphasize the ways in which their authors disagree. Everyone agrees that human activities are having a huge impact on the environment and that the impact could be substantially reduced by various remedial actions. The articles in these three parts emphasize different aspects of the problem. The orthodox belief of the majority of climate experts is "climate alarmism." Climate alarmists say that climate change is mainly caused by humans' burning of fossil fuels and that our present patterns of fuel burning are already leading us to disaster. Elizabeth Kolbert's two pieces in Part Four are strong statements of the climate-alarmist position. The articles in Part Five do not concern themselves with global climate; they describe local environmental problems that may have local remedies. "The Monkey and the Fish" gives us a wonderfully vivid picture of an intractable environmental situation in Mozambique. Finally, Part Six pays attention to climate problems but asks new questions that the orthodox climate alarmists have ignored. Richard Manning's piece, "Graze Anatomy," is to me the most illuminating of the whole collection.

Before I discuss Manning's piece in detail, I must first declare my own interest in climate and the environment. Thirty years ago, it was already clear that fossil-fuel burning would cause climate change and that this was an important problem. It was also clear that fossil-fuel burning would have large effects on the growth of vegetation. Carbon dioxide is an excellent fertilizer for agricultural crops and for natural forests. Commercial fruit growers were enriching the air in greenhouses with carbon dioxide in order to accelerate the growth of fruit. From the experience of greenhouse growers, we can calculate that the carbon dioxide put into the atmosphere by fossil-fuel burning has increased the worldwide yield of agricultural crop plants by roughly 15 percent in the last fifty years. In addition, when there is more carbon dioxide in the atmosphere, plants will put more growth into roots and less into aboveground stems and leaves. These effects of carbon dioxide on vegetation might in turn cause large effects on topsoil. After they decay, roots add carbon to the soil, while stems and leaves mostly return carbon to the atmosphere. The plowing of fields by farmers all over the world then exposes topsoil to the air and increases the loss of carbon from soil to atmosphere. The flows of carbon among soil and vegetation and atmosphere may be as important as the flows between fossil fuels and atmosphere.

Thirty years ago, the place where all these ecological effects of fuel burning were studied was the Oak Ridge National Laboratory in Tennessee. I went to Oak Ridge to work as a consultant, and I listened to the experts. They understood fluid dynamics and climate modeling, but they also knew a lot about forestry and soil science, agriculture and ecology. I learned two basic facts from them. First, the natural environment contains five reservoirs of carbon of roughly equal size: fossil fuels, the atmosphere, the upper level of the ocean, land vegetation, and topsoil. Second, these five reservoirs are tightly coupled together. Anything we do to change any one of them has important effects on all of them. The carbon that we add to the atmosphere by burning fossil fuels has major effects on the growth of food crops and forests. The carbon that we subtract from the atmosphere by building up topsoil has major effects on climate.

The orthodox climate-alarmist view describes the problem of climate change as involving only two reservoirs of carbon, fossil fuels and the atmosphere, ignoring the other three. This simplification of the problem makes predictions seem more certain and more dire. Nothing is said about the large fertilizing effects of carbon in the atmosphere and in topsoil upon food crops. Nothing is said about the large fluxes of carbon into the atmosphere caused by the plowing of soil. For reasons that are not clear to me, the public debate about the environment is dominated by climate scientists who are expert in fluid dynamics, while experts in soil and land management remain silent. The problems of climate change become much more tractable if we look at them through a broader lens.

Having lived for thirty years with these unorthodox opinions about the climate-change debate, I was amazed and delighted to read Manning's article. Here is a story about two farmers in Minnesota who actually make a living by raising beef on grass instead of on feedlots. This is just what I have been hoping for the last thirty years. The prevailing method of raising beef is to keep the animals in feedlots and to grow corn and soybeans to feed them. This method is prevalent partly because it is profitable and partly because it is subsidized by the United States government. It has at least six seriously harmful effects on the environment. First, it requires massive amounts of fertilizer to keep the corn growing, and the fertilizer carried off by rainwater causes excessive growth of green algae in rivers and lakes, using up the oxygen in the water and finally killing fish in the Gulf of Mexico. Second, it decreases the ability of the land to retain water and increases the frequency of serious flooding. Third, it increases the erosion of topsoil. Fourth, it destroys habitat for birds and other wildlife. Fifth, it raises the price of corn for poor countries that need corn to feed humans. Sixth, it is cruel to the animals and creates a stinking atmosphere for human farm workers and their neighbors. When the Minnesota farmers switch from feedlots to grass, all six environmental insults disappear. The grass is efficiently fertilized by the animals, the rainwater mostly stays in the ground instead of running off, and the erosion is reduced to zero.

These two farmers are not the only ones. It turns out that many others in different parts of the country are doing similar things. This might be a growing trend, and it might have a major effect on the environment. Raising beef on grass without plowing means reversing the flow of carbon out of the soil into the atmosphere. It means pushing big quantities of carbon down into the roots of the grass and turning a substantial fraction of it into topsoil. Instead of shouting, "Stop burning coal!" the climate alarmists might shout, "Stop plowing soil!" The effects on climate of plowing less soil might be as large as the effects of burning less coal, while the economic costs might be smaller and the ancillary ecological benefits might be greater.

For a farmer, it is not enough to be environmentally virtuous. A farm must be financially profitable, and it must be economical in its use of land. Manning brings us the splendid news that the farmers who switched from feedlots to grass are doing well. On the average, they are making net profits about eight times larger than the government subsidies that they received for their feedlots. Since the quality of their beef is superior, they have no difficulty selling it for good prices to food stores and restaurants. In addition, they are using less land in grass than they used for the same number of animals in feedlots. They can raise roughly two steer per acre on grass instead of one steer per acre on corn and feedlot.

We do not know whether the switch from feedlot to grass could be practical for a majority of Midwestern farmers. Farming on grass requires skills and motivation that an average farmer may not possess. It is at least possible that a massive switch to grass farming may be practical and profitable, with or without a change in government subsidies. Until we explore these questions, we cannot say that reducing consumption of coal is the only remedy for climate change. Richard Manning estimates that switching the entire American Midwest from feedlot to grass would remove from the atmosphere to topsoil about one quarter of the total greenhouse emissions of the United States. Raising beef on grass will not solve all our environmental problems, but it might give us a powerful push in the right direction. Even the reddest-blooded Americans do not live on beef alone. Additional environmental benefits will come from raising pigs and chickens or vegetable crops on unplowed land in other parts of the country.

I find another feature of Manning's story attractive. The key to the efficient raising of beef on grass is low-tech rather than high-tech. No genetic engineering or other controversial biotechnology is required. The key technical innovation is polywire, a simple and cheap electric-fencing material. Polywire makes it possible to move the animals frequently from place to place by moving fences, so that they eat the grass more uniformly. This simple technology will be easily adaptable to big and small farms and to rich and poor countries. It will not raise religious or ideological opposition. I also find attractive the fact that the switch to grass came from the bottom up and not from the top down. Social changes that come from the bottom up are usually more solid and more durable. In the eyes of most ordinary citizens, Minnesota farmers have more credibility than professors of economics.

The last two stories in Part Six are staged in India and China. They reinforce the evidence that Manning's story brings from Minnesota. India and China are now the center of gravity of the world's population and of the world's environmental problems. The fate of the planet, from an ecological point of view, is being decided by India and China and not by the United States. These two stories, one in India and one in China, bring us good news. Neither India nor China is about to stop burning coal, but both countries are taking environmental problems seriously. Each in its own way is putting big efforts into the healing of nature's wounds. Indian entrepreneurs and Chinese government officials are like Minnesota farmers. When they see something obviously wrong, they are willing to take responsibility and work hard to put it right. They take a long view of the future and try to solve only one problem at a time. They do not despair. They are happy if they leave their piece of the planet a little healthier than they found it. The lesson that I learn from these stories is that our future is in good hands.

FREEMAN DYSON

The Best American Science
and Nature Writing 2010

PART ONE

Visions of Space

ANDREW CORSELLO

The Believer

FROM *GQ*

ONCE IN A WHILE, this planet gives birth to a child with freakish talent—freakish not only because it is vast but because it is ready upon arrival, with batteries included and no assembly required. One need only open the box and step back.

In this case, the talent belongs to a six-year-old boy with a rather odd name. The year is 1977, and Elon (pronounced *Ee-lon*) Musk lives in the most odious country in the world: South Africa. It's summertime, and Elon and his kid brother and sister and their cousins have been playing outside their grandmother's suburban Pretoria home for hours. Now it's getting dark. The other children head for the house. *Come on, Elon. Let's go.*

But Elon doesn't want to go inside and doesn't understand why the others do. It's beautiful out here in the dark.

Elon and his siblings and cousins start to argue. *Come on, Elon. No! Come, Elon! I won't! Please, Elon.*

Tosca, the three-year-old, starts to yell, then cry. Then she blurts out what the other children are thinking.

"Elon, I'm *scared!*"

Tosca's mummy has come outside to see what the tears are about. Huddled there on the porch are Tosca and Kimbal—the middle sibling, fifteen months Elon's junior—and the cousins. And there at the tree line is Elon. The light has mostly waned, but Elon, he's so *white*, skin as pale as a fish's belly, and Maye Musk can see his face so clearly. Beaming. Euphoric. Because he *knows*.

Elon hasn't been bickering with his sister and brother; he has been evangelizing. And now he raises both arms to make sure they can see, as well as hear, the good news.

"Do not be scared of the darkness!" Elon Musk calls out to them from the wilderness. "There is nothing to fear—it is merely the absence of light!"

Though Elon has been issuing such pronouncements for several years, it seems to Maye Musk that the distinct way her son has of inspecting the world around him—so precise, so sober—was fully formed even before he could speak. A carefulness was evident, a stillness. Now, at six, he is creative and imaginative, but not in a fanciful way. Other than a fondness for comic books and Tolkien, he doesn't engage in make-believe, doesn't make things up. There are no imaginary friends—a surprise, since he doesn't have many real ones—or monsters in the closet. Elon simply isn't interested in things that are not there. Only in things that are, or plausibly could be. Facts. Elon needs facts the way he needs air.

And so he reads. Four, five hours a day, even as a first-grader. He forgets nothing he reads. Tosca will say, "I wonder how high up in the sky the moon is!" and Kimbal will respond, "A billion kilometers!" And Elon, smiling, sharing, will say, "Actually, it is 384,400 kilometers away." His siblings will stop and look at him then, and Elon, interpreting the silence as an invitation, will add, "On average."

Just the facts. They're all Elon needs. What he doesn't seem to need is a mentor, or even encouragement. Sometimes he fires questions at his father, an electrical and mechanical engineer. Problem is, many of his questions involve computers, which Errol Musk dismisses as "toys that will amount to nothing." His son calls this opinion "very silly" and, at the age of ten, buys his first computer and begins teaching himself how to program it. Two years later, he sells his first piece of software—a video game called Blastar—for $500.

Intelligence like Elon's—self-originating, self-sustaining, seemingly parentless—provokes a reflexive question from everyone who encounters it. *Where does such a child come from?* It's also a rhetorical question. The better thing to ask is: *Where does such a child go?*

This is the more relevant question not only because it is answerable but because it can and must be asked and answered *now.* Now—when we are more uncertain about one another and about ourselves and about our direction than we have been in decades—it is important for us to hear a story like Elon Musk's. As a reminder. And as a bracing slap to the face.

Because when children like Elon Musk attain the kind of self-awareness that leads to questions about environment — *Where in the world can I go for the license and the room to do what I must do? Where in the world are my peers?* — they always, and *still,* come to the same conclusion.

Elon Musk knew when he was a child. A remarkable conviction for a child to have, and all the more so because there was no specific dream attached to it. There was no "to build rocket ships" or "to make millions" or "to design computer software." Instead, Elon had this thought, consciously, literally, at the age of ten: *America is where people like me need to go. That is where people like me have always gone.* A place that was the photographic negative of apartheid South Africa, a place less encumbered than any in the world, ever, by fear.

"It is as true now as it has always been," says Elon Musk, the man who is endeavoring — as preposterously as he is credibly — to give the human race its biggest upgrade since the advent of consciousness. "Funny how people seem to have forgotten that. But almost all innovation in the world takes place in the United States."

By the time he's ten he's reading eight to ten hours a day. Elon reads and Elon retains, and his retention armors him. When the negative injunctions *You can't* and *You won't* come at Elon the way they come at all children, tens of thousands of times and in every conceivable form, sometimes overt and hard, sometimes insidious and soft, he simply doesn't hear them. Another couple of decades will pass before his biography fills in the specifics, but Elon Musk — the metamorphic intellect, the stuntman brazenness, the aura of immanence — is already *there.* The twenty-four-year-old physics Ph.D. candidate at Stanford who drops the program after forty-eight hours to become a software programmer who sells his first venture, a media-software company called Zip2, for $307 million? *There.* The propulsive personality who, within weeks of that sale, starts X.com, an online-banking company that morphs into PayPal before being sold to eBay in 2002 for $1.5 billion? *There.* The thirty-year-old autodidact who then dispenses with digital ephemera in order to become a man, a rocket man, a rocket *scientist,* and creates Space Exploration Technologies (SpaceX), a company whose short-term purpose is to commercialize an endeavor — orbital rocketry — that has previously been the province of a handful of nations and huge aerospace concerns (Northrop Grumman, Lockheed Martin, Boeing, etc.) and whose long-term aim is, yup, a mission to Mars? *There.* The thirty-two-year-old entrepreneur who decides it's time to gin up some *ambition* already and wean America off the teat of foreign oil while combating global warming and in 2004 makes himself the controlling shareholder and, eventually, CEO of Tesla Motors, manufacturer of the world's first all-electric sports car? *There.* The thirty-four-year-old penitent who realizes he's just not doing his part, greenhouse-gas-wise, and becomes the chairman and controlling shareholder of SolarCity, turning the company into one of the nation's biggest installers of solar panels? *There.*

The above reads like a chronology, which it is, but much of it is also a simultaneity: Elon Musk is currently helming three companies, all of them start-ups, each of them created to address an intractable global problem, two of them on the cutting edge of entirely different engineering technologies, and none of them in even remotely related industries. He is doing so as a businessman (he devised the business plan for SolarCity, which is run by his two cousins, and is the chief executive of the other two) and as a financier (having put more than $100 million of his own money into SpaceX and $55 million into Tesla), and that's something.

But really, it's nothing, because he's not just the vision guy or the money guy or the marketing guy, although he's all of those, too. He's also *designing* the stuff. At each of his companies, he knows what the engineers — chemical, mechanical, electrical, structural — know and what the software programmers know, and he does what they do. When the brushed-aluminum pedal of the Tesla Roadster is floored, unleashing 650 amps and 14,000 rpm from the car's 6,831 lithium-ion cells and launching it from zero to 60 miles per hour in 3.9 seconds, the engine roars like . . . a cell phone on vibrate mode — a phenomenon made possible not only by Elon Musk's money but by his mind. Likewise, the "CTO" in his official SpaceX title is descriptive, not ceremonial: Elon Musk taught himself how to design and build rockets. "I'd never seen anything like it," says Chris Thompson, explaining what persuaded him to leave a senior position at Boeing to oversee "structures" at SpaceX. "He was the quickest learner I've ever come across. You had this guy who knew everything from a business point of view, but who was also clearly capable of knowing everything from a technical point

of view—and the place he was creating was a blank sheet of paper."
Musk says (as do the rocket scientists he works with) that after
founding SpaceX, it took him "about two years to get up to speed."
How is such a thing possible?

Books. They did for Elon what they'd always done. They gave
him what he needed—facts. And, less obvious but just as crucial,
they took away what he didn't need: fear, or even any kind of hesi-
tation.

Did you see what Elon did this fall? It was big. It almost, but not
quite, made up for what he did in August, when one of his Falcon 1
rockets failed to make orbit and ended up dumping James Doo-
han's ashes—*Scotty's ashes*—in the Pacific Ocean. The achievement might have slipped under your radar, though, because it
came at the very end of September, just before the bottom truly
fell out. Most of us had already assumed the fetal position by then,
thrust our eyes into the softs of our elbows, anything to avoid looking at our latest 401(k) statements. So know this: on September 28,
Elon Musk did something that had never been done before, and
which experts had repeatedly said could never be done: launched
a privately funded rocket built from scratch into Earth orbit. Previously, only nine nations (and the European Space Agency) had independently done such a thing—each after decades of trial and
error, dozens of failed launches, and billions (of dollars, rubles,
francs . . .) invested. Musk's Falcon 1 rocket, built for $100 million
by a company with fewer than 150 employees, succeeded on only
its fourth attempt.

Walk into the giant hangar housing the offices of SpaceX and you
will immediately find your eye drawn to a large glass-walled space
named after Wernher von Braun, the Teutonic creator-god of rocketry and, like Elon Musk, a naturalized American citizen. Spend
an hour or two at the company and you'll realize that *von Braun*
refers less to a room than a state of mind. "We'll take it to the von
Braun"—that's the argot, the invocation, in the face of any conflict that requires immediate resolution. Engineers at SpaceX talk
about takin' it to the von Braun the way toughs in dive bars talk
about takin' it *outside*.

It's early November, late in the afternoon on election eve, as the
dozen men who make up SpaceX's senior design team file into the
von Braun Room. (Yes, they're all men, ranging from their early

twenties to early fifties. Two wear wedding rings; all but Elon sport metal watches chunky enough to deflect gunfire.) There is every reason to believe that this will be a truly terrible meeting for Elon Musk. Actually, meetings are terrible almost by definition in Musk's view. Meetings, he's fond of saying, are what happens when people aren't working.

But this Monday afternoon is special, thanks to Tesla. October has just proven to be the single worst month for the auto industry in twenty-five years. Despite being a new kind of company making a new kind of car, Tesla isn't immune from what is ailing Detroit. People aren't buying cars, *period*, much less $109,000 electric sports cars with a 244-mile range — a fact not lost on the venture capitalists Tesla relies on for financing. In recent weeks, Musk has had to close Tesla's engineering office in Michigan, lay off 20 percent of the company's staff (mostly from the Michigan office but also from the Silicon Valley headquarters), and announce a significant production delay in Tesla's Model S — the $57,000 sedan that Musk (and those venture capitalists) have been hoping will broaden the company's client base.

Yet more: that announcement about the S has nearly coincided with another, on the blog of Elon's wife, the fantasy novelist Justine Musk, that he has left her and their five boys (four-year-old twins and two-year-old triplets) for a twenty-three-year-old English actress named Talulah Riley. ("By all accounts she is bright and sweet and of course beautiful, and about as personally responsible for the death of my marriage as she is for the dynamic that played out inside it. In other words, not very," Justine wrote. "Also, she is not blonde, and I do find this refreshing.") And about a week after *that,* a Tesla employee leaked information to a popular Silicon Valley blog about how low morale at Tesla had sunk and revealing the proprietary fact that the company — which has taken more than a thousand deposits from buyers who haven't yet received their Roadsters — was down to its last $9 million in liquid reserves. The same day the blog item appeared, Musk issued a statement confirming the $9-million figure while announcing his intention to bolster Tesla's cash with at least $20 million in additional financing. Then, in search of the leaker, he sent a computer-forensics team to seize and search the computers of various employees. The only redeeming pieces of news about Tesla? Leonardo DiCaprio,

Matt Damon, and George Clooney are all having their Roadsters delivered this week.

Today, Elon and his SpaceX engineers are takin' it to the von Braun to discuss a fine point of reentry physics, per an exchange in one of the day's earlier meetings.

> Engineer #1: Would you VPPA?
> Engineer #2: [*lustily*] *Naaaaaah*, I'd probably go to soft plasma.
> Elon: You always get misplaced diameters with that.
> Engineer #2: What if the heat shield attached to the Dragon's base . . .
> [*A prolonged exchange of glances; a clear consensus that there are sometimes feelings for which there can be no words.*]
> Elon: We'll take it to the von Braun.
> [*Satisfied nods from all. Exeunt, pursued by a bear.*]

Now Musk sits, his engineers loosely grouped around him, waiting for one of them to begin a PowerPoint presentation. He just misses being extremely handsome, and somehow, by just missing the extreme of handsomeness, he also just misses being merely handsome. Yet Elon Musk draws eyes the way an extremely handsome man does, for two reasons. The first is that he is unusual-looking, in a boyish and pleasant way. The second is that physically, Elon Musk is a very, very still human being, and there is something arresting about that. Or as one Silicon Valley blog recently put it, "The liquored-up consensus at San Francisco watering hole Joey & Eddie's last night: Tesla Motors CEO Elon Musk is actually kind of hot."

"The economy is *shit*," he says, apropos of nothing and everything. Though Elon Musk almost never raises his voice and doesn't now, his tone is unmistakably . . . chipper. "Do you *realize* what that's going to do to the value of secondhand machines? They'll be in the toilet! We can get an EB welder on-site!"

It's the damnedest thing. The world is shit. *Elon's* world is shit. Yet when Elon asks, "Should we buy a welder?" he seems to be doing so in the same way a ten-year-old asks, "Should we ride the roller coaster now?" Here in the von Braun, everything that comes out of his mouth, whether in the form of a question or comment, is about building, hiring, investing. If and when the present woes of the world are acknowledged—the economy is shit!—the point is to exult in how *easy* that's going to make things.

But then the fun time is over and the meeting begins in earnest. The issue at hand involves the physics of reentry on the Falcon 9 —the larger and more ambitious successor to the Falcon 1 that SpaceX put into orbit in September. SpaceX is all about making orbital rockets that are both cheap and *reusable*—in other words, rockets that can survive the hellfires of atmospheric reentry. The adherence to one-time-use technology is *the* reason space exploration has always been the province of governments and their contractors. (Even the space shuttle is only partly reusable—and still costs $450 million per launch.) The commercial viability of SpaceX is therefore less about getting rockets into Earth orbit than reliably getting them back. Without reusability, there can be no economy of scale, and if there can be no economy of scale, there can be no SpaceX.

Though SpaceX already has several Falcon 1 contracts lined up with government and private satellite makers, it is the Falcon 9—scheduled to be test-launched from Cape Canaveral this summer—that could transform the aerospace industry. With its single engine, the 70-foot Falcon 1 can send payloads of 1,400 pounds or less into low Earth orbit (up to about 1,200 miles above the planet's surface). The 180-foot F9, with its nine engines and its Dragon capsule, is designed to catapult five tons of cargo 22,000 miles into the sky. What that will cost, at least initially: about $40 million per launch—somewhere between a third and a half of what NASA is accustomed to paying. The F9 is now vying for a contract to resupply the International Space Station. If NASA goes with SpaceX, the Dragon—which is both free-flying and reusable—will be able to supply the station not only with cargo but with people.

The PowerPoint presenter mentions reentry temperatures and the need for "restoring the pitch moment." All par for the course. But then he says it.

"So we want to make sure there's room enough for the avionics . . ."

"There is shitloads of fucking room there," Elon says quietly. Then, for clarity, he adds, "There is shitloads of fucking room there." He's not angry. This is just what Elon does; as nature hates a vacuum, Elon hates an inaccuracy. Without a trace of defensiveness, the engineer whom Elon has corrected explains the calculations underlying his previous statement. There's some back-and-forth on that. On paper, the language can look a bit violent ("If people do anything that contributes to this [rocket] stage not being recoverable," Elon says at one point, "they will find their work undone"). Spoken, it's a purely informational exchange.

Will the molten slag cause any problems?

Nah, it'll all be blowin' off so hard . . .

If we think of this as an upside-down Dragon . . .

If we need three inches of cork, how the fuck will the inflatable survive?

And so it goes. It's quite something to see a group of human beings offering themselves up in the service of *facts* the way these men are. Here in the von Braun, it is possible to comprehend the *facts*—all those mind-blowing technological facts—as molds into which these men have poured their lives, and not the other way around.

Something remains to be said about the nature of Elon Musk's ambition, something that makes the man either a sublime or a comic figure. Or perhaps both. Because according to Musk, the point of SpaceX is not to make money. The point is that mission to Mars. And that mission, as well as the Martian colonization to which it leads, still isn't the end-all Elon has in mind. There is a larger imperative.

According to Elon Musk, there are such things as "epochal moments." As he defines them, epochal moments are not moments "in" human history. They're larger. Rarer. Musk says the first epochal moment in the grand human ascent was "the advent of the single-celled organism. The next was the emergence of multicelled life. Then the differentiation into plants and animals. Then the move onto land. Then mammals. Then consciousness." To Musk, epochal moments are make-or-break moments; either there is a great leap forward or there is extinction. One does not fuck with epochal moments. And now, Elon says, for the first time since we humans began peering at our own reflections and wondering *Who?*, another epochal moment is upon us: we've got to get off this rock—soon—or face oblivion.

"I founded SpaceX and put much of my fortune in it because I really believe it is a matter of when and not if, and that when is probably a lot sooner than most of us are comfortable thinking about," Musk says of the end of life—all life, not just human—on

Earth. There is a constant motion in his eyes as he says this. Actually, the motion is present even when he's not prophesying the end-time. While his gaze remains fixed, the irises and pupils of his eyes—small, gray, wide-set—never stop moving. Very rapidly, almost imperceptibly, horizontally. *Vibrate* and even *shiver* overplay it, make it seem like a leer or tic when it is neither, and unappealing, which it isn't. When Elon Musk speaks, the colored parts of his eyes *shimmer* with attentiveness. He seems less like a person who is speaking than one who is listening. "I'm not saying we'll do it, to be sure. The odds are we *won't* succeed. But if something is important enough, then you should do it anyway.

"Things have happened quickly," he continues. "It took us millions of years to evolve into what we are, but in the last sixty years, with atomic weaponry, we've created the potential to extinguish ourselves. And if it's not us, it will inevitably be something else. If not a meteorite in the relative short term, then the expansion of the sun's corona. It *will* happen."

Yeah, man, the *corona*. Guy needs to lay off the comic books or go into movies, right? (Oh, wait—he already did that when he and two of his PayPal co-founders were producers on *Thank You for Smoking*.)

But he's not done.

"It's important enough to be on the scale of life itself, and therefore goes beyond the parochial concerns of humanity," Musk says of our interplanetary destiny. "We're all focused on our little things that are of concern to humanity itself. People think of curing AIDS or cancer as being very important, and they are—within the context of humanity. But curing all forms of cancer would improve the average life span by only two to three years. That's it."

In other words, while eradicating disease is a worthy pursuit, and would extend the lives of individual human beings, my life's work is extending the life span of life itself.

How does a person say the kinds of things Elon Musk is fond of saying and not conjure up a man stroking an albino show cat while reclining in the control room of his volcanic lair? Yet he doesn't. This neutral, disarming tone of his has always been part of his gift, his ability to make grandiose pronouncements without coming across as arrogant or show-offy (as an adult) or obnoxious or precious (when he was a child). Part of this may be physical. In addition to being soft-voiced, Musk has an unusually small mouth that barely moves, if it moves at all, when he speaks. There's a vaguely ventriloquistic effect. To listen to Elon Musk speak is to encounter words that somehow feel displaced from motive; he never seems to be attempting to prove anything about himself. There is a drawing power in that.

People follow Elon Musk. Many things about the man amaze, but none so much as the way he gets people to follow him. The pattern is constant: Elon declares his intentions and asks people to join him. Without exception, those intentions are ludicrous; without exception, the people he's wooing are bright, older, and more experienced than he is and safely ensconced in lives they find fulfilling. Yet these people forsake the secure and the known to follow Elon Musk. It has been this way since he was a teenager.

Only a few months after Musk left South Africa at the age of seventeen, his sister, Tosca, announced to their mother that she intended to join him in Canada. (Elon was unable to immigrate straight to the United States—only to Canada, where his mother had been born and still had citizenship.)

"Elon is eighteen, you're *fifteen*," said Maye Musk, pointing out the obvious.

"Elon will look after me," Tosca told her mother.

Tosca, it turned out, had absorbed more than a little of what her older brother had told her years before about the absence of light; Maye Musk returned from a trip to find that her fifteen-year-old daughter had not only managed to put their house on the market but had sold it. "I think we must leave now," Tosca explained. Elon was expecting them.

Nearly a decade earlier, Maye Musk had taken her three children and, as she now puts it, "run away" from her husband. (Elon says he and his father have been in "limited touch" ever since. "Quite an astute engineer, although he's gone a little crazy later in life. I don't think he has all his cookies in the jar.") In the years following her divorce, Maye had built dual careers as a model and a dietitian. If she chose to emigrate, the South African government would freeze all her assets—even the money from the house sale. The choice filled her with panic. Don't worry, Tosca assured her mother, "Elon knows everything." How could Maye argue with that? Elon really did know everything, which meant there was nothing to fear. Which was why Maye Musk packed a bag and, with her two younger children, followed her eldest child across an ocean.

Twelve years later, when Musk started SpaceX and began recruiting, it was the same. Some of those who came gave up senior, secure, lucrative positions at places like TRW and Boeing to come work at SpaceX with Elon Musk—an Internet entrepreneur with no background in rocket science. "It was a can-do attitude combined with the fact that he knew what he was talking about and was happy to be corrected if he didn't," says Tom Mueller, the man who now oversees all "propulsion" at SpaceX. "You couldn't say no to it." Others left places like Google and Microsoft to come work at SpaceX and, later, at Tesla because Musk liked the way their minds worked and convinced them that like him, they didn't need physics Ph.D.'s to build rockets and cars.

Perhaps the most remarkable thing about Elon Musk's drawing power is that, strange as it may seem, he is not a particularly charming or charismatic person. Charm and charisma require, first, the desire to be charming and charismatic and, second, the ability to execute; both of these require energy. And Elon, a man deeply attuned to the physics of his energy—its caloric and intellectual and financial and historical value—simply isn't interested in using it to light up a room. And yet there is some . . . *thing* whereby Elon Musk enters a room and both the space and the other people in it suddenly seem distilled. What *is* that—and how does it obviate the questions about charm and charisma and even likability?

Here's how: Elon Musk's leadership, his ability to inspire and motivate the people who work for him, derives completely, and only, from his *knowledge*. The knowledge—the millions upon millions of facts that he has not forgotten—is itself a form of charisma, that infectious "thing you just can't put your finger on" that we normally expect to see in the form of "star power." Ironic, that the man with the intelligence and the will to take the human species to Mars and beyond would lack star power. And strangely assuring that he doesn't need it. What he needs is to persuade the smartest engineers in the world to come work for him. Star power can't do that. Facts can.

In the end—although it is patently ridiculous to speak of any kind of "end," seeing as how Elon Musk is still three years shy of

forty—all those facts add up to a kind of dazzling mosaic. Of . . . ?
Well, that depends on who you are. If you are Elon or one of the
people working with him to build his cars and his rockets, it is an
image that convinces—an image without metaphysical content: he
doesn't have *faith* that he can do it; he knows. Because Elon knows
everything.

But if you are just . . . a citizen, and especially if you are an *American*
can citizen weary of soul after eight years of a national leadership
not only incurious about but hostile toward science, even if this
man fails to midwife the next epochal moment in the great human
ascension, or even to deliver an affordable all-electric car that helps
kill this country's oil addiction, the mosaic image of Elon Musk's
life is one that galvanizes: *this* guy had to come *here* to attempt *that.*

Brothers, sisters, fellow earthbound humans, if that doesn't wake
you up, you're already extinct.

TOM WOLFE

One Giant Leap to Nowhere

FROM *The New York Times*

WELL, LET'S SEE NOW . . . That was a small step for Neil Armstrong, a giant leap for mankind, and a real knee in the groin for NASA.

The American space program, the greatest, grandest, most Promethean—OK if I add "godlike"?—quest in the history of the world, died in infancy at 10:56 P.M. New York time on July 20, 1969, the moment the foot of *Apollo 11*'s Commander Armstrong touched the surface of the moon.

It was no ordinary dead-and-be-done-with-it death. It was full-blown purgatory, purgatory being the holding pen for recently deceased but still restless souls awaiting judgment by a Higher Authority.

Like many another youngster at that time, or maybe retro-youngster in my case, I was fascinated by the astronauts after *Apollo 11*. I even dared to dream of writing a book about them someday. If anyone had told me in July 1969 that the sound of Neil Armstrong's small step plus mankind's big one was the shuffle of pallbearers at graveside, I would have averted my eyes and shaken my head in pity. Poor guy's bucket's got a hole in it.

Why, putting a man on the moon was just the beginning, the prelude, the prologue! The moon was nothing but a little satellite of Earth. The great adventure was going to be the exploration of the planets . . . Mars first, then Venus, then Pluto. Jupiter, Mercury, Saturn, Neptune, and Uranus? NASA would figure out their slots in the schedule in due course. In any case, we Americans wouldn't stop until we had explored the entire solar system. And after that . . . the galaxies beyond.

NASA had long since been all set to send men to Mars, starting with manned flybys of the planet in 1975. Wernher von Braun, the German rocket scientist who had come over to our side in 1945, had been designing a manned Mars project from the moment he arrived. In 1952 he published his Mars Project as a series of graphic articles called "Man Will Conquer Space Soon" in *Collier's* magazine. It created a sensation. He was front and center in 1961 when NASA undertook Project Empire, which resulted in working plans for a manned Mars mission. Given the epic, the saga, the triumph of Project Apollo, Mars would naturally come next. All NASA and von Braun needed was the president's and Congress's blessings and the great adventure was a Go. Why would they so much as blink before saying the word?

Three months after the landing, however, in October 1969, I began to wonder . . . I was in Florida, at Cape Kennedy, the space program's launching facility, aboard a NASA tour bus. The bus's spielmeister was a tall-fair-and-handsome man in his late thirties . . . and a real piece of lumber when it came to telling tourists on a tour bus what they were looking at. He was so bad, I couldn't resist striking up a conversation at the end of the tour.

Sure enough, it turned out he had not been put on Earth for this job. He was an engineer who until recently had been a NASA heat-shield specialist. A baffling wave of layoffs had begun, and his job was eliminated. It was so bad he was lucky to have gotten this standup spielmeister gig on a tour bus. Neil Armstrong and his two crewmates, Buzz Aldrin and Mike Collins, were still on their triumphal world tour . . . while back home, NASA's irreplaceable team of highly motivated space scientists—irreplaceable!—there were no others! . . . anywhere! . . . You couldn't just run an ad saying, "Help Wanted: Experienced heat-shield expert" . . . The irreplaceable team was breaking up, scattering in nobody knows how many hopeless directions.

How could such a thing happen? In hindsight, the answer is obvious. NASA had neglected to recruit a corps of philosophers.

From the moment the Soviets launched *Sputnik I* into orbit around Earth in 1957, everybody from Presidents Eisenhower, Kennedy, and Johnson on down looked upon the so-called space race as just one thing: a military contest. At first there was alarm over the Soviets' seizure of the "strategic high ground" of space. They were already up there—right above us! They could now hurl thunderbolts down whenever and wherever they wanted. And what could we do about it? Nothing. *Ka-boom!* There goes Bangor . . . *Ka-boom!* There goes Boston . . . *Ka-boom!* There goes New York . . . Baltimore . . . Washington . . . St. Louis . . . Denver . . . San Jose—blown away!—just like that.

Physicists were quick to point out that nobody would choose space as a place from which to attack Earth. The spacecraft, the missile, Earth itself, plus Earth's own rotation, would be traveling at wildly different speeds upon wildly different geometric planes. You would run into the notorious "three-body problem" and then some. You'd have to be crazy. The target would be untouched and you would wind up on the floor in a fetal ball, twitching and gibbering. On the other hand, the rockets that had lifted the Soviets' five-ton manned ships into orbit were worth thinking about. They were clearly powerful enough to reach any place on Earth with nuclear warheads.

But that wasn't what was on President Kennedy's mind when he summoned NASA's administrator, James Webb, and Webb's deputy, Hugh Dryden, to the White House in April 1961. The president was in a terrible funk. He kept muttering: "If somebody can just tell me how to catch up. Let's find somebody—anybody . . . There's nothing more important." He kept saying, "We've got to catch up." Catching up had become his obsession. He never so much as mentioned the rockets.

Dryden said that, frankly, there was no way we could catch up with the Soviets when it came to orbital flights. A better idea would be to announce a crash program on the scale of the Manhattan Project, which had produced the atomic bomb. Only the aim this time would be to put a man on the moon within the next ten years.

Barely a month later Kennedy made his famous oration before Congress: "I believe that this nation should commit itself to achieving the goal, before this decade is out, of landing a man on the moon and returning him safely to Earth." He neglected to mention Dryden.

Intuitively, not consciously, Kennedy had chosen another form of military contest, an oddly ancient and archaic one. It was called "single combat."

The best known of all single combats was David versus Goliath. Before opposing armies clashed in all-out combat, each would send forth its "champion," and the two would fight to the death, usually with swords. The victor would cut off the head of the loser and brandish it aloft by its hair.

The deadly duel didn't take the place of the all-out battle. It was regarded as a sign of which way the gods were leaning. The two armies then had it out on the battlefield . . . unless one army fled in terror upon seeing its champion slaughtered. There you have the Philistines when Little David killed their giant, Goliath . . . and cut his head off and brandished it aloft by its hair (1 Samuel 17:1–58). They were overcome by a mad desire to be somewhere else. (The Israelites pursued and destroyed them.)

More than two millenniums later, the mental atmosphere of the space race was precisely that. The details of single combat were different. Cosmonauts and astronauts didn't fight hand to hand and behead one another. Instead, each side's brave champions, including one woman (Valentina Tereshkova), risked their lives by sitting on top of rockets and having their comrades on the ground light the fuse and fire them into space like the human cannonballs of yore.

The Soviets rocketed off to an early lead. They were the first to put an object into orbit around Earth (*Sputnik*), the first to put an animal into orbit (a dog), the first to put a man in orbit (Yuri Gagarin). No sooner had NASA put two astronauts (Gus Grissom and Alan Shepard) into fifteen-minute suborbital flights to the Bahamas— *the Bahamas!—fifteen minutes!—two miserable little mortar lobs!* —than the Soviets put a second cosmonaut (Gherman Titov) into orbit. He stayed up there for twenty-five hours and went around the globe seventeen times. Three times he flew directly over the United States. The gods had shown which way they were leaning, all right!

At this point, the mental atmospheres of the rocket-powered space race of the 1960s and the sword-clanking single combat of ancient days became so similar you had to ask: Does the human beast ever really change—or merely his artifacts? The Soviet cosmo-champions beat our astro-champions so handily, gloom spread like a gas. Every time you picked up a newspaper you saw headlines with the phrase, SPACE GAP . . . SPACE GAP . . . SPACE GAP . . . The

Soviets had produced a generation of scientific geniuses—while we slept, fat and self-satisfied! Educators began tearing curriculums apart as soon as *Sputnik* went up, introducing the New Math and stressing another latest thing, the Theory of Self-Esteem.

At last, in February 1962, NASA managed to get a man into Earth orbit, John Glenn. You had to have been alive at that time to comprehend the reaction of the nation, practically all of it. He was up for only five hours, compared to Titov's twenty-five, but he was our . . . Protector! Against all odds he had risked his very hide for . . . us!—protected us from our mortal enemy!—struck back in the duel in the heavens!—showed the world that we Americans were born fighting and would never give up! John Glenn made us whole again!

During his ticker-tape parade up Broadway, you have never heard such cheers or seen so many thousands of people crying. Big Irish cops, the classic New York breed, were out in the intersections in front of the world, sobbing, blubbering, boo-hooing, with tears streaming down their faces. John Glenn had protected all of us, cops too. All tears have to do with protection . . . but I promise not to lay that theory on you now. John Glenn, in 1962, was the last true national hero America has ever had.

There were three more Mercury flights, and then the Gemini series of two-man flights began. With Gemini, we dared to wonder if perhaps we weren't actually pulling closer to the Soviets in this greatest of all single combats. But we held our breath, fearful that the Soviets' anonymous Chief Designer would trump us again with some unimaginably spectacular feat.

Sure enough, the CIA brought in sketchy reports that the Soviets were on the verge of a moon shot.

NASA entered into the greatest crash program of all time, Apollo. It launched five lunar missions in one year, December 1968 to November 1969. With *Apollo 11,* we finally won the great race, landing a man on the moon before the end of this decade and returning him safely to Earth.

Everybody, including Congress, was caught up in the adrenal rush of it all. But then, on the morning after, congressmen began to wonder about something that hadn't dawned on them since Kennedy's oration. What was this single-combat stuff—they didn't use the actual term—really all about? It had been a battle for morale at home and image abroad. Fine, OK, we won, but it had no tactical military meaning whatsoever. And it had cost a fortune, $150 billion or so. And this business of sending a man to Mars and whatnot? Just more of the same, when you got right down to it. How laudable . . . how far-seeing . . . but why don't we just do a Scarlett O'Hara and think about it tomorrow?

And that NASA budget! Now there was some prime pork you could really sink your teeth into! And they don't need it anymore! Game's over, NASA won, congratulations. Who couldn't use some of that juicy meat to make the people happy? It had an ambrosial aroma . . . made you think of reelection . . .

NASA's annual budget sank like a stone from $5 billion in the mid-1960s to $3 billion in the mid-1970s. It was at this point that NASA's lack of a philosopher corps became a real problem. The fact was, NASA had only one philosopher, Wernher von Braun. Toward the end of his life, von Braun knew he was dying of cancer and became very contemplative. I happened to hear him speak at a dinner in his honor in San Francisco. He raised the question of what the space program was really all about.

It's been a long time, but I remember him saying something like this: here on Earth we live on a planet that is in orbit around the sun. The sun itself is a star that is on fire and will someday burn up, leaving our solar system uninhabitable. Therefore we must build a bridge to the stars, because as far as we know, we are the only sentient creatures in the entire universe. When do we start building that bridge to the stars? We begin as soon as we are able, and this is that time. We must not fail in this obligation we have to keep alive the only meaningful life we know of.

Unfortunately, NASA couldn't present as its spokesman and great philosopher a former high-ranking member of the Nazi Wehrmacht with a heavy German accent.

As a result, the space program has been killing time for forty years with a series of orbital projects . . . Skylab, the Apollo-Soyuz joint mission, the International Space Station, and the space shuttle. These programs have required a courage and an engineering brilliance comparable to the manned programs that preceded them. But their purpose has been mainly to keep the lights on at the Kennedy Space Center and Houston's Johnson Space Center — by removing manned flight from the heavens and bringing it very much down to Earth. The shuttle program, for example, was actually supposed to appeal to the public by offering orbital tourist rides, only to end in the *Challenger* disaster, in which the first such passenger, Christa McAuliffe, a schoolteacher, perished.

Forty years! For forty years, everybody at NASA has known that the only logical next step is a manned Mars mission, and every overture has been entertained only briefly by presidents and Congress. They have so many more luscious and appealing projects that could make better use of the close to $10 billion annually the Mars program would require. There is another overture even at this moment, and it does not stand a chance in the teeth of Depression II.

"Why not send robots?" is a common refrain. And once more it is the late Wernher von Braun who comes up with the rejoinder. One of the things he most enjoyed saying was that there is no computerized explorer in the world with more than a tiny fraction of the power of a chemical analog computer known as the human brain, which is easily reproduced by unskilled labor.

What NASA needs now is the power of the Word. On Darwin's tongue, the Word created a revolutionary and now well-nigh universal conception of the nature of human beings or, rather, human beasts. On Freud's tongue, the Word means that at this very moment there are probably several million orgasms occurring that would not have occurred had Freud never lived. Even the fact that he is proved to be a quack has not diminished the power of his Word.

July 20, 1969, was the moment NASA needed, more than anything else in this world, the Word. But that was something NASA's engineers had no specifications for. At this moment, that remains the only solution to recovering NASA's true destiny, which is, of course, to build that bridge to the stars.

STEVEN WEINBERG

The Missions of Astronomy

FROM *The New York Review of Books*

A FEW YEARS AGO, I decided that I needed to know more about the history of science, so naturally I volunteered to teach the subject. In working up my lectures, I was struck by the fact that in the ancient world, astronomy reached what from a modern perspective was a much higher level of accuracy and sophistication than any other science.[1] One obvious reason for this is that visible astronomical phenomena are much simpler and easier to study than the things we can observe on Earth's surface. The ancients did not know it, but Earth and the moon and planets all spin at nearly constant rates, and they travel in their orbits under the influence of a single dominant force, that of gravitation.

In consequence, the changes in what is seen in the sky are simple and periodic: the moon regularly waxes and wanes, the sun and moon and stars seem to revolve once a day around the celestial pole, and the sun traces a path through the same constellations of stars every year, those of the zodiac.[2] Even with crude instruments, these periodic changes could be and were studied with a fair degree of mathematical precision, much greater than was possible for things on Earth like the flight of a bird or the flow of water in a river.

But there was another reason why astronomy was so prominent in ancient and medieval science. It was useful in a way that the physics and biology of the time were not. Even before history began, people must have used the apparent motion of the sun as at least a crude clock, calendar, and compass. These functions became much more precise with the introduction of what may have been the first scientific instrument, the gnomon, attributed by the Greeks variously to Anaximander or to the Babylonians.

The gnomon is simply a straight pole, set vertically in a flat, level patch of ground open to the sun's rays. When during each day the gnomon's shadow is shortest, that is noon. At noon the gnomon's shadow anywhere in the latitude of Greece or Mesopotamia points due north, so all the points of the compass can be permanently and accurately marked out on the ground around the gnomon. Watching the shadow from day to day, one can note the days when the noon shadow is shortest or longest. That is the summer or the winter solstice. From the length of the noon shadow at the summer solstice one can calculate the latitude. The shadow at sunset points somewhat south of east in the spring and summer, and somewhat north of east in the fall and winter; when the shadow at sunset points due east, that is the spring or fall equinox.[3]

Using the gnomon as a calendar, the Athenian astronomers Meton and Euctemon made a discovery around 430 BC that was to trouble astronomers for two thousand years: the four seasons, whose beginnings and endings are precisely marked by the solstices and equinoxes, have slightly different lengths. This ruled out the possibility that the sun travels around Earth (or Earth around the sun) with constant velocity in a circle, for in that case the equinoxes and solstices would be evenly spaced throughout the year. This was one of the reasons that Hipparchus of Nicaea, the greatest observational astronomer of the ancient world, found it necessary around 150 BC to introduce the idea of epicycles, the idea that the sun (and planets) move on circles whose centers themselves move on circles around Earth, an idea that was picked up and elaborated three centuries later by Claudius Ptolemy.

Even Copernicus, because he was committed to orbits composed of circles, retained the idea of epicycles. It was not until the early years of the seventeenth century that Johannes Kepler finally explained what Hipparchus and Ptolemy had attributed to epicycles. Earth's orbit around the sun is not a circle but an ellipse; the sun is not at the center of the ellipse but at a point called the focus, off to one side; and Earth's speed is not constant but faster when it is near the sun and slower when farther away.

*

For the human uses I have been discussing, the sun has its limitations. The sun can of course be used to tell time and directions only during the day, and before the introduction of the gnomon its annual motions gave only a crude idea of the time of year. From earliest recorded times, the stars were put to use to fill these gaps. Homer knew of the stars' use at night as a compass. In the *Odyssey*, Calypso gives Odysseus instructions on how to go from her island eastward toward Ithaca: he is told to keep the Bear on his left. The Bear, of course, is Ursa Major, aka the Big Dipper, a constellation near the North Pole of the sky (called the celestial pole), which in the latitude of the Mediterranean never sets beneath the horizon (or, as Homer says, never bathes in the ocean). With north on his left, Odysseus would be sailing east, toward home.[4]

The stars were also put to use as a calendar. The Egyptians very early appear to have anticipated the flooding of the Nile by observing the rising of the star Sirius. Around 700 BC the Greek poet Hesiod in *Works and Days* advised farmers to plow at the cosmical setting of the Pleiades constellation — that is, on the day in the year on which the Pleiades star cluster is first seen to set before the sun comes up.

Observing the stars for these reasons, it was noticed in many early civilizations that there are five "stars," called planets by the Greeks, that in the course of a year move against the background of all the other stars, staying pretty much on the same path along the zodiac as the sun, but sometimes seeming to reverse their course. The problem of understanding these motions perplexed astronomers for millennia and finally led to the birth of modern physics with the work of Isaac Newton.

The usefulness of astronomy was important not only because it focused attention on the sun and stars and planets and thereby led to scientific discoveries. Utility was also important in the development of science, because when one is actually using a scientific theory rather than just speculating about it, there is a large premium on getting things right. If Calypso had told Odysseus to keep the moon on his left, he would have gone around in circles and never reached home. In contrast, Aristotle's theory of motion survived through the Middle Ages because it was never put to practical use in a way that could reveal how wrong it was. Astronomers did try to use Aristotle's theory of the planetary system (due originally to

Plato's pupil Eudoxus and his pupil Callippus), in which the sun
and moon and planets ride on coupled transparent spheres centered on Earth, a theory that (unlike the epicycle theory) was consistent with Aristotle's physics.

They found that it did not work—for instance, Aristotle's theory could not account for the changes in brightness of the planets
over time, changes that Ptolemy understood to be due to the fact
that each planet is not always at the same distance from Earth. Because of the prestige of Aristotle's philosophy, some philosophers
and physicians (but few working astronomers) continued through
the ancient world and the Middle Ages to adhere to his theory of
the solar system, but by the time of Galileo it was no longer taken
seriously. When Galileo wrote his *Dialogue Concerning the Two Chief
Systems of the World,* the two systems that Galileo considered were
those of Ptolemy and Copernicus, not Aristotle's.

There was one more reason that the usefulness of astronomy was
important to the advance of science: it promoted government support of scientific research. The first great example was the Museum
of Alexandria, established by the Greek kings of Egypt early in the
Hellenistic era, around 300 BC. This was not a museum in the
modern sense, a place where visitors can come to look at fossils and
pictures, but a research institution, devoted to the Muses, including Urania, the muse of astronomy. The kings of Egypt supported
studies in Alexandria, probably at the museum, of the construction
of catapults and other artillery and of the flights of projectiles, but
the museum also provided salaries to Aristarchus, who measured
the distances and sizes of the sun and moon, and to Eratosthenes,
who measured the circumference of Earth.

The museum was the first of a succession of government-sup-
ported centers of research, including the House of Wisdom, established around 830 AD by the caliph al-Mamun in Baghdad, and
Tycho Brahe's observatory Uraniborg, on an island given to Brahe
by the Danish king Frederick II in 1576. The tradition of govern-
ment-supported research continues in our day, at particle physics
laboratories like CERN and Fermilab and on unmanned observatories like Hubble and WMAP and Planck, put into space by NASA
and the European Space Agency.

In fact, in the past astronomy benefited from an overestimate of
its usefulness. The legacies of the Babylonians to the Hellenistic

world included not only a large body of accurate astronomical observations (and perhaps the gnomon) but also the pseudoscience of astrology. Ptolemy was the author not only of a great astronomical treatise, the *Almagest,* but also of a book on astrology, the *Tetrabiblos.* Much of the royal support for compiling tables of astronomical data in the medieval and early modern periods was motivated by the use of these tables by astrologers. This appears to contradict what I said about the importance in applications of getting the science right, but the astrologers did generally get the astronomy right, at least as to the apparent motions of the planets and stars, and they could hide their failure to account for human affairs in the obscurity of their predictions.

Not everyone has been enthusiastic about the utilitarian side of astronomy. In Plato's *Republic* there is a discussion of the education to be provided for future philosopher kings. Socrates suggests that astronomy ought to be included, and his stooge Glaucon hastily agrees, because "it's not only farmers and others who need to be sensitive to the seasons, months, and phases of the year; it's just as important for military commanders as well." Poor Glaucon—Socrates calls him naive and explains that the real reason to study astronomy is that it forces the mind to look upward and think of things that are nobler than our everyday world.

Although surprises are always possible, my own main research area, elementary particle physics, has no direct applications that anyone can foresee,[5] so it gives me little joy to note the importance of utility to the historical development of science. By now pure sciences like particle physics have developed standards of verification that make applications unnecessary in keeping us honest (or so we like to think), and their intellectual excitement incites the efforts of scientists without any thought of practical use. But research in pure science still has to compete for government support with more immediately useful sciences, like chemistry and biology.

Unfortunately for the ability of astronomy to compete for support, the uses of astronomy that I have discussed so far have largely become obsolete. We now use atomic clocks to tell time so accurately that we can measure tiny changes in the length of the day and year. We can look up today's date on our watches or computer screens. And recently the stars have even lost their importance for navigation.

In 2005 I was on the bark *Sea Cloud,* cruising the Aegean Sea. One evening I fell into a discussion about navigation with the ship's captain. He showed me how to use a sextant and chronometer to find positions at sea. Measuring the angle between the horizon and the position of a given star with the sextant at a known chronometer time tells you that your ship must lie somewhere on a particular curve on the map of Earth. Doing the same with another star gives another curve, and where they intersect, there is your position. Doing the same with a third star and finding that the third curve intersects the first two at the same point tells you that you have not made a mistake. After demonstrating all this, my friend the captain of the *Sea Cloud* complained that the young officers coming into the merchant marine could no longer find their position with chronometer and sextant. The advent of global positioning satellites had made celestial navigation unnecessary.

One use remains to astronomy: it continues to have a crucial part in our discovery of the laws of nature. As I mentioned, it was the problem of the motion of the planets that led Newton to the discovery of his laws of motion and gravitation. The fact that atoms emit and absorb light at only certain wavelengths, which in the twentieth century led to the development of quantum mechanics, was discovered in the early nineteenth century in observations of the spectrum of the sun. Later in the nineteenth century these solar observations revealed the existence of new elements, such as helium, that were previously unknown on Earth. Early in the twentieth century, Einstein's General Theory of Relativity was tested astronomically, at first by comparison of his theory's predictions with the observed motion of the planet Mercury and then through the successful prediction of the deflection of starlight by the gravitational field of the sun.

After the confirmation of General Relativity, for a while the source of the data that inspired progress in fundamental physics switched away from astronomy, first toward atomic physics and then in the 1930s toward nuclear and particle physics. But progress in particle physics has slowed since the formulation of the Standard Model of elementary particles in the 1960s and 1970s, which accounted for all the data about elementary particles that was then available. The only things discovered in recent years in particle physics that go beyond the Standard Model are the tiny masses of the various kinds of neutrinos, and these first showed up in a sort of astronomy, the search for neutrinos from the sun.

Meanwhile, we are now in what it has become trite to call a golden age of cosmology. Astronomical observation and cosmological theory have invigorated each other, to the point that we can now say with a straight face that the universe in its present phase of expansion is 13.73 billion years old, give or take 0.16 billion years. This work has revealed that only about 4.5 percent of the energy of the universe is in the form of ordinary matter — electrons and atomic nuclei. Some 23 percent of the total energy is in the masses of particles of "dark matter," particles that do not interact with ordinary matter or radiation, and whose existence is so far known only through observations of effects of the gravitational forces they exert on ordinary matter and light. The greatest part of the energy budget of the universe, about 72 percent, is a "dark energy" that does not reside in the masses of any sort of particle but in space itself, and that is causing the present expansion of the universe to accelerate. The explanation of dark energy is now the deepest problem facing elementary particle physics.

Exciting as all this is, both astronomy and particle physics have increasingly had to struggle for government support. In 1993 Congress canceled a program to build an accelerator, the Superconducting Super Collider, that would have greatly extended the range of masses of new particles that might be created, perhaps including the particles of dark matter. The European consortium CERN has picked up this task, but its new accelerator, the Large Hadron Collider, will be able to explore only about a third of the range of masses that could have been reached by the Super Collider, and support for the next accelerator after the Large Hadron Collider seems increasingly in doubt. In astronomy NASA has cut back on the Beyond Einstein and Explorer programs, major programs of astronomical research of the sort that has made possible the great progress of recent years in cosmology.

Of course, there are many worthy calls on government funds. What particularly galls many scientists is the existence of a vastly expensive NASA program that often masquerades as science.[6] I refer, of course, to the manned space flight program. In 2004 President Bush announced a "new vision" for NASA, a return of astronauts to the moon followed by a manned mission to Mars. A few days later the NASA Office of Space Science announced cuts in its unmanned Beyond Einstein and Explorer programs, with the explanation that they did not support the president's new vision.

Astronauts are not effective in scientific research. For the cost of taking astronauts safely to the moon or planets and bringing them back, one could send many hundreds of robots that could do far more in the way of exploration. Astronauts in orbiting astronomical observatories would create vibrations and radiate heat, which would foul up sensitive astronomical observations. All of the satellites like Hubble or COBE or WMAP or Planck that have made possible the recent progress in cosmology have been unmanned. No important science has been done at the manned International Space Station, and it is hard to imagine any significant future work that could not be done more cheaply on unmanned facilities.

It is often said that manned space flight is necessary for science because without it the public would not support any space programs,[7] including unmanned missions like Hubble and WMAP that do real science. I doubt this. I think that there is an intrinsic excitement to astronomy in general and cosmology in particular, quite apart from the spectator sport of manned space flight. As illustration, I will close with a verse of Claudius Ptolemy:

> I know that I am mortal and the creature of a day; but when I search out the massed wheeling circles of the stars, my feet no longer touch the Earth, but, side by side with Zeus himself, I take my fill of ambrosia, the food of the gods.

Notes

1. This article is based on a talk given on September 25, 2009, at the Harry Ransom Center for Humanistic Studies of the University of Texas at Austin, to commemorate its exhibition "Other Worlds: Rare Astronomical Works," on view September 8, 2009–January 3, 2010.

2. Of course the stars are not visible during the day, but some of them can be seen just after sunset, when the sun's position in the sky is still known.

3. A gnomon is different from a sundial because the pole that casts a shadow on a sundial is not vertical but set at an angle chosen so that the pole's shadow follows about the same path during each day of the year.

This makes the sundial more useful as a clock but less useful as a calendar.

4. It may be wondered why Calypso did not tell Odysseus to keep the North Star on his left. The reason is that in Homer's time the star Polaris, which is now the North Star, was not at the North Pole of the sky. This is not because of any motion of Polaris itself but because of a phenomenon known as the precession of the equinoxes, discovered by Hipparchus. In modern terms, the axis of Earth's rotation does not keep a fixed direction in the sky, but precesses like the axis of a spinning top, making a full circle every 25,727 years. It is a measure of the accuracy of Greek astronomy that the data of Hipparchus indicated a period of 28,000 years.

5. I say "direct" applications because experimental and theoretical work in particle physics that pushes technology and mathematics to their current limits occasionally spins off new technology or mathematics of great practical importance. One celebrated example is the World Wide Web. This can provide a valid argument for government support, but it is not why we do the research.

6. I have written about this at greater length in "The Wrong Stuff," *New York Review of Books*, April 8, 2004.

7. This opinion was most recently expressed by Giovanni Bignami, the head of the European Space Agency's Science Advisory Committee, in "Why We Need Space Travel," *Nature*, July 16, 2009.

TIMOTHY FERRIS

Cosmic Vision

FROM *National Geographic*

WHEN YOU START STARGAZING with a telescope, two experiences typically ensue. First, you are astonished by the view—Saturn's golden rings, star clusters glittering like jewelry on black velvet, galaxies aglow with gentle starlight older than the human species—and by the realization that we and our world are part of this gigantic system. Second, you soon want a bigger telescope.

Galileo, who first trained a telescope on the night sky four hundred years ago this fall, pioneered this two-step program. First, he marveled at what he could see. Galileo's telescope revealed so many previously invisible stars that when he tried to map all of those in just one constellation, Orion, he gave up, confessing that he was "overwhelmed by the vast quantity of stars." He saw mountains on the moon—in contradiction of the prevailing orthodoxy, which declared that all celestial objects were made of an unearthly "ether." He charted four bright satellites as they bustled around Jupiter like planets in a miniature solar system, something that critics of the Copernican sun-centered cosmology had dismissed as physically impossible. Evidently Earth was a small part of a big universe, not a big part of a small one.

And soon, sure enough, Galileo went to work making bigger and better telescopes. Large light-gathering lenses were not yet available, so he concentrated on making longer telescopes, which produced higher magnifying powers and reduced the halos of spurious colors that afflicted glass lenses in those days. Subsequent observers took the design of glass-lensed, refracting telescopes to great lengths, sometimes literally so. In Danzig, Johannes Hevelius deployed a telescope 150 feet long; hung by ropes from a pole, it undulated in the slightest breeze. In the Netherlands, the Huygens brothers unveiled lanky telescopes that had no tubes at all: the objective lens was perched on a high platform in a field, while an observer up to 200 feet away aligned a magnifying eyepiece and peered through it. Such instruments proffered fleeting glimpses of planets and stars, which, like the dance of the seven veils, only aroused a burning desire to see more.

The reflecting telescope, pioneered by Isaac Newton, made it practical to gratify such desires: mirrors required that only one surface be ground to gather and reflect starlight to a focal point, and since the mirror was supported from behind, it could be quite large without sagging under its own weight, as large lenses tended to do. William Herschel discovered the planet Uranus with a handmade reflecting telescope—he cast his metal mirrors in his garden and basement and once had to flee from a coursing river of molten metal after the horse-dung mold fractured. Spiral-armed galaxies were first glimpsed through a massive reflecting telescope with a six-foot-diameter primary mirror that Lord Rosse constructed on his estate in Ireland.

Today's largest telescopes have mirrors up to some 10 meters (33 feet) in diameter, with quadruple the light-gathering power of the legendary 5-meter Hale Telescope at Palomar Observatory in southern California. Looming as large as office buildings, some of these giants are so highly automated that they can dust off their optics at sundown, open the dome, sequence and carry out observations throughout the night, and shut down come threatening weather, all with little or no human intervention. Yet humans, being human, still intervene a lot, if only to make sure nothing goes awry: to lose just one night's work at a big telescope these days is to squander as much as $100,000 in operating costs.

Three of today's largest telescopes—Gemini North, Subaru, and Keck—stand within hailing distance of one another atop the nearly 14,000-foot peak of Hawaii's Mauna Kea, an inactive volcano. The altitude puts them above 40 percent of Earth's atmosphere—and most of its water vapor, which is opaque to the infrared wavelengths the astronomers like to study—but also makes it difficult for the astronomers and engineers who work there to breathe and think. Many wear clear plastic oxygen tubes in their nostrils as routinely as we might wear eyeglasses. Others rely on the body's ability to adapt but worry about making what they call a CLM, or "career-limiting mistake." "At altitude, we don't improvise; that would be a disaster," says the Gemini astronomer Scott Fisher. "We're kind of trained monkeys up here. The real thinking goes on at sea level."

These big Mauna Kea observatories are similarly smart and costly, yet each exudes a distinct personality. The 8.1-meter Gemini telescope is housed in an onion-shaped silver dome ringed by a set of shutters that, when closed during the day, make the observatory look as ungainly as a fat man in an inner tube. But the shutters open at dusk to create an enormous set of windows, three stories tall and stretching nearly three-quarters of the way around the observatory, that let in the night air and happen to afford a panorama of the blue Pacific all the way to Maui and beyond. Gemini's four main digital detectors—cameras and spectrometers, as heavy as cars and costing around $5 million each—are attached to a carousel surrounding the telescope's focal point, where they can be rotated into place in minutes. Computers run the telescope by night, shuffling requested observations to make the most of every minute. "We're all about nighttime efficiency," says Fisher.

The Subaru telescope's instruments are housed in alcoves like jeroboams of champagne in a heavenly wine cellar. (The comparison is not entirely fanciful; one leading Japanese astronomer propitiates the gods at the start of each Subaru observing run by pouring vintage sake on the ground outside the dome at the four points of the compass.) When a particular instrument is required, a robotic yellow trolley makes its way to the alcove, picks up the detector, ferries it to the bottom of the massive telescope, and locks it in place, attaching the data cables and the plumbing for the detector's refrigeration system. Subaru happens to be one of the few giant telescopes that anybody has ever actually looked through. For its inauguration in 1999, an eyepiece was attached so that Princess Sayako of Japan could have a look through the scope, and for several nights thereafter eager Subaru staffers did the same. "Everything you can see in the Hubble Space Telescope photos—the colors, the knots in the clouds—I could see with my own eyes, in stunning Technicolor," one recalled.

Keck consists of two identical telescopes. Both have 10-meter mirrors made of thirty-six segments; with its support structure, each segment weighs close to a thousand pounds, costs close to a million dollars, and would suffice to create a fine, university-grade telescope on its own. The telescopes' "tubes" are spindly steel skeletons that look as delicate as spiders' webs but are more precisely configured than a racing sloop's rigging. "We use the telescope's mission to motivate ourselves," one Keck astronomer told me. "If a little wire or something is found intruding into the optical path, we think, If the light has been traveling through space for 90 percent of the history of the universe, and it got this close to the telescope, we'd better make sure it gets the rest of the way."

Few of the astronomers awarded time on the big telescopes actually go there to observe anymore. Most submit their requests electronically—on a recent night at Gemini, the scheduled projects ranged from "Primordial Solar System Masses" to "Magnetic Activity in Ultracool Dwarfs"—and the results are sent back to them. Geoff Marcy, a modern-day Prince Henry the Navigator, whose team has discovered more than 150 planets orbiting stars other than our sun, gets more observing time than most at Keck but has not been there for years. Instead, his extrasolar-planet team observes from a remote operating facility at UC Berkeley. During observing runs, Marcy reports, "we settle into a routine of working all night. We have all our books and other resources here at hand, plus enough normal life so our spouses don't forget us."

In addition to their unprecedented light-gathering power, today's big telescopes benefit from their adaptive optics (AO) systems, which compensate for atmospheric turbulence. The turbulence is what makes stars glitter; telescopes magnify every twinkle. A typical AO system fires a laser beam into a thin layer of sodium atoms 56 miles high in the atmosphere, causing them to glow. By monitoring this artificial star, the system determines how the air is churning and adjusts the telescope's optics more than a thousand times each second to compensate. Gemini pays a pair of students ten dollars an hour to sit outside the dome all night, walkie-talkies in hand, ready to warn the astronomers to turn off the laser should an airplane approach. "It's incredible to see in practice," says Scott Fisher. "When the AO system is off, you see a nice, pretty star that looks a little fuzzy. Turn the AO on, and the star just goes *phonk!* and collapses to a tiny point."

Objects in the night sky are measured in degrees, the full moon spanning about one half of a degree. Without AO, a powerful telescope on a fine night can perceive objects separated from each other by as little as one 3,600th of a degree, or one arc second. Thanks to Keck's AO system, UCLA astronomer Andrea Ghez was able to make a motion picture of seven bright stars whirling around the invisible black hole at the center of our galaxy over a period of fourteen years: the entire movie takes place *inside* a box measuring only one arc second on a side. Based on the frenzy of the stars in the grip of the black hole, Ghez calculated that it has the mass of 4 million suns, generating enough gravitational force to sling-shot some stars that pass too close right out of our galaxy. Several such hypervelocity stars have been located, speeding off toward the depths of intergalactic space like party crashers ejected from an exclusive nightclub.

What's next? Even bigger telescopes, of course, with the capability to shoot cosmic pictures faster, wider, and in even greater detail. Among the behemoths due to come on line within a decade are the Giant Magellan Telescope, the Thirty Meter Telescope, and the 42-meter European Extremely Large Telescope—a scaled-down version of the 100-meter Overwhelmingly Large Telescope, which was tabled at the planning stage when its projected budget turned out to be overwhelming too.

Particularly innovative is the Large Synoptic Survey Telescope, or LSST, whose 8.4-meter primary mirror was cast last August in a spinning furnace under the stands of the University of Arizona Wildcats' football stadium in Tucson. (The rotation technique produces a mirror blank that is already concave, reducing the amount of glass that must be ground away to bring the mirror to a proper figure.) Conventional telescopes have narrow fields of view, typically spanning no more than half a degree on a side—much too narrow to take in the enormous patterns that grew out of the big bang. The LSST will have a field of view covering 10 square degrees, the area of fifty full moons. From its site in the Chilean Andes, it will be able to image galaxies far across the universe in exposures of just 15 seconds each, capturing fleeting events to distances of over 10 billion light-years, 70 percent of the way across the observable universe. "Since we'll have a big field of view, we can take a whole lot of short exposures and—*bang, bang, bang, bang*—cover the entire visible sky every several nights, and then repeat," says LSST Director Tony Tyson. "If you keep doing that for ten years, you have a movie—the first movie of the universe."

The LSST's fast, wide-angle imaging could help answer two of the biggest questions confronting astronomers today: the nature of dark matter and the nature of dark energy. Dark matter makes its presence known by its gravitational attraction—it explains the rotation speed of galaxies—but it emits no light, and its constitution is unknown. Dark energy is the name given to the mysterious phenomenon that for the past 5 billion years has been accelerating the rate at which the universe expands. "It's a little bit scary," says Tyson, "as if you were flying an airplane and suddenly something unknown took over the controls."

The LSST could help solve these immense riddles thanks in part, oddly enough, to the science of acoustics. The big bang was *noisy*. Although sound cannot propagate through the vacuum of today's space—as pedants are fond of reminding the directors of science-fiction films—the early universe was a thick plasma and as alive with sound as a drummers' convention. Certain tones resonated in the primordial plasma like the tones of struck wineglasses, and these harmonies, etched into sheets of galaxies that today shamble across billions of light-years, contain precise information about the nature of dark matter and dark energy. If astronomers can map these large-scale structures accurately, they should be able to identify the signatures of dark matter and dark energy in the big bang's harmonics. The Sloan Digital Sky Survey, a pioneering wide-angle study, captured some of this information when it mapped the sky from 1999 through 2008. The LSST is designed to go much deeper into cosmic space. It may not resolve the mysteries, but, predicts Tyson, "it will go a long way toward showing what dark energy and dark matter aren't."

The LSST's photographic "speed" will also give astronomers a better look at events too short-lived to be readily studied today. Most astronomers, even amateurs using backyard telescopes and off-the-shelf digital cameras, regularly record fleeting events of unknown origin. You take a series of digital exposures, and in one of them a spot of light appears where none was before or after. It may have been a cosmic ray hitting the light-detection chip, a high-velocity asteroid hurtling through the field of view, or a blue flare on the surface of a dim red star. You just don't know, so you shrug and move on. Because the LSST will take so many repeat exposures of the entire sky, it could resolve many such riddles.

Tomorrow's enormous telescopes will do as much in one night as today's do in a year, but that will not necessarily render the older telescopes obsolete. When the giants come on line, says Scott Fisher, "the Geminis of today will become the telescopes that get to go out and do the surveys," finding interesting phenomena for the largest scopes to investigate in detail. "It's like a pyramid, and it feeds both ways: when a really big telescope finds something exciting that we can't spend every night observing, the astronomers can apply for time on a smaller telescope to, say, check it out every clear night for a year and see how it changes over time."

Orbiting space telescopes are opening up another dimension. NASA's Kepler satellite, which launched in March 2009, is methodically imaging the constellation Cygnus, looking for the slight dimming of light caused when planets—some perhaps Earthlike —transit in front of their stars; Geoff Marcy's team will then use Keck to scrutinize stars flagged by Kepler to confirm that they have planets. In the future, pairs of mirrors deployed in orbit and linked by laser-ranging systems could attain the resolving power of telescopes measuring thousands of meters across. One day, observatories sitting in craters on the far side of the moon may probe the universe from surroundings ideally quiet, dark, and cold. The coming combination of smart satellites talking to big, increasingly automated ground telescopes, themselves linked together by fiber-optic networks and employing artificial intelligence systems to search out patterns in the torrents of data, suggests a process as much biological as mechanical, akin to the evolution of global eyes, optic nerves, and brains.

Film directors like to say that each movie is really two movies—the one you make and the one you say you're going to make while raising the money. The point is that nobody can accurately predict the outcome of any genuinely creative venture. The same is true of scientific discovery: scientists can explain what they expect to accomplish with bigger and better telescopes, but such predictions are mostly just extrapolations from the past. "If you're going to Washington to seek funding for a new telescope and you make a list of what you'll see through this new window on the universe, you know that the most interesting thing it will discover is probably not on your list," says Tyson. "It's likely to be something totally new, some out-of-the-box physics that's going to blow our minds."

The marvelous model of the big-bang universe pieced together in the twentieth century arose largely from just such unanticipated discoveries. Edwin Hubble discovered the expansion of the universe accidentally, at the telescope: cosmic expansion had been implied by Einstein's general theory of relativity, but Hubble knew nothing of the prediction, and not even Einstein had taken it seriously. Dark matter was discovered accidentally; so was dark energy. A telescope doesn't just show you what's out there; it impresses upon you how little you know, opening your imagination to wonders as big as all outdoors. "The spyglass is very truthful," said Galileo.

TIMOTHY FERRIS

Seeking New Earths

FROM *National Geographic*

IT TOOK HUMANS thousands of years to explore our own planet and centuries to comprehend our neighboring planets, but nowadays new worlds are being discovered every week. To date, astronomers have identified more than 370 "exoplanets," worlds orbiting stars other than the sun. Many are so strange as to confirm the biologist J. B. S. Haldane's famous remark that "the universe is not only queerer than we suppose, but queerer than we *can* suppose." There's an Icarus-like "hot Saturn" 260 light-years from Earth, whirling around its parent star so rapidly that a year there lasts less than three days. Circling another star 150 light-years out is a scorched "hot Jupiter," whose upper atmosphere is being blasted off to form a gigantic, cometlike tail. Three benighted planets have been found orbiting a pulsar—the remains of a once mighty star shrunk to a spinning atomic nucleus the size of a city—while untold numbers of worlds have evidently fallen into their suns or been flung out of their systems to become "floaters" that wander in eternal darkness.

Amid such exotica, scientists are eager for a hint of the familiar: planets resembling Earth, orbiting their stars at just the right distance—neither too hot nor too cold—to support life as we know it. No planets quite like our own have yet been found, presumably because they're inconspicuous. To see a planet as small and dim as ours amid the glare of its star is like trying to see a firefly in a fireworks display; to detect its gravitational influence on the star is like listening for a cricket in a tornado. Yet by pushing technology to the limits, astronomers are rapidly approaching the day when they can find another Earth and interrogate it for signs of life.

Only eleven exoplanets, all of them big and bright and conveniently far away from their stars, have as yet had their pictures taken. Most of the others have been detected by using the spectroscopic Doppler technique, in which starlight is analyzed for evidence that the star is being tugged ever so slightly back and forth by the gravitational pull of its planets. In recent years astronomers have refined the Doppler technique so exquisitely that they can now tell when a star is pulled from its appointed rounds by only one meter a second—about human walking speed. That's sufficient to detect a giant planet in a big orbit or a small one if it's very close to its star, but not an Earth at anything like our Earth's 93-million-mile distance from its star. Earth tugs the sun around at only one-tenth walking speed, or about the rate that an infant can crawl; astronomers cannot yet prize out so tiny a signal from the light of a distant star.

Another approach is to watch a star for the slight periodic dip in its brightness that will occur should an orbiting planet circle in front of it and block a fraction of its light. At most a tenth of all planetary systems are likely to be oriented so that these mini-eclipses, called transits, are visible from Earth, which means that astronomers may have to monitor many stars patiently to capture just a few transits. The French COROT satellite, now in the third and final year of its prime mission, has discovered seven transiting exoplanets, one of which is only 70 percent larger than Earth.

The United States' Kepler satellite is COROT's more ambitious successor. Launched from Cape Canaveral last March, Kepler is essentially just a big digital camera with a .95-meter aperture and a 95-megapixel detector. It makes wide-field pictures every thirty minutes, capturing the light of more than 100,000 stars in a single patch of sky between the bright stars Deneb and Vega. Computers on Earth monitor the brightness of all those stars over time, alerting humans when they detect the slight dimming that could signal the transit of a planet.

Because that dimming can be mimicked by other phenomena, such as the pulsations of a variable star or a large sunspot moving across a star's surface, the Kepler scientists won't announce the presence of a planet until they have seen it transit at least three times—a wait that may be only a few days or weeks for a planet rapidly circling close to its star but years for a terrestrial twin. By combining Kepler results with Doppler observations, astronomers expect to determine the diameters and masses of transiting planets. If they manage to discover a rocky planet roughly the size of Earth orbiting in the habitable zone—not so close to the star that the planet's water has been baked away nor so far out that it has frozen into ice—they will have found what biologists believe could be a promising abode for life.

The best hunting grounds may be dwarf stars, smaller than the sun. Such stars are plentiful (seven of the ten stars nearest to Earth are M dwarfs), and they enjoy long, stable careers, providing a steady supply of sunlight to any life-bearing planets that might occupy their habitable zones. Most important for planet hunters, the dimmer the star, the closer in its habitable zone it lies—dim dwarf stars are like small campfires, where campers must sit close to be comfortable—so transit observations will pay off more quickly. A close-in planet also exerts a stronger pull on its star, making its presence easier to confirm with the Doppler method. Indeed, the most promising planet yet found—the "super Earth" Gliese 581 d, seven times Earth's mass—orbits in the habitable zone of a red dwarf star only a third the mass of the sun.

Should Earthlike planets be found within the habitable zones of other stars, a dedicated space telescope designed to look for signs of life there might one day take a spectrum of the light coming from each planet and examine it for possible biosignatures such as atmospheric methane, ozone, and oxygen, or for the "red edge" produced when chlorophyll-containing photosynthetic plants reflect red light. Directly detecting and analyzing the planet's own light, which might be one ten-billionth as bright as the star's, would be a tall order. But when a planet transits, starlight shining through the atmosphere could reveal clues to its composition that a space telescope might be able to detect.

While grappling with the daunting technological challenge of performing a chemical analysis of planets they cannot even see, scientists searching for extraterrestrial life must keep in mind that it may be very different from life here at home. The lack of the red edge, for instance, might not mean a terrestrial exoplanet is lifeless: life thrived on Earth for billions of years before land plants appeared and populated the continents. Biological evolution is so inherently unpredictable that even if life originated on a planet identical to Earth at the same time it did here, life on that planet today would almost certainly be very different from terrestrial life.

As the biologist Jacques Monod once put it, life evolves not only through necessity—the universal workings of natural law—but also through chance, the unpredictable intervention of countless accidents. Chance has reared its head many times in our planet's history, dramatically so in the many mass extinctions that wiped out millions of species and, in doing so, created room for new life forms to evolve. Some of these baleful accidents appear to have been caused by comets or asteroids colliding with Earth—most recently the impact, 65 million years ago, that killed off the dinosaurs and opened up opportunities for the distant ancestors of human beings. Therefore scientists look not just for exoplanets identical to the modern Earth but for planets resembling Earth as it used to be or might have been. "The *modern* Earth may be the worst template we could use in searching for life elsewhere," notes Caleb Scharf, head of Columbia University's Astrobiology Center.

It was not easy for earlier explorers to plumb the depths of the oceans, map the far side of the moon, or discern evidence of oceans beneath the frozen surfaces of Jovian moons, and it will not be easy to find life on the planets of other stars. But we now have reason to believe that billions of such planets must exist and that they hold the promise of expanding not only the scope of human knowledge but also the richness of the human imagination.

For thousands of years we humans knew so little about the universe that we were apt to celebrate our imaginations and denigrate reality. (As the Spanish philosopher Miguel de Unamuno wrote, the mysticism of the religious visionaries of old arose from an "intolerable disparity between the hugeness of their desire and the smallness of reality.") Now, with advances in science, it has become gallingly evident that nature's creativity outstrips our own. The curtain is going up on countless new worlds with stories to tell.

PART TWO

Neurology Displacing Molecular Biology

JONAH LEHRER

Don't!

FROM *The New Yorker*

IN THE LATE NINETEEN SIXTIES, Carolyn Weisz, a four-year-old with long brown hair, was invited into a "game room" at the Bing Nursery School, on the campus of Stanford University. The room was little more than a large closet containing a desk and a chair. Carolyn was asked to sit down in the chair and pick a treat from a tray of marshmallows, cookies, and pretzel sticks. Carolyn chose the marshmallow. (Although she's now forty-four, Carolyn still has a weakness for those air-puffed balls of corn syrup and gelatin. "I know I shouldn't like them," she says. "But they're just so delicious!") A researcher then made Carolyn an offer: she could either eat one marshmallow right away or, if she was willing to wait while he stepped out for a few minutes, she could have two marshmallows when he returned. He said that if she rang a bell on the desk while he was away, he would come running back, and she could eat one marshmallow but would forfeit the second. Then he left the room.

Although Carolyn has no direct memory of the experiment, and the scientists would not release any information about the subjects, she strongly suspects that she was able to delay gratification. "I've always been really good at waiting," Carolyn told me. "If you give me a challenge or a task, then I'm going to find a way to do it, even if it means not eating my favorite food." Her mother, Karen Sortino, is still more certain: "Even as a young kid, Carolyn was very patient. I'm sure she would have waited." But her brother Craig, who also took part in the experiment, displayed less fortitude. Craig, a year older than Carolyn, still remembers the torment of trying to wait. "At a certain point, it must have occurred to me that I was all by myself," he recalls. "And so I just started taking all the candy." According to Craig, he was also tested with little plastic toys—he could have a second one if he held out—and he broke into the desk, where he figured there would be additional toys. "I took everything I could," he says. "I cleaned them out. After that, I noticed the teachers encouraged me to not go into the experiment room anymore."

Footage of these experiments, which were conducted over several years, is poignant, as the kids struggle to delay gratification for just a little bit longer. Some cover their eyes with their hands or turn around so that they can't see the tray. Others start kicking the desk, or tug on their pigtails, or stroke the marshmallow as if it were a tiny stuffed animal. One child, a boy with neatly parted hair, looks carefully around the room to make sure that nobody can see him. Then he picks up an Oreo, delicately twists it apart, and licks off the white cream filling before returning the cookie to the tray, a satisfied look on his face.

Most of the children were like Craig. They struggled to resist the treat and held out for an average of less than three minutes. "A few kids ate the marshmallow right away," Walter Mischel, the Stanford professor of psychology in charge of the experiment, remembers. "They didn't even bother ringing the bell. Other kids would stare directly at the marshmallow and then ring the bell thirty seconds later." About 30 percent of the children, however, were like Carolyn. They successfully delayed gratification until the researcher returned, some fifteen minutes later. These kids wrestled with temptation but found a way to resist.

The initial goal of the experiment was to identify the mental processes that allowed some people to delay gratification while others simply surrendered. After publishing a few papers on the Bing studies in the early seventies, Mischel moved on to other areas of personality research. "There are only so many things you can do with kids trying not to eat marshmallows."

But occasionally Mischel would ask his three daughters, all of whom attended the Bing, about their friends from nursery school. "It was really just idle dinnertime conversation," he says. "I'd ask them, 'How's Jane? How's Eric? How are they doing in school?'" Mischel began to notice a link between the children's academic performance as teenagers and their ability to wait for the second marshmallow. He asked his daughters to assess their friends academically on a scale of zero to five. Comparing these ratings with the original data set, he saw a correlation. "That's when I realized I had to do this seriously," he says. Starting in 1981, Mischel sent out a questionnaire to all the reachable parents, teachers, and academic advisers of the 653 subjects who had participated in the marshmallow task, who were by then in high school. He asked about every trait he could think of, from their capacity to plan and think ahead to their ability to "cope well with problems" and get along with their peers. He also requested their SAT scores.

Once Mischel began analyzing the results, he noticed that low delayers, the children who rang the bell quickly, seemed more likely to have behavioral problems, both in school and at home. They got lower SAT scores. They struggled in stressful situations, often had trouble paying attention, and found it difficult to maintain friendships. The child who could wait fifteen minutes had an SAT score that was, on average, 210 points higher than that of the kid who could wait only thirty seconds.

Carolyn Weisz is a textbook example of a high delayer. She attended Stanford as an undergraduate, and got her Ph.D. in social psychology at Princeton. She's now an associate psychology professor at the University of Puget Sound. Craig, meanwhile, moved to Los Angeles and has spent his career doing "all kinds of things" in the entertainment industry, mostly in production. He's currently helping to write and produce a film. "Sure, I wish I had been a more patient person," Craig says. "Looking back, there are definitely moments when it would have helped me make better career choices and stuff."

Mischel and his colleagues continued to track the subjects into their late thirties. Ozlem Ayduk, an assistant professor of psychology at the University of California at Berkeley, found that low-delaying adults have a significantly higher body-mass index and are more likely to have had problems with drugs—but it was frustrating to have to rely on self-reports. "There's often a gap between what people are willing to tell you and how they behave in the real world," he explains. And so last year Mischel, who is now a professor at Columbia, and a team of collaborators began asking the original Bing subjects to travel to Stanford for a few days of experiments in an fMRI machine. Carolyn says she will be participating in the scanning experiments later this summer; Craig completed a survey several years ago but has yet to be invited to Palo Alto. The scientists are hoping to identify the particular brain regions that allow some people to delay gratification and control their temper. They're also conducting a variety of genetic tests, as they search for the hereditary characteristics that influence the ability to wait for a second marshmallow.

If Mischel and his team succeed, they will have outlined the neural circuitry of self-control. For decades, psychologists have focused on raw intelligence as the most important variable when it comes to predicting success in life. Mischel argues that intelligence is largely at the mercy of self-control: even the smartest kids still need to do their homework. "What we're really measuring with the marshmallows isn't will power or self-control," Mischel says. "It's much more important than that. This task forces kids to find a way to make the situation work for them. They want the second marshmallow, but how can they get it? We can't control the world, but we can control how we think about it."

Walter Mischel is a slight, elegant man with a shaved head and a face of deep creases. He talks with a Brooklyn bluster, and he tends to act out his sentences, so that when he describes the marshmallow task he takes on the body language of an impatient four-year-old. "If you want to know why some kids can wait and others can't, then you've got to think like they think," Mischel says.

Mischel was born in Vienna in 1930. His father was a modestly successful businessman with a fondness for café society and Esperanto, while his mother spent many of her days lying on the couch with an ice pack on her forehead, trying to soothe her frail nerves. The family considered itself fully assimilated, but after the Nazi annexation of Austria in 1938, Mischel remembers being taunted in school by the Hitler Youth and watching as his father, hobbled by childhood polio, was forced to limp through the streets in his pajamas. A few weeks after the takeover, while the family was burning evidence of their Jewish ancestry in the fireplace, Walter found a long-forgotten certificate of U.S. citizenship issued to his maternal grandfather decades earlier, thus saving his family.

The family settled in Brooklyn, where Mischel's parents opened up a five-and-dime. Mischel attended New York University, studying poetry under Delmore Schwartz and Allen Tate, and taking studio-art classes with Philip Guston. He also became fascinated by psychoanalysis and new measures of personality, such as the Rorschach test. "At the time, it seemed like a mental X-ray machine," he says. "You could solve a person by showing them a picture." Although he was pressured to join his uncle's umbrella business, he ended up pursuing a Ph.D. in clinical psychology at Ohio State.

But Mischel noticed that academic theories had limited application, and he was struck by the futility of most personality science. He still flinches at the naiveté of graduate students who based their diagnoses on a battery of meaningless tests. In 1955 Mischel was offered an opportunity to study the "spirit possession" ceremonies of the Orisha faith in Trinidad, and he leapt at the chance. Although his research was supposed to involve the use of Rorschach tests to explore the connections between the unconscious and the behavior of people when possessed, Mischel soon grew interested in a different project. He lived in a part of the island that was evenly split between people of East Indian and of African descent; he noticed that each group defined the other in broad stereotypes. "The East Indians would describe the Africans as impulsive hedonists who were always living for the moment and never thought about the future," he says. "The Africans, meanwhile, would say that the East Indians didn't know how to live and would stuff money in their mattress and never enjoy themselves."

Mischel took young children from both ethnic groups and offered them a simple choice: they could have a miniature chocolate bar right away or, if they waited a few days, they could get a much bigger chocolate bar. Mischel's results failed to justify the stereotypes—other variables, such as whether or not the children lived with their father, turned out to be much more important—but they did get him interested in the question of delayed gratification. Why did some children wait and not others? What made waiting possible? Unlike the broad traits supposedly assessed by personality tests, self-control struck Mischel as potentially measurable.

In 1958 Mischel became an assistant professor in the Department of Social Relations at Harvard. One of his first tasks was to develop a survey course on "personality assessment," but Mischel quickly concluded that while prevailing theories held personality traits to be broadly consistent, the available data didn't back up this assumption. Personality, at least as it was then conceived, couldn't be reliably assessed at all. A few years later, he was hired as a consultant on a personality assessment initiated by the Peace Corps. Early Peace Corps volunteers had sparked several embarrassing international incidents—one mailed a postcard on which she expressed disgust at the sanitary habits of her host country—so the Kennedy administration wanted a screening process to eliminate people unsuited for foreign assignments. Volunteers were tested for standard personality traits, and Mischel compared the results with ratings of how well the volunteers performed in the field. He found no correlation; the time-consuming tests predicted nothing. At this point, Mischel realized that the problem wasn't the tests—it was their premise. Psychologists had spent decades searching for traits that exist independently of circumstance, but what if personality can't be separated from context? "It went against the way we'd been thinking about personality since the four humors and the ancient Greeks," he says.

While Mischel was beginning to dismantle the methods of his field, the Harvard psychology department was in tumult. In 1960 the personality psychologist Timothy Leary helped start the Harvard Psilocybin Project, which consisted mostly of self-experimentation. Mischel remembers graduate students' desks giving way to mattresses, and large packages from Ciba chemicals, in Switzerland, arriving in the mail. Mischel had nothing against hippies, but he wanted modern psychology to be rigorous and empirical. And so, in 1962, Walter Mischel moved to Palo Alto and went to work at Stanford.

There is something deeply contradictory about Walter Mischel—a psychologist who spent decades critiquing the validity of personality tests—inventing the marshmallow task, a simple test with impressive predictive power. Mischel, however, insists there is no contradiction. "I've always believed there are consistencies in a person that can be looked at," he says. "We just have to look in the right way." One of Mischel's classic studies documented the aggressive behavior of children in a variety of situations at a summer camp in New Hampshire. Most psychologists assumed that aggression was a stable trait, but Mischel found that children's responses depended on the details of the interaction. The same child might consistently lash out when teased by a peer but readily submit to adult punishment. Another might react badly to a warning from a counselor but play well with his bunkmates. Aggression was best assessed in terms of what Mischel called "if-then patterns." If a certain child was teased by a peer, then he would be aggressive.

One of Mischel's favorite metaphors for this model of personality, known as interactionism, concerns a car making a screeching noise. How does a mechanic solve the problem? He begins by trying to identify the specific conditions that trigger the noise. Is there a screech when the car is accelerating, or when it's shifting gears, or turning at slow speeds? Unless the mechanic can give the screech a context, he'll never find the broken part. Mischel wanted psychologists to think like mechanics and look at people's responses under particular conditions. The challenge was devising a test that accurately simulated something relevant to the behavior being predicted. The search for a meaningful test of personality led Mischel to revisit, in 1968, the protocol he'd used on young children in Trinidad nearly a decade earlier. The experiment seemed especially relevant now that he had three young daughters of his own. "Young kids are pure id," Mischel says. "They start off unable to wait for anything—whatever they want they need. But then, as I watched my own kids, I marveled at how they gradually learned how to delay and how that made so many other things possible."

A few years earlier, in 1966, the Stanford psychology department had established the Bing Nursery School. The classrooms were designed as working laboratories, with large one-way mirrors that allowed researchers to observe the children. This past February, Jennifer Winters, the assistant director of the school, showed me around the building. While the Bing is still an active center of research—the children quickly learn to ignore the students scribbling in notebooks—Winters isn't sure that Mischel's marshmallow task could be replicated today. "We recently tried to do a version of it, and the kids were very excited about having food in the game room," she says. "There are so many allergies and peculiar diets today that we don't do many things with food."

Mischel perfected his protocol by testing his daughters at the kitchen table. "When you're investigating will power in a four-year-old, little things make a big difference," he says. "How big should the marshmallows be? What kind of cookies work best?" After several months of patient tinkering, Mischel came up with an experi-

mental design that closely simulated the difficulty of delayed gratification. In the spring of 1968, he conducted the first trials of
his experiment at the Bing. "I knew we'd designed it well when a
few kids wanted to quit as soon as we explained the conditions to
them," he says. "They knew this was going to be very difficult."

At the time, psychologists assumed that children's ability to wait
depended on how badly they wanted the marshmallow. But it soon
became obvious that every child craved the extra treat. What, then,
determined self-control? Mischel's conclusion, based on hundreds
of hours of observation, was that the crucial skill was the "strategic
allocation of attention." Instead of getting obsessed with the marshmallow—the "hot stimulus"—the patient children distracted
themselves by covering their eyes, pretending to play hide-and-seek
underneath the desk, or singing songs from *Sesame Street*. Their desire wasn't defeated—it was merely forgotten. "If you're thinking
about the marshmallow and how delicious it is, then you're going
to eat it," Mischel says. "The key is to avoid thinking about it in the
first place."
    In adults, this skill is often referred to as metacognition, or
thinking about thinking, and it's what allows people to outsmart
their shortcomings. (When Odysseus had himself tied to the ship's
mast, he was using some of the skills of metacognition: knowing he
wouldn't be able to resist the Sirens' song, he made it impossible
to give in.) Mischel's large data set from various studies allowed
him to see that children with a more accurate understanding of
the workings of self-control were better able to delay gratification.
"What's interesting about four-year-olds is that they're just figuring
out the rules of thinking," Mischel says. "The kids who couldn't
delay would often have the rules backward. They would think that
the best way to resist the marshmallow is to stare right at it, to keep
a close eye on the goal. But that's a terrible idea. If you do that,
you're going to ring the bell before I leave the room."
    According to Mischel, this view of will power also helps explain
why the marshmallow task is such a powerfully predictive test. "If
you can deal with hot emotions, then you can study for the SAT instead of watching television," Mischel says. "And you can save more
money for retirement. It's not just about marshmallows."
    Subsequent work by Mischel and his colleagues found that these
differences were observable in subjects as young as nineteen

months. Looking at how toddlers responded when briefly separated from their mothers, they found that some immediately burst into tears or clung to the door, but others were able to overcome their anxiety by distracting themselves, often by playing with toys. When the scientists set the same children the marshmallow task at the age of five, they found that the kids who had cried also struggled to resist the tempting treat.

The early appearance of the ability to delay suggests that it has a genetic origin, an example of personality at its most predetermined. Mischel resists such an easy conclusion. "In general, trying to separate nature and nurture makes about as much sense as trying to separate personality and situation," he says. "The two influences are completely interrelated." For instance, when Mischel gave delay-of-gratification tasks to children from low-income families in the Bronx, he noticed that their ability to delay was below average, at least compared with that of children in Palo Alto. "When you grow up poor, you might not practice delay as much," he says. "And if you don't practice, then you'll never figure out how to distract yourself. You won't develop the best delay strategies, and those strategies won't become second nature." In other words, people learn how to use their mind just as they learn how to use a computer: through trial and error.

But Mischel has found a shortcut. When he and his colleagues taught children a simple set of mental tricks—such as pretending that the candy is only a picture, surrounded by an imaginary frame —he dramatically improved their self-control. The kids who hadn't been able to wait sixty seconds could now wait fifteen minutes. "All I've done is given them some tips from their mental user manual," Mischel says. "Once you realize that will power is just a matter of learning how to control your attention and thoughts, you can really begin to increase it."

Marc Berman, a lanky graduate student with an easy grin, speaks about his research with the infectious enthusiasm of a freshman taking his first philosophy class. Berman works in the lab of John Jonides, a psychologist and neuroscientist at the University of Michigan, who is in charge of the brain-scanning experiments on the original Bing subjects. He knows that testing forty-year-olds for self-control isn't a straightforward proposition. "We can't give these people marshmallows," Berman says. "They know they're part of a long-term study that looks at delay of gratification, so if you give them an obvious delay task they'll do their best to resist. You'll get a bunch of people who refuse to touch their marshmallow."

This meant that Jonides and his team had to find a way to measure will power indirectly. Operating on the premise that the ability to delay eating the marshmallow had depended on a child's ability to banish thoughts of it, they decided on a series of tasks that measure the ability of subjects to control the contents of working memory—the relatively limited amount of information we're able to consciously consider at any given moment. According to Jonides, this is how self-control "cashes out" in the real world: as an ability to direct the spotlight of attention so that our decisions aren't determined by the wrong thoughts.

Last summer the scientists chose fifty-five subjects, equally split between high delayers and low delayers, and sent each one a laptop computer loaded with working-memory experiments. Two of the experiments were of particular interest. The first is a straightforward exercise known as the "suppression task." Subjects are given four random words, two printed in blue and two in red. After reading the words, they're told to forget the blue words and remember the red words. Then the scientists provide a stream of "probe words" and ask the subjects whether the probes are the words they were asked to remember. Though the task doesn't seem to involve delayed gratification, it tests the same basic mechanism. Interestingly, the scientists found that high delayers were significantly better at the suppression task: they were less likely to think that a word they'd been asked to forget was something they should remember.

In the second, known as the Go/No Go task, subjects are flashed a set of faces with various expressions. At first, they are told to press the space bar whenever they see a smile. This takes little effort, since smiling faces automatically trigger what's known as "approach behavior." After a few minutes, however, subjects are told to press the space bar when they see frowning faces. They are now being forced to act against an impulse. Results show that high delayers are more successful at not pressing the button in response to a smiling face.

When I first started talking to the scientists about these tasks last summer, they were clearly worried that they wouldn't find any behavioral differences between high and low delayers. It wasn't until early January that they had enough data to begin their analysis (not surprisingly, it took much longer to get the laptops back from the low delayers), but it soon became obvious that there were provocative differences between the two groups. A graph of the data shows that as the delay time of the four-year-olds decreases, the number of mistakes made by the adults sharply rises.

The big remaining question for the scientists is whether these behavioral differences are detectable in an fMRI machine. Although the scanning has just begun—Jonides and his team are still working out the kinks—the scientists sound confident. "These tasks have been studied so many times that we pretty much know where to look and what we're going to find," Jonides says. He rattles off a short list of relevant brain regions, which his lab has already identified as being responsible for working-memory exercises. For the most part, the regions are in the frontal cortex—the overhang of brain behind the eyes—and include the dorsolateral prefrontal cortex, the anterior prefrontal cortex, the anterior cingulate, and the right and left inferior frontal gyri. While these cortical folds have long been associated with self-control, they're also essential for working memory and directed attention. According to the scientists, that's not an accident. "These are powerful instincts telling us to reach for the marshmallow or press the space bar," Jonides says. "The only way to defeat them is to avoid them, and that means paying attention to something else. We call that will power, but it's got nothing to do with the will."

The behavioral and genetic aspects of the project are overseen by Yuichi Shoda, a professor of psychology at the University of Washington, who was one of Mischel's graduate students. He's been following these "marshmallow subjects" for more than thirty years: he knows everything about them, from their academic records and their social graces to their ability to deal with frustration and stress. The prognosis for the genetic research remains uncertain. Although many studies have searched for the underpinnings of personality since the completion of the Human Genome Project in 2003, many of the relevant genes remain in question. "We're incredibly complicated creatures," Shoda says. "Even the simplest aspects of personality are driven by dozens and dozens of different genes." The scientists have decided to focus on genes in the dopamine pathways, since those neurotransmitters are believed to regulate both motivation and attention. However, even if minor

coding differences influence delay ability—and that's a likely possibility—Shoda doesn't expect to discover these differences: the
sample size is simply too small.

In recent years, researchers have begun making house visits to
many of the original subjects, including Carolyn Weisz, as they try
to better understand the familial contexts that shape self-control.
"They turned my kitchen into a lab," Carolyn told me. "They set up
a little tent where they tested my oldest daughter on the delay task
with some cookies. I remember thinking, I really hope she can
wait."

While Mischel closely follows the steady accumulation of data from
the laptops and the brain scans, he's most excited by what comes
next. "I'm not interested in looking at the brain just so we can use a
fancy machine," he says. "The real question is what can we do with
this fMRI data that we couldn't do before?" Mischel is applying for
an NIH grant to investigate various mental illnesses, like obsessive-
compulsive disorder and attention-deficit disorder, in terms of the
ability to control and direct attention. Mischel and his team hope
to identify crucial neural circuits that cut across a wide variety of
ailments. If there is such a circuit, then the same cognitive tricks
that increase delay time in a four-year-old might help adults deal
with their symptoms. Mischel is particularly excited by the example
of the substantial subset of people who failed the marshmallow
task as four-year-olds but ended up becoming high-delaying adults.
"This is the group I'm most interested in," he says. "They have substantially improved their lives."

Mischel is also preparing a large-scale study involving hundreds
of schoolchildren in Philadelphia, Seattle, and New York City to see
if self-control skills can be taught. Although he previously showed
that children did much better on the marshmallow task after being
taught a few simple "mental transformations," such as pretending
the marshmallow was a cloud, it remains unclear if these new skills
persist over the long term. In other words, do the tricks work only
during the experiment or do the children learn to apply them at
home, when deciding between homework and television?

Angela Lee Duckworth, an assistant professor of psychology at
the University of Pennsylvania, is leading the program. She first
grew interested in the subject after working as a high school math
teacher. "For the most part, it was an incredibly frustrating experi-

ence," she says. "I gradually became convinced that trying to teach a teenager algebra when they don't have self-control is a pretty futile exercise." And so, at the age of thirty-two, Duckworth decided to become a psychologist. One of her main research projects looked at the relationship between self-control and grade point average. She found that the ability to delay gratification — eighth-graders were given a choice between a dollar right away or two dollars the following week — was a far better predictor of academic performance than IQ. She said that her study shows that "intelligence is really important, but it's still not as important as self-control."

Last year, Duckworth and Mischel were approached by David Levin, the cofounder of KIPP, an organization of sixty-six public charter schools across the country. KIPP schools are known for their long workday — students are in class from 7:25 A.M. to 5 P.M. — and for dramatic improvement of inner-city students' test scores. (More than 80 percent of eighth-graders at the KIPP academy in the South Bronx scored at or above grade level in reading and math, which was nearly twice the New York City average.) "The core feature of the KIPP approach is that character matters for success," Levin says. "Educators like to talk about character skills when kids are in kindergarten — we send young kids home with a report card about 'working well with others' or 'not talking out of turn.' But then, just when these skills start to matter, we stop trying to improve them. We just throw up our hands and complain."

Self-control is one of the fundamental "character strengths" emphasized by KIPP — the KIPP academy in Philadelphia, for instance, gives its students a shirt emblazoned with the slogan "Don't Eat the Marshmallow." Levin, however, remained unsure about how well the program was working — "We know how to teach math skills, but it's harder to measure character strengths," he says — so he contacted Duckworth and Mischel, promising them unfettered access to KIPP students. Levin also helped bring together additional schools willing to take part in the experiment, including Riverdale Country School, a private school in the Bronx; the Evergreen School for gifted children, in Shoreline, Washington; and the Mastery Charter Schools, in Philadelphia.

For the past few months, the researchers have been conducting pilot studies in the classroom as they try to figure out the most effective way to introduce complex psychological concepts to young children. Because the study will focus on students between the ages of four and eight, the classroom lessons will rely heavily on peer modeling, such as showing kindergartners a video of a child successfully distracting herself during the marshmallow task. The scientists have some encouraging preliminary results—after just a few sessions, students show significant improvements in the ability to deal with hot emotional states—but they are cautious about predicting the outcome of the long-term study. "When you do these large-scale educational studies, there are ninety-nine uninteresting reasons the study could fail," Duckworth says. "Maybe a teacher doesn't show the video, or maybe there's a field trip on the day of the testing. This is what keeps me up at night."

Mischel's main worry is that even if his lesson plan proves to be effective, it might still be overwhelmed by variables the scientists can't control, such as the home environment. He knows that it's not enough just to teach kids mental tricks—the real challenge is turning those tricks into habits, and that requires years of diligent practice. "This is where your parents are important," Mischel says. "Have they established rituals that force you to delay on a daily basis? Do they encourage you to wait? And do they make waiting worthwhile?" According to Mischel, even the most mundane routines of childhood—such as not snacking before dinner, or saving up your allowance, or holding out until Christmas morning—are really sly exercises in cognitive training: we're teaching ourselves how to think so that we can outsmart our desires. But Mischel isn't satisfied with such an informal approach. "We should give marshmallows to every kindergartner," he says. "We should say, 'You see this marshmallow? You don't have to eat it. You can wait. Here's how.'"

KATHLEEN McGOWAN

Out of the Past

FROM *Discover*

RITA MAGIL WAS DRIVING down a Montreal boulevard one
sunny morning in 2002 when a car came blasting through a red
light straight toward her. "I slammed the brakes, but I knew it was
too late," she says. "I thought I was going to die." The oncoming
car smashed into hers, pushing her off the road and into a building with large cement pillars in front. A pillar tore through the car,
stopping only about a foot from her face. She was trapped in the
crumpled vehicle, but to her shock, she was still alive.

The accident left Magil with two broken ribs and a broken col-
larbone. It also left her with posttraumatic stress disorder (PTSD)
and a desperate wish to forget. Long after her bones healed, Magil
was plagued by the memory of the cement barriers looming toward her. "I would be doing regular things—cooking something,
shopping, whatever—and the image would just come into my
mind from nowhere," she says. Her heart would pound; she would
start to sweat and feel jumpy all over. It felt visceral and real, like
something that was happening at that very moment.

Most people who survive accidents or attacks never develop
PTSD. But for some, the event forges a memory that is patho-
logically potent, erupting into consciousness again and again.
"PTSD really can be characterized as a disorder of memory," says
the McGill University psychologist Alain Brunet, who studies and
treats psychological trauma. "It's about what you wish to forget and
what you cannot forget." This kind of memory is not misty and wa-
tercolored. It is relentless.

More than a year after her accident, Magil saw Brunet's ad for an
experimental treatment for PTSD, and she volunteered. She took

a low dose of a common blood-pressure drug, propranolol, that reduces activity in the amygdala, a part of the brain that processes emotions. Then she listened to a taped re-creation of her car accident. She had relived that day in her mind a thousand times. The difference this time was that the drug broke the link between her factual memory and her emotional memory. Propranolol blocks the action of adrenaline, so it prevented her from tensing up and getting anxious. By having Magil think about the accident while the drug was in her body, Brunet hoped to permanently change how she remembered the crash. It worked. She did not forget the accident but was actively able to reshape her memory of the event, stripping away the terror while leaving the facts behind.

Brunet's experiment emerges from one of the most exciting and controversial recent findings in neuroscience: that we alter our memories just by remembering them. Karim Nader of McGill — the scientist who made this discovery — hopes it means that people with PTSD can cure themselves by editing their memories. Altering remembered thoughts might also liberate people imprisoned by anxiety, obsessive-compulsive disorder, even addiction. "There is no such thing as a pharmacological cure in psychiatry," Brunet says. "But we may be on the verge of changing that."

These recent insights into memory are part of a larger about-face in neuroscience research. Until recently, long-term memories were thought to be physically etched into our brain, permanent and unchanging. Now it is becoming clear that memories are surprisingly vulnerable and highly dynamic. In the lab they can be flicked on or dimmed with a simple dose of drugs. "For a hundred years, people thought memory was wired into the brain," Nader says. "Instead, we find it can be rewired — you can add false information to it, make it stronger, make it weaker, and possibly even make it disappear." Nader and Brunet are not the only ones to make this observation. Other scientists probing different parts of the brain's memory machinery are similarly finding that memory is inherently flexible.

Someday this new science of memory could cure PTSD and other mental traumas. Already it corrodes our trust in what we know and how we know it. It pokes holes in eyewitness testimony, in memoirs, in our most intimate records of truth. Every time we remember, it seems, we add new details, shade the facts, prune and tweak. Without realizing it, we continually rewrite the stories of our lives. Memory, it turns out, has a surprising amount in common with imagination, conjuring worlds that never existed until they were forged by our minds.

On the Trail of the Memory Meme

Neuroscientists have long viewed memory as a kind of neural architecture, a literal physical reshaping of the microstructure of the brain. In the nineteenth century, the pioneering neuroanatomist Santiago Ramón y Cajal theorized that information was processed in our heads each time an electrical impulse traveled across a synapse, the gap between one nerve cell and the next. Memories were made or altered, he proposed, when structures near the synapse changed.

More than a century later, the textbook description of episodic memory (conscious knowledge of an event) is a more sophisticated version of that same basic idea. Memory formation requires an elaborate chemical choreography of more than a hundred proteins, but the upshot is that sensory information, coded as electrical pulses, zips through neural networks of the brain. The impulses cause glutamate (one of the brain's main neurotransmitters) to pop out of one nerve cell and travel across the synapse to activate the next by binding to its receptors, chemically active signaling stations on the cell surface. Ultimately the electrical and chemical signals reach the centers of memory, the almond-size amygdala and the banana-shaped hippocampus, adjacent structures buried on either side of the brain.

Neuroscientists believe that memory forms when neurons in these key brain structures are simultaneously activated by glutamate and an electrical pulse, a result of everyday sensory experience. The experience triggers a biochemical riot, causing a specialized glutamate receptor, called NMDA, to spring open and allow calcium ions to flood the cells. The ions stimulate dozens of enzymes that reshape the cells by opening up additional receptors and by prompting the formation of more synapses and new protrusions that contain still more receptors and synapses. In aggregate, these changes make neurons more sensitive to each other and put the anatomical scaffold of a memory in place.

Enacting all these changes takes time, and for up to a few hours the memory is like wet concrete — solidifying but not quite set, still open to interference. Once the process is over, though, the memory is said to be "consolidated." In the textbook description, neuroscientists talk of memory the way geoscientists describe mountains—built through a dynamic process, but once established almost impossible to reshape quickly except by extraordinary means. By the late 1990s, this explanation of memory was so widely accepted by neuroscientists that its major author, the Columbia University neuroscientist Eric Kandel, was awarded the Nobel Prize. It seemed that the most important questions about memory had been answered.

No wonder, then, that Nader—at the time a young postdoc studying the neurobiology of fear at New York University—was electrified when he attended one of Kandel's lectures. "It was so beautiful and so convincing," Nader says. But he began to wonder: What actually happens when we recall the past? Does the very act of remembering undo what happened? Does a memory have to go through the consolidation process again? Nader asked his adviser, the noted fear researcher Joseph LeDoux, if he could study these questions. LeDoux says his initial response was "Don't waste our time and money," but Nader talked him into it, little suspecting just how far this line of research would go.

Meanwhile, doubts about the standard theory of memory were piling up in the world outside the neuroscience lab. In the early 1990s many people began reporting what seemed to be long-buried memories of childhood sexual abuse. These traumatic recollections frequently surfaced with the help of recovered-memory therapy techniques like hypnosis and guided imagery, in which patients are encouraged to visualize terrible experiences. Cognitive scientists suspected that some of these memories were bogus, the unwitting product of suggestion by the therapist. In support of this view, the psychologist Elizabeth Loftus, then of the University of Washington, proved how easy it is to implant a false memory, especially one that is plausible. In a famous experiment, she gave volunteers a booklet narrating three true stories of events from their own childhood along with an invented tale that described their getting lost in the mall at age five. When prompted later to write down all they could remember about the events, 25 percent were sure that all four events had actually happened to them.

Spurred on by the controversy over recovered memory, other cognitive scientists found that false memory is a normal phenomenon. David Rubin, who studies autobiographical memory at Duke University, observed that adult twins often disagree over who experienced something in childhood. Each might believe, for example, that he was the one to get pushed off his bike by a neighbor at age eight. Apparently, even the most basic facts about a past event (such as who experienced it) could be lost.

Even harrowing memories—the so-called flashbulb memories that feel as if they have been permanently seared into the brain —are not as accurate as we think. Less than a year after a cargo plane crashed into an Amsterdam apartment building in 1992, 55 percent of the Dutch population said they had watched the plane hit the building on TV. Many of them recalled specifics of the crash, such as the angle of descent, and could report whether or not the plane was on fire before it hit. But the event had not been caught on video. The "memory" shared by the majority was a hallucination, a convincing fiction pieced together out of descriptions and pictures of the event.

By the late 1990s, hundreds of psychology experiments suggested that the description of memory as a neurally encoded recapitulation of the past was so oversimplified as to completely miss the point. Instead of being a perfect movie of the past, psychologists found, memory is more like a shifting collage, a narrative spun out of scraps and constructed anew whenever recollection takes place. The science of memory was conflicted, with the neurobiological and psychological versions at odds. If a memory is wired into brain cells—a literal engraving of information—then why is it so easy to alter many years after the fact? It took an outsider to connect the dots.

Rewriting the Past

In the hierarchy of memory science, Karim Nader hardly ranked— a lowly postdoc, only thirty-three years old, and not even a memory researcher. But in 1999, inspired by Kandel's talk, he set out to satisfy his big questions about how we recall and forget through a simple experiment. Nader tweaked a standard method used in fear research, in which rats are trained to associate a tone with an electric shock to the foot. The animals quickly learn that the sound is bad news. If they hear it weeks later, they freeze in fear. It is an easy way for the experimenter to know that they remember what took place.

Nader trained some rats, then played the tone again fourteen days later, prompting them to remember. He also simultaneously injected them with a protein-synthesis inhibitor, which prevents new memory from forming by prohibiting alteration at the synapses. According to the standard model of memory, the chemical should have no effect, since the memory of the tone has already consolidated. In reality, the treated rats' memory disappeared. When Nader sounded the tone again later, the animals did not freeze. LeDoux was won over by this simple but powerful demonstration. In 2000 Nader's paper on reconsolidation sparked a commotion in the world of memory research. He showed that reactivating a memory destabilizes it, putting it back into a flexible, vulnerable state.

Immediately "reconsolidation" became a fighting word. The gossip Nader heard terrified him; some of the biggest bigwigs of memory research thought he had made a ludicrous mistake. "I had no idea how much of a backlash there was going to be," he says. Even so, Nader kept at his experiments, and in the fall of 2001, he was scheduled to present his research at a huge Society for Neuroscience meeting. It would be his moment of truth, his one chance to persuade the field to take his finding seriously. "I knew the old guard was saying, 'This sucks; it's all crap,'" he says. "I knew if I didn't hit a grand slam, this thing was dead." The talk drew an overflow crowd of more than a thousand, including the legend himself, Eric Kandel. ("I really wanted to die," Nader says.)

That day, by addressing the major criticisms of his research, Nader managed to convince his colleagues that memory reconsolidation was at least worth a serious look. Various labs took on the challenge, soon repeating his findings and discovering that many types of memory in many different species reconsolidate. Other groups began teasing out the reconsolidation process molecule by molecule. Nader's group found that the NMDA glutamate receptor—which solidifies memory—also is involved in destabilizing it. A group led by Sue-Hyun Lee at Seoul National University demonstrated that proteins must be actively dismantled to destabilize a memory, more evidence that the old memory is actually changed as it is recalled.

Brain researchers are still grappling with the implications of this idea, trying to figure out exactly how malleable memory really is. "People are willing to say we have to go back to the drawing board," says LeDoux, whose group has also continued to study reconsolidation. At the 2008 Society for Neuroscience meeting in Washington, D.C., forty-three presentations focused on reconsolidation, and Nader was besieged by students and young researchers eager to talk.

With this new understanding of memory has come the even more startling possibility of new ways to control it: the era of memory treatment has arrived. For Rita Magil, who got just two doses of propranolol over the course of a single day, the results were encouraging. Her heart rate and muscle tension eased while the drug was in her body. She sensed the difference too. "I felt more detached from it," she says. "I felt that I was relating a narrative rather than describing something right in front of me right now." After the study was over, the flashbacks returned, though with less intensity. For her, the only real cure was time.

Six-session treatments with a total of twelve doses of propranolol have shown better results. Collaborating with the Harvard psychiatrist Roger Pitman, who was the first to try propranolol for posttraumatic stress, the McGill group has treated about forty-five PTSD patients, ranging from soldiers to rape victims. Most had been suffering for years. But after the longer treatment, their symptoms declined by half and stayed that way even six months afterward. They still remember what happened, but it is less disturbing. "They say: 'I'm not thinking about it as much. It just doesn't bother me as much anymore,'" Brunet says. As a group, they are considered to be in remission.

The researchers must still prove that the improvement will last. If it does, it could offer rare hope to millions of people with PTSD, a disorder from which only a third completely recover.

Brunet hopes that similar treatments can address other psychiatric problems, too. Anxiety, acquired phobias, and addiction are increasingly described as disorders of emotional memory. An overly powerful fear memory, for example, can crystallize into a phobia, in which a relatively safe experience like flying in a plane is inextricably linked to a feeling of extreme danger. No matter how the phobic person tries, his emotional memory refuses to update itself to incorporate reassuring information. A treatment that restores his emotional memory to a flexible state could help him cope.

Addiction is another kind of pathological remembering, but in this case the memory is pleasurable. Just as adrenaline sears emotional memories into the brain with the help of the amygdala, drugs of abuse enlist the amygdala and the brain's reward centers to forge unforgettable memories of pleasure. Anything connected to the bliss reawakens the memory, in the form of craving. "When you see someone with a beer and a smoke and you get a craving, you are suffering from reminiscence, from an emotional memory," Brunet says. Adapting experimental methods of forgetting to addiction might make it easier to quit.

The Reconsolidated Life

While neuroscientists were skeptical of Nader's findings, cognitive scientists were immediately fascinated that memory might be constantly revamped. It certainly seemed to explain their observations: The home run you hit in Little League? Your first kiss? As you replay these memories, you reawaken and reconsolidate them hundreds of times. Each time, you replace the original with a slightly modified version. Eventually you are not really remembering what happened; you are remembering your story about it. "Reconsolidation suggests that when you use a memory, the one you had originally is no longer valid or maybe no longer accessible," LeDoux says. "If you take it to the extreme, your memory is only as good as your last memory. The fewer times you use it, the more pristine it is. The more you use it, the more you change it." We've all had the experience of repeating a dramatic story so many times that the events seem dead, as if they came from a novel rather than real life. This might be reconsolidation at work.

Reconsolidation research has helped foster a growing sense that the flexibility of memory might be functional—an advantage rather than a bug in the brain. Reconsolidation might be how we update our store of knowledge, by making old memories malleable in response to new information. "When you encounter a familiar experience, you are remembering the original memory at the same time, and the new experience somehow gets blended in," says Jonathan Lee of the University of Birmingham in England, who recently found evidence for this effect in animals. "That is essentially what reconsolidation is." The evident purpose of episodic memory, after all, is to store facts in the hope of anticipating what might happen next. From the perspective of survival, constructive memory is an asset. It allows you to pull together scraps of information to simulate the future on the fly.

"The brain knows there is a future," says the neuroscientist Yadin Dudai, head of the department of neurobiology at the Weizmann Institute of Science in Israel, who collaborates with Nader and LeDoux. Facing something new, we want to link the novel information with memories to better interpret the situation. If the side effect is a few mistakes, that is probably a small price to pay. "Having a memory that is too accurate is not always good," he says.

Put another way, memory and imagination are two sides of the same coin. Like memory, imagination allows you to put yourself in a time and place other than the one you actually occupy. This isn't just a clever analogy: in recent neuroimaging studies, the Harvard psychologist Daniel Schacter has shown that remembering and imagining mobilize many of the same brain circuits. "When people are instructed to imagine events that might happen in their personal future and then to remember actual events in the past, we find extensive and very striking overlap in areas of brain activation," he says. Other researchers have found that people with severe amnesia lose their ability to imagine. Without memory, they can barely picture the future at all.

The Spotless Mind

Reconsolidation modifies old memories, but other new research points the way toward erasing them wholesale. One technique for blanking out the past, developed by Joe Tsien at the Medical College of Georgia, flows from his studies of memory formation. When calcium floods a neuron as a memory is formed, it turns on an enzyme called CaMkII (calcium/calmodulin-dependent protein kinase). Among many other things, the enzyme responds to signals from NMDA receptors, leading to more receptor activity and stronger signaling throughout the network of cells.

You would think, therefore, that the more CaMkII, the more robust a memory would be. But in experiments with mice, Tsien has
found there is a limit. If he drives CaMkII above the normal limit
while the animal is actively remembering an experience, the memory simply vaporizes, as the connections between the cells suddenly
weaken. The effect happens within minutes, and it is permanent
and selective, affecting the recalled memory while leaving the others unchanged. Indeed, when Tsien trained a mouse to fear both
an unfamiliar cage and a particular tone, then pumped up CaMkII
while the mouse was in the cage, it forgot the cage-fear memory
but not the tone-fear memory. "At the time the memory was retrieved, it disappeared," he says. "It erases the memory being recalled. It is feasible that by manipulating specific molecules, we can
selectively alter memories in the brain."

Todd Sacktor, a professor of physiology, pharmacology, and neurology at the State University of New York Downstate Medical Center in Brooklyn, has found a blunter but more powerful technique
that can eradicate whole categories of memory. He studies protein
kinase M-zeta (PKM-zeta), which is involved in memory maintenance. As calcium rushes into a memory neuron, PKM-zeta is synthesized, linking up with spare glutamate receptors and dragging
them to the synapse, where memory construction occurs. With
more receptors at the synapse, signals are boosted and amplified
and the memory persists.

When Sacktor deactivated PKM-zeta with a compound called
zeta-inhibitory peptide (ZIP), he got a spectacular response: total
amnesia for one type of memory. Rats that had learned a day or a
month before to avoid part of a platform that was rigged with electric shock forgot everything they knew about the location generating the jolt. "You inhibit the PKM-zeta and those glutamate receptors float away very, very fast," he says. "As a result, the memory is
lost — very, very fast."

Certain types of memory are encoded in different brain areas,
and depending on where Sacktor injects the inhibitor in his animals, he can zap away different categories of memory. In the hippocampus, he erases memory for spatial locations like the platform; in the amygdala, fear memories; in the insular cortex,
memories of nauseating taste. Very rarely, Sacktor says, neurosur-
geons remove nerve clusters to help disturbed psychiatric patients

who do not respond to any other treatment. His research may eventually provide a way to erase memory without causing damage.

The implications are staggering. If stored memories were inscribed in the brain, it is hard to imagine how flipping one chemical switch could erase them so quickly. "It really is a paradigm shift in how people think about long-term memories," Sacktor says. In the old view, erasure should cause permanent brain damage as the synapses are ripped apart. Instead, Sacktor's rats' brains remain intact. Once the ZIP treatment wears off, the animals behave and even learn normally again. "It's like wiping a hard disk," he says.

ZIP is nowhere near ready for human use. First, the compound would have to be made activity-dependent in order to target specific memories. You would also have to find a way to get it to the right spot in the brain without using a needle. People are clamoring to be test subjects anyway. When Sacktor's study first came out in 2006, people, especially rape survivors, tracked him down, imploring him to eradicate their painful memories. "They were suffering," he says. "They couldn't work or have relationships. Some of them wanted everything erased." They didn't care that it would also vaporize all they had ever known.

Benevolent Forgetting

If you feel that you've heard this story before, there's a reason. Moviemakers love the idea of erasing memory, and they work a consistent theme: if you try to undo the past, you pay the price. Nader's research supposedly inspired the 2004 movie *Eternal Sunshine of the Spotless Mind,* in which Jim Carrey and Kate Winslet both pay to have memories of their painful love affair obliterated. Needless to say, it makes them both miserable. But not as miserable as Arnold Schwarzenegger's character in *Total Recall,* from 1990, in which he learns that his real memories have been erased, that his life is a fake, and that his faux wife, played by Sharon Stone, is trying to kill him.

You don't have to be a rape survivor or a soldier to have memories you would rather forget. For most people, though, unpleasant memories also serve as a guide. Indeed, some fear the consequences of undermining appropriately bad memories—say, allowing a murderer to forget what he did. Members of the President's Council on Bioethics warn that altering the memory of a violent crime could unleash moral havoc by lifting the repercussions of malice. "Perhaps no one has a greater interest in blocking the painful memory of evil than the evildoer," their report cautions.

Beyond all this, memory is the essence of who we are. *Eternal Sunshine of the Spotless Mind* is difficult to watch as Carrey's character flails around in confusion and loss. His fear and desperation may be a realistic portrayal of what it would be like to erase your memory: basically, a waking nightmare. Memory is how you know who you are, how you point yourself toward a destination. We already know that people with Alzheimer's disease do not feel liberated. They feel utterly lost.

Thankfully, Nader and Brunet's studies suggest much more benevolent possibilities. If he had received reconsolidation therapy, Carrey's character would not have forgotten Winslet's. He simply wouldn't care that much about her anymore. He would be able to look at his failed relationship as if through the wrong end of a telescope. What is on the other side is still visible, but it is tiny and far away.

That is basically what all these scientists hope to do. Nader, Brunet, and Pitman are now expanding their PTSD study with a new, $6.7 million grant from the U.S. Army, looking for drugs that go beyond propranolol. They are increasingly convinced that reconsolidation will prove to be a powerful and practical way to ease traumatic memories. Sacktor also believes that some version of the techniques they apply in the lab will eventually be used to help people. Most recently, LeDoux's lab has figured out a way to trigger reconsolidation without drugs to weaken memory, simply by carefully timing the sessions of remembering. "The protocol is ridiculously simple," LeDoux says.

None of these researchers are looking to create brain-zapped, amoral zombies—or even amnesiacs. They are just trying to take control of the messy, fragile biological process of remembering and rewriting and give it a nudge in the right direction. Brunet's patients remember everything that happened, but they feel a little less tortured by their own pathological powers of recollection. "We're turning traumatic memories into regular bad memories," Brunet says. "That's all we want to do."

JOHN COLAPINTO

Brain Games

FROM *The New Yorker*

ONE MORNING IN JANUARY, a tall, gray-haired man whom I will call Arthur Jamieson arrived at the Mandler Hall psychology building at the University of California, San Diego, in La Jolla. Jamieson is seventy years old and lives in the Midwest. He is a physician and an amateur cellist and has been married for forty-seven years. He also suffers from a rare and bewildering condition called apotemnophilia, the compulsion to have a perfectly healthy limb amputated — in his case, the right leg, at midthigh. He had come to La Jolla not to be cured of his desire (like most people with the syndrome, he believed that relief would come only with the removal of the limb) but to gain insight into its cause. To that end, he had scheduled a meeting with Dr. Vilayanur S. Ramachandran, an Indian-born behavioral neurologist who is the director of the Center for Brain and Cognition at UCSD and who has a reputation among his peers for being able to solve some of the most mystifying riddles of neuroscience.

Ramachandran, who is fifty-seven, has held prestigious fellowships at All Souls College in Oxford and at the Royal Institution in London. His 1998 book, *Phantoms in the Brain,* about rare neurological disorders, was adapted as a miniseries on BBC television, and the Indian government recently accorded him the title Padma Bhushan, the country's third-highest civilian honor. But it is the awe that he inspires in his scientific colleagues that best illuminates his position in neuroscience, where the originality of his thinking and the simple elegance of his experiments give him a unique status. "Ramachandran is a latter-day Marco Polo, journeying the silk road of science to strange and exotic Cathays of the mind,"

Richard Dawkins once wrote. Eric Kandel, the Columbia University neuroscientist whose work on the physiological basis of learning and memory earned him a Nobel Prize in 2000, invoked two pioneering brain scientists to describe Ramachandran's contribution to the field: "He is a continuation of a tradition in neurology that goes back to the nineteenth century, to giants like Broca and Wernicke, who gave us, from studying clinical material, enormous insights into the functioning of the human mind."

Ramachandran, who has dark skin, curly black hair, and a mustache, cultivates a slightly rebellious image, often wearing dark polo shirts and a black leather jacket. However, when he meets with patients he tends to dress more conservatively. The day that he met with Jamieson, he was wearing a wool blazer and a tie. He greeted Jamieson in his office, whose décor reflects Ramachandran's many interests outside neurology: Darwinian evolution, plate tectonics, Indian art, Victorian medicine, paleontology, optical illusions. A four-foot stone sculpture of the god Shiva stood behind his desk. On one wall, there was a 300-million-year-old fossil of a mesosaur, a freshwater reptile found only in South America and Africa (and which, as Ramachandran likes to explain, is a central piece of evidence in the theory of continental drift). On a side table was an array of antique scientific items: a brass Gilbert telescope, a hand-cranked electrical machine for curing "nervous diseases," a box of glass tubes containing Victorian homeopathic medicines. Another table held what appeared to be a smoothly sanded wooden sculpture of a woman's pelvis. Ramachandran often tells visitors that the object is a Henry Moore before revealing, with a booming laugh, that it is actually a specimen of the world's largest seed, from the coco-de-mer palm.

Ramachandran listened closely as Jamieson talked about his condition. In a specialty that today relies chiefly on the power of multi-million-dollar imaging machines to peer deep inside the brain, Ramachandran is known for his low-tech method, which often involves little more than interviews with patients and a few hands-on tests — an approach that he traces to his medical education in India, in the 1970s, when expensive diagnostic machines were scarce. "The lack of technology actually forces you to be ingenious," he told me. "You have to rely on your clinical acumen. You have to use your Sherlock Holmes–like deductive abilities to figure things out."

Ramachandran suspected that apotemnophilia was a neurological disorder and not, as Freudians have theorized, a psychological syndrome associated with repressed sexual desires. After interviewing several apotemnophiliacs—Jamieson is the fifth person with the disorder whom he has studied—Ramachandran was struck by the fact that all of them said they became aware of the compulsion in early childhood, that it centered on a particular limb (or limbs), that they could draw a line at the exact spot where they wanted the amputation to occur, and that they attached little or no erotic significance to the condition. Furthermore, none rejected the limb as "not belonging" to them, as some stroke victims do in the case of a paralyzed arm or leg, and as Ramachandran had predicted they might. Instead, they said that the limb *over*-belonged to them: it felt intrusive. "If you talk to independent apotemnophiliacs, they say the same bloody things," Ramachandran told me. "'The line for cutting is here.' 'It started in early childhood.' 'It's over-present.' They're not crazy."

Jamieson, who was born and raised in New York City, first remembers having an unusual relationship with his right leg when, at around the age of seven, he was waiting for a bus. He found himself thinking that if he stuck out his leg it would be crushed and severed by the bus. "What came to me was not 'No, I don't want to do that' but 'How would I ever explain this?'" he told Ramachandran. In recounting his childhood memories, he said, "One of the things that's astonishing to me is how clear these recollections are."

"These things are very salient," Ramachandran said in a resonant baritone, which carries a British-inflected Indian accent. "It's interesting to contrast these very clear-cut descriptions with these vague, Freudian notions about this whole phenomenon—that it's primarily connected with sexual stuff."

"Yeah," Jamieson said with disgust. "I've got no desire to cozy up to anyone with a stump. It's psychobabble."

Asked where he would make the cut line for the amputation, Jamieson unhesitatingly drew an index finger across the middle of his right thigh. As to whether he felt that his leg didn't "belong" to him, Jamieson was emphatic. "Somehow, for me, that just doesn't compute, that kind of language," he said. "I have always been fascinated by amputation and wished that I had one. Why? Who the hell knows?"

*

Ramachandran is one of a dozen or so scientists and doctors who, in the past thirty years, have revolutionized the field of neurology by overturning a paradigm that dates back more than a hundred years: that of the brain as an organ with discrete modules (for vision, touch, pain, language, memory, and so on) that are fixed early in life and immutable. Neurological syndromes, such as paralysis from stroke, forms of mental illness, and the perception of pain in an amputated limb (a phenomenon known as phantom-limb pain), were considered largely untreatable. But Ramachandran and other researchers have shown that the brain is what scientists call "plastic"—it can reorganize itself. Not only are different regions of the brain engaged in ongoing communication with one another, with the body, and with the surrounding world; these relationships can be manipulated in ways that can reverse damage or dysfunction previously believed to be permanent. Ramachandran's work with patients at UCSD has led to one of the most effective treatments for chronic phantom-limb pain and to a new therapy for paralysis resulting from a stroke. (In both instances, his treatment involves only a five-dollar household mirror.) It has also provided suggestive insights into the physiological cause of such mystifying syndromes as autism.

Until the mid-1990s, Ramachandran's specialty was visual perception, but he had been interested in brain science since his days as a medical student in India. He made the switch to neurology in midcareer. "A scientist with that kind of creativity—it's rare," says Michael Merzenich, a neuroscientist at the University of California, San Francisco, whose experiments with monkeys in the 1980s provided much of the groundwork for understanding brain plasticity. "It's usually not allowed, in some sense. You're not supposed to be a butterfly like that."

Little about Ramachandran's scientific career has been conventional. He was born in Tamil Nadu, in southern India, to a Hindu family of the Brahman caste. His grandfather, Alladi Krishnaswamy Iyer, was the attorney general of Madras and a framer of India's constitution. Ramachandran's father was a diplomat in the United Nations. However, science ran in the family. His mother had a master's degree in mathematics; one uncle was a professor of optics at the University of Sydney; another was an expert in theoretical physics and relativity.

At around the age of nine, Ramachandran began collecting fossils and seashells and became fascinated by taxonomy and evolution. He wrote to a conchologist at the American Museum of Natural History. "Here's this little kid from India sending him sketches of shells and asking, 'Are these new species?'" Ramachandran said. "And he is writing back saying, 'A, B, C, and D are well-known species; E is very rare and has not been reported from your locality and is very interesting.' So for a while I was the only conchologist in India!" Ramachandran continues to collect fossils and has gone on digs in South Dakota, where he has found specimens of trilobites and a 30-million-year-old oreodon, a sheeplike creature. His most notable find, however, was not in the field but at the annual Tucson Gem and Mineral Show, in 2004, when he noticed on a table, amid heaps of bones and rocks, a skull that he thought could be a new species of ankylosaur, a herbivorous dinosaur from the Jurassic and Cretaceous periods. Ramachandran's friend Cliff Miles, a paleontologist, was with him and suggested that Ramachandran buy the fossil so that it could be studied and described. In January of this year, Miles and his brother Clark, also a paleontologist, announced the discovery of a new species of ankylosaur from the Upper Cretaceous period: *Minotaurasaurus ramachandrani.*

In his early teens, Ramachandran began conducting experiments in chemistry and biology in a makeshift laboratory under the staircase in the family's house in Bangkok, where his father was stationed. He also read books on the history of science and was struck by the role of intuition and play in many important discoveries: Galileo adapting a child's spyglass and discovering the moons of Jupiter, which led him to challenge the geocentric model of the universe; Faraday tinkering with a magnet and coil and discovering electromagnetism. Ramachandran often recounts these anecdotes to his students. "These stories are inspirational and fun," he told me. "But they're also telling you about how to do science."

Ramachandran's father discouraged him from pursuing a career as a researcher. "My father was intensely pragmatic," Ramachandran said. "He told me, 'Forget about chemistry and biology and all that. I know it's fun, but you're not going to make a living out of this.'" He urged his son to become a doctor, and Ramachandran duly enrolled at Stanley Medical College, in Madras. But he continued to read British and American science journals, and in his second year, he devised an experiment that was inspired in part by conversations he had had as a child with his uncle the optics professor. The experiment addressed a question debated by experts since the time of Hermann von Helmholtz, in the late nineteenth century, about how the brain harmonizes the two slightly different images seen by each eye. For years, scientists believed that when the eyes are given conflicting information — for instance, a green image in front of one eye and a red one in front of the other — the brain accepts input from one retina at a time. Ramachandran, using an old-fashioned stereoscope and volunteers from his medical-school class, found that, when presented with a pattern that was colored differently for each eye, his test subjects continued to see in three dimensions. He concluded that a neural channel was still active in the "shut-down" eye — even though his subjects were consciously seeing only one eye's color at a time. "This suggests that concepts of retinal rivalry need drastic revision," Ramachandran wrote in a report of the experiment.

He sent the report to *Nature* in December 1971, a few months after his twentieth birthday. "To my astonishment, it was published without revision," Ramachandran told me. Soon he published a more ambitious paper, "The Role of Contours in Stereopsis," which explored ideas about visual processing that became influential decades later. Ramachandran also wrote to one of the foremost vision scientists at the time, Dr. William Rushton, a professor of physiology at Trinity College, Cambridge, describing several original experiments that he was eager to try. The letter was passed to Oliver Braddick, a psychology lecturer who worked on vision. "The letter was obviously the product of a very fertile young mind," Braddick, who is now a professor of experimental psychology at Oxford, told me. "Perhaps a little kind of spinning off in all directions. But he had all these great ideas."

Braddick and another researcher, Fergus Campbell, invited Ramachandran to visit Cambridge for a month, at the university's expense, to conduct experiments. The results of one experiment, on which Braddick collaborated, were published as "Orientation-Specific Learning in Stereopsis," in the journal *Perception*. "Maybe fifteen years later, various people started publishing in this area of how specific developments of perceptual skills could be highly related," Braddick told me.

Ramachandran returned to Madras to complete his medical degree, and in the fall of 1974 he enrolled at Trinity College to begin a Ph.D. in visual perception. "I thought they'd all be like Faraday and the great Renaissance scientists," he said of the researchers he met in England. "Ninety percent of them are like Indian scientists, or scientists here, for that matter, or anywhere — it's a nine-to-five job. They're not moved by the great romantic spirit of science, and they're not great Renaissance people. So I was a bit disillusioned. Then I ran into Richard Gregory soon after I arrived, and I said, 'Well, at least there's some of them here!'"

Gregory was a professor of neuropsychology at the University of Bristol, the author of several best-selling books about science, and an expert in visual perception who had a special interest in optical illusions. Typical of his approach was a demonstration involving a Charlie Chaplin mask on a rotating axle, in which he shows how the brain uses prior knowledge of shape, shading, and other light effects to make sense of visual information and assemble a coherent representation of the world. Gregory's playful style irritated some of his colleagues, but Ramachandran found it electrifying. "He came to Cambridge to give a lecture," Ramachandran recalls. "He was like a magician! He is truly one of the five most amazing men I have met in my life."

During his four years at Cambridge, Ramachandran commuted regularly to Bristol to design experiments with Gregory. They have since written a number of scientific papers together, including groundbreaking work on the blind spot, the region at the back of the eyeball where the retina's photoreceptors are interrupted by the optic nerve. This region creates a gap in our vision the size of a palm held at arm's length, but, owing to several strategies of the brain, we never perceive it. Using optical illusions to trick the eyes and the brain, Ramachandran and Gregory determined how the brain "fills in" the gap, and published influential articles on stroke victims suffering from scotoma — a particularly large blind spot sometimes caused by a focal lesion in the visual cortex.

In the mid-nineties, Gregory visited Ramachandran at UCSD to undertake further experiments on scotoma, but they were unable to find a patient with a focal lesion. Instead, they spent Gregory's weeklong visit investigating a phenomenon that had long fascinated Ramachandran: the reported ability of flounder to camouflage itself against patterned backgrounds. Leading ichthyologists disagreed about whether the fish changed its appearance or whether the camouflage effect was an illusion. Ramachandran's local pet store had no cold-water flounder, so he bought five peacock flounder, a related species that lives in tropical coral reefs. The men placed the fish on the bottom of four small tanks against various backgrounds: widely spaced polka dots, a neutral gray, and two checkerboard patterns. The fish, whose natural tendency is to lie flat on the sea bottom, precisely matched on their bodies the patterns at the bottom of the tanks—and they did so within two to eight seconds, far faster than the hours and, in some cases, days reported by researchers using cold-water flounder. Ramachandran and Gregory surmised that the rapid change was an adaptive mechanism, since the species lived among bright colors and patterns. The experiment, which they meticulously documented in photographs and on videotape, effectively ended the debate on flounder camouflage—and, incidentally, throws an instructive sidelight on visual processing in human beings. Even though the fish sees the background close up and in a distorted, slanted perspective, it recreates the pattern on its body with perfect fidelity, as viewed from directly above. Human beings, Ramachandran points out, visually process the world in the same way. "Your eyeball distorts the image—it's curved," he says. "Your lens inverts it—it's upside down. And your two eyes double it. The brain *interprets* the image."

When they wrote up the results of the experiment, Ramachandran and Gregory laced their paper with puns. In a caption for a photograph showing one fish on a polka-dot background, they wrote, "Spot the flounder," and they said that they had conducted the experiments "just for the halibut."

"So we sent this off to *Nature*," Ramachandran told me, "and back come the referees' comments: 'Brilliant paper, publish it right away, but remove all the puns.'" He laughed. The paper, "Rapid Adaptive Camouflage in Tropical Flounders," was published in a 1996 issue of *Nature*. "Since then," Ramachandran said, "I get papers on octopuses and squids and fish—because they all think I'm an expert on ichthyology!"

In 1983 Ramachandran joined the psychology department at UCSD as an assistant professor working on visual perception. In

1991 he became interested in the work of Tim Pons, a neuroscientist at the National Institute of Mental Health, who had been investigating the ability of neurons in the sensory cortex to adapt to change.

The sensory cortex is in the deeply ridged tissue that makes up the outermost layer of the brain. Until recently, much of what was known about it was the result of the work of Wilder Penfield, a neurosurgeon in Montreal who, beginning in the 1930s, had conducted a series of extraordinary experiments while performing open-skull operations on cancer and epilepsy patients. Seeking to distinguish healthy tissue from diseased tissue, Penfield touched the surface of his patients' brains with an electric probe, and, because the brain lacks pain receptors, the patients were fully conscious and able to talk to him about what they felt. As he stimulated different areas of the brain, his patients reported feeling touch sensations in specific parts of their bodies. In this way, over several decades and hundreds of operations, Penfield mapped areas of the brain according to their corresponding body parts. The "Penfield homunculus," as it came to be called, is oriented upside down: the areas corresponding to the feet and the legs are at the top of the brain, the arms and the hands are in the middle, and the face is near the bottom. Body parts with the greatest sensitivity—lips, fingertips—take up a far larger area of the cortical surface than less sensitive areas.

The regions representing separate body parts on the Penfield homunculus, like the brain centers, were believed to be unchangeable. This view came under challenge as the technology for mapping the brain improved. Whereas Penfield had used a large electrode that affected thousands of neurons at a time, brain researchers in the fifties began to use tiny microelectrodes, which could be inserted into the brains of animals to record the firing of single neurons and, thus, communication among them. In the seventies, Michael Merzenich became expert at using microelectrodes to map the sensory cortex of monkeys. In one experiment, he mapped a monkey's hand area in the brain, then amputated its middle finger. Some months later, he remapped the monkey's hand and discovered that the brain map for the missing finger had vanished and been replaced by maps for the two adjacent fingers, which had spread to fill the gap. The results, published in the *Jour-*

nal of Comparative Neurology in 1984, were decisive proof that the brain can reorganize itself—at least across very short distances of one to two millimeters.

Pons, at NIMH, was curious to know whether the brain could accomplish more dramatic reorganizations, across greater distances. He wondered what happened in the brains of monkeys that had lost brain input from an entire hand and arm, and he thought that he could procure some animals to test. In 1981 a member of PETA had infiltrated a Maryland lab where a researcher studying stroke paralysis had severed the sensory nerves in a group of macaque monkeys that connected the animals' arms to their spinal cords —a procedure known as deafferentation. PETA released photographs of the monkeys, and the animals were seized and placed in the custody of the National Institutes of Health. By 1990 the monkeys had grown old and were about to be euthanized. Pons successfully appealed to the NIH to allow him to conduct a final experiment on four of them.

Pons anesthetized the first animal, opened its skull, and inserted electrodes into the brain-map area for the deafferented arm. He stroked the corresponding limb. As expected, the brain electrodes recorded no activity, since no messages were being sent to the brain from the arm. But when Pons stroked the monkey's face, the neurons in the map of the deafferented arm began to fire. The experiment showed that the neurons in the face map had invaded the area of the hand-and-arm map, which had been inactive for twelve years. Fourteen millimeters of the monkey's arm map had been reorganized to process sensory input from the face. Pons repeated the experiment on three more monkeys and published the results in *Science* in 1991, as a paper titled "Massive Cortical Reorganization After Sensory Deafferentation in Adult Macaques."

Ramachandran read Pons's paper and wondered whether it could help solve the long-standing medical puzzle of phantom limbs. Many amputees continue to experience sensations—often painful—from a missing limb, and the phenomenon has baffled scientists since it was first reported, in the sixteenth century, by the French surgeon Ambroise Paré. Ramachandran says that his interest in phantom limbs was a natural extension of his work in visual processing. "I was interested in the 'filling in' of the blind spot and other holes in the visual field; how the brain deals with undersampled regions—gaps," he said. "This resulted in my asking, 'How do you "fill in" a missing limb?'" Pons's monkeys seemed to offer a clue.

"Often, the best experiments begin as jokes," Ramachandran told me. "I joked with my students. I said, 'Hey, this means that if I touch the monkey's face the monkey should feel it in the hand.' And they all laughed, and I said, 'Hey, why not?' Then they said, 'Well, how do you train a monkey to tell you what it's feeling?' And I said, 'Why do you need a monkey? Let's try it on a person.'"

Ramachandran arranged to examine a seventeen-year-old boy whom he calls Tom, who had recently lost his left arm, just above the elbow, in a car crash. In a basement lab at Mandler Hall, Ramachandran lightly stroked Tom's cheek with a Q-Tip. Tom said that he felt the touch in his cheek but also in his phantom thumb. A touch on the lip he felt on his phantom index finger, a touch on the lower jaw in his phantom pinkie. Ramachandran realized that every time Tom moved his face and his lips—smiling, talking, frowning—the nerve impulses from his face activated the "hand" area in his cortex. "Stimulated by all these spurious signals," he later wrote, "Tom's brain literally hallucinates his arm."

Ramachandran immediately telephoned his wife, Diane Rogers-Ramachandran, and told her, "Come in right now. You've got to see this guy."

Rogers-Ramachandran is also a scientist, specializing in vision and experimental psychology. She and Ramachandran met in the late 1970s at a vision conference in Florida. She was then a graduate student at the University of North Carolina, Chapel Hill. They married in 1987. (They have two boys: Chandramani, who is nineteen, and Jaya, fourteen.) Rogers-Ramachandran rushed from their home in nearby Del Mar to watch the experiment. In the course of a few hours, she and her husband mapped Tom's phantom hand on his face. In a later experiment, they applied warm water to Tom's cheek. He felt heat in his phantom hand. When the water trickled down his cheek, he felt it running down his phantom arm. Ramachandran and his wife published their findings in 1992 in *Science*.

Rogers-Ramachandran, a vivacious woman with bright blue eyes, continues to collaborate with her husband on papers, and they write a regular science column for *Scientific American Mind*. She says that it has sometimes been a challenge to be married to a man of Ramachandran's mental energy and intellectual curiosity. "Like, when we got married," she said one evening, over dinner at a restaurant with her husband and Jaya, "we went to England for our honeymoon and spent the whole time going to bookstores and collecting prints, books, scientific instruments. Never went to a play! None of those things! The collecting! He went from scientific instruments to fossils, to learning about his Indian heritage, to art. You say, 'Well, can't we just go walk on the beach?'"

She mentioned Ramachandran's abstracted air—it's as if he were constantly mulling over an abstruse neurological conundrum. I knew something about this. On the first day of my visit to UCSD, Ramachandran was unable to remember where in the parking lot he had left his car and finally had to activate the alarm on the remote control to locate it. His embarrassment suggested that this was the first time such a thing had happened. Yet during the six days that I spent with him, it happened every time. When I told this story to Diane at dinner, she snorted.

"When we leave a place, he'll go into the parking lot, and a lot of time he'll just start walking," she said. "He has no idea where he's going. He just walks. One time I picked him up from a trip—"

"Oh, don't tell him that," Ramachandran said.

But Diane went on. "He reached in his pocket and he said, 'Oh, my God, I had a rental car in that city! I completely forgot! I have the keys and I didn't turn the car in!' Another time," she continued, "I got a call from Sears and a woman said, 'There's a man here who says he's your husband and he's trying to purchase something on this credit card.' I said, 'Ye-e-e-s.' And she said, 'We're kind of concerned if it's really your husband, because he doesn't know your birth date.' I said, 'Oh, *that's* my husband!'"

"Ha-ha-ha-ha-ha!" Ramachandran boomed. "That is a good story."

I could not resist asking whether Ramachandran had since learned Diane's birthday. They have been married for twenty-two years.

"I know she's a Leo," he said slowly, eying her from across the table.

"I'm not a Leo," Diane said. "*You're* a Leo."

"No," he corrected himself. "Virgo! Virgo!"

"Yup," she said.

"August eighteenth," he said with confidence.

"No," Diane said. Then she turned to me. "See, he gets the month, because it's the same as his."

"It's not the eighteenth?" Ramachandran asked.

"No."

"Twenty-second?" he offered.

"No."

At this point, Jaya asked, "Do you know *my* birthday?"

Ramachandran looked helplessly at his son and shrank into his seat. "It doesn't mean I don't love you," he said.

In 1994 Ramachandran published a paper in *Nature* that is now considered a landmark in the field of neuroplasticity. He described experiments that he had conducted with UCSD's multimillion-dollar magnetoencephalography machine, which records the changing magnetic fields caused by brain activity. (Though he calls himself a "technophobe," Ramachandran occasionally uses high-tech gadgetry, chiefly as a means to support his hunches.) The high-resolution MEG scans clearly showed that in the brains of arm amputees the area associated with the face had invaded the area associated with the missing arm — "the first direct demonstration of massive reorganization of sensory maps in the adult human brain," Ramachandran wrote.

His most startling revelation about the brain's capacity for reorganizing itself was yet to come. It emerged from his efforts to address phantom-limb pain, which afflicts up to 90 percent of amputees. Some report feeling that they are clenching their phantom fist so hard that their phantom fingernails are digging into their phantom palm. Phantom-limb pain can be so agonizing that some sufferers commit suicide.

For more than a century, doctors theorized that the pain was psychological or originated in the stump — in swollen nerve endings called neuromas. Some resorted to repeated amputations, making the stump shorter and shorter. When this didn't work, they tried severing the nerves at the spinal cord and even disabling parts of the thalamus, an organ at the base of the brain that processes pain. All to no avail. "They can chase the phantom farther and farther into the brain, but of course they'll never find it," Ramachandran once wrote. The phantoms, as he had shown, are *produced* in the sensory cortex, where neurons for the face have invaded territory once reserved for the arm.

Ramachandran posited that the phantom sensations are also created by higher brain centers, produced by a complex interplay among the sensory cortex, the motor cortex in the frontal lobes, and a "body image" map in the right superior parietal lobule, a section of the cerebral cortex just above the right ear. One of the main tasks of the right superior parietal lobule is to assemble a coherent body image from touch signals ("I feel my fingers touch the cup"), visual signals ("I see my hand reaching for the cup"), and nerve signals from the muscles, joints, and tendons ("I feel my arm extending toward the cup"). Even though amputees no longer received these signals from the nonexistent limb, Ramachandran believed that memories of these inputs remained in the nervous system and the brain.

Reviewing the histories of amputees, Ramachandran noticed that many who suffered from cramping or clenching spasms had experienced, before their amputations, a period during which the limb was immobilized, sometimes for months, in a sling or a cast. He theorized that a kind of "learned paralysis" was burned into the brain's circuitry, as repeated commands from the patients' brains to move the limb were met with touch, visual, and nerve evidence that the limb could not move. When the limb was later amputated, the patient was stuck with a revised body-image map, which included a paralyzed phantom whose neural pathways retained a memory of pain signals that could not be shut off. Ramachandran wondered what would happen if such a patient was presented with evidence that the phantom could move ("I see my hand reaching for the cup"). If the brain could be tricked into thinking that the phantom was moving, would the cramping sensations cease?

His first test subject was a young man who a decade earlier had crashed his motorcycle and torn from his spinal column the nerves supplying his left arm. After keeping the useless arm in a sling for a year, the man had the arm amputated above the elbow. Ever since, he had felt unremitting cramping in the phantom limb, as though it were immobilized in an awkward position.

In his office in Mandler Hall, Ramachandran positioned a twenty-inch-by-twenty-inch drugstore mirror upright and perpendicular to the man's body and told him to place his intact right arm on one side of the mirror and his stump on the other. He told the man to arrange the mirror so that the reflection created the illusion that his intact arm was the continuation of the amputated one. Then Ramachandran asked the man to move his right and left arms simultaneously, in synchronous motions — like a conductor — while keeping his eyes on the reflection of his intact arm. "Oh, my God!" the man began to shout. "Oh, my God, Doctor, this is unbelievable." For the first time in ten years, the patient could feel his phantom limb "moving," and the cramping pain was instantly relieved. After the man had used the mirror therapy ten minutes a day for a month, his phantom limb shrank — "the first example in medical history," Ramachandran later wrote, "of a successful 'amputation' of a phantom limb."

Ramachandran conducted the experiment on eight other amputees and published the results in *Nature* in 1995. In all but one patient, phantom hands that had been balled into painful fists opened, and phantom arms that had stiffened into agonizing contortions straightened. "People always ask, 'How did you think of the mirror?'" Ramachandran told me. "And I say, 'I don't know!' There was a mirror in the lab, so that must have been in my mind, and I said, 'Let's try it.' It's not any more mysterious than if you say something 'popped into' your mind."

Dr. Jack Tsao, a neurologist for the U.S. Navy, was doing graduate work in physiology at Oxford University when he read Ramachandran's *Nature* paper on mirror therapy for phantom-limb pain. "I said, 'Why the heck should this work? It doesn't make sense,'" Tsao told me. Several years later, in 2004, Tsao began working at Walter Reed Military Hospital, where he saw hundreds of soldiers with amputations returning from Iraq and Afghanistan. Ninety percent of them had phantom-limb pain, and Tsao, noting that the painkillers routinely prescribed for the condition were ineffective, suggested mirror therapy. "We had a lot of skepticism from the people at the hospital, my colleagues as well as the amputee subjects themselves," Tsao said. But in a clinical trial of eighteen service members with lower-limb amputations, in which six were given mirror therapy and the twelve others were evenly divided between two control therapies (a covered mirror and mental visualization), the six who used the mirror reported that their pain decreased (and, in some cases, disappeared altogether). In the two control groups, only three patients reported pain relief, and others found that their pain increased. Tsao published his results in the *New England Journal of Medicine* in 2007. "The people who really got completely pain-free remain so, two years later," said Tsao, who is currently conducting a study involving mirror therapy on upper-limb amputees at Walter Reed.

Buoyed by these successes, in the mid-nineties Ramachandran abandoned his work in visual perception to devote himself to neurology. "Vision was getting overcrowded," he told me. Neurology seemed like virgin territory. Much of the specialty was concerned with describing strange syndromes rather than with explaining their cause or alleviating symptoms. "You've got a hundred papers saying, 'My God, they can move their phantom' — but it stayed at that level, a descriptive level," Ramachandran said. "We said, 'Look, we can do experiments. What if you do *this* to the patient?' And I took that same style to other syndromes. Then the sky was the limit. No one was studying these things."

Gradually, Ramachandran began to specialize in rare conditions and disorders, including the Capgras delusion, in which an otherwise lucid victim of a head injury insists that close loved ones (spouses, parents, children) are impostors. Freudians had theorized that Capgras patients were suffering from unbearable Oedipal desires aroused by the blow to the head, but Ramachandran demonstrated that severed neural pathways between the facial-recognition areas of the visual cortex and the emotional centers of the brain were responsible for the disorder. He also investigated post-stroke syndromes, in which patients deny that a paralyzed limb has become immobile or, in a more severe version, insist that the paralyzed arm or leg belongs to someone else. Ramachandran traced the delusion to damage in the right superior parietal lobule, the body-map region, where the discrepancy between the absence of signals from the limb to the brain and the presence of the limb on the body results in a defensive rationalization that the arm or leg must be someone else's. A few years ago, Ramachandran began studying apotemnophilia, the compulsion to amputate a healthy limb. He is, he said, "ninety-five percent sure" that he has figured out the cause of the disorder. His consultation with Arthur Jamieson strengthened this conviction.

After interviewing Jamieson in his office, Ramachandran led him to a lab for a galvanic skin response, or GSR, test, which would reveal how Jamieson's legs reacted to a mild pain stimulus. He escorted Jamieson into a small room that held only a table, a desktop computer, and two chairs. He asked Jamieson to sit with his back to the computer. Then David Brang, one of Ramachandran's graduate students, attached a sensor to the middle two fingers of Jamieson's right hand using a Velcro strap. The sensor would measure the reaction of Jamieson's sympathetic nervous system by monitoring the sweat on his fingers. With a sterilized pin, Brang pricked Jamieson's legs at random points, waiting a few seconds between each prick. A scrolling graph on the computer screen registered Jamieson's responses.

The unaffected leg—the left one—and the right leg above where he wished to have it amputated showed a normal response: the graph at first shot upward with each prick, but with further pricks it ceased to rise, then began to flatten out, indicating that Jamieson's nervous system was getting used to the stimulus. But when Brang pricked Jamieson anywhere on the leg *below* the amputation line, his nervous system responded with increasing distress, the graph climbing higher and higher with each prick.

The experiment seemed to support Ramachandran's theory about the disorder. He believed that people with apotemnophilia had a deficit in the right superior parietal lobule, where the body-image map is assembled. According to this notion, Jamieson was missing the neurons in the map that corresponded to his right leg from the midthigh down. He had normal sensation in the unwanted part of his leg—he felt the pin prick. But when the pain signal traveled to the right superior parietal lobule, there was nothing in the body-image map to receive it.

"So there's a big discrepancy—a *clash*—and the brain doesn't like discrepancies," Ramachandran said. "When a discrepancy comes in, it says, 'Shit! What the hell is going on here?' and it kicks in and sends a message to the insular part of the brain, which is involved in emotional reactions—so you're getting this crazy GSR." In apotemnophilia sufferers, the discrepancy causes a feeling of distress that is no less agonizing for being below the level of conscious awareness.

In the past two years, Ramachandran has tested four other apotemnophiliacs using MEG brain scans. "You touch them anywhere in the body and the right superior parietal lobule lights up, as you would expect," Ramachandran said. "But if you touch him here"—he gestured to a point on Jamieson's leg below the amputation line—"nothing happens." Ramachandran said that the experiment needed to be repeated by other researchers, but, he added, "This takes a spooky psychological phenomenon and, as Shakespeare said, gives it a 'habitation and a name.'" Furthermore, the findings suggested to Ramachandran a possible method for alleviating the oppressive sensations in the unwanted limb.

Later, he asked Jamieson to stand in a corner of his office and placed a three-foot-high mirror in front of him in such a way that in place of his right leg Jamieson saw his left, which he held bent at the knee. Jamieson gazed into the mirror. "Astonishing," he said. For a moment, the leg looked "right."

The mirror was a less risky kind of sham amputation than the method that Jamieson had recently adopted: injecting anesthetic to block the sciatic nerve of his right leg, shutting down the touch sensation. (As a physician, Jamieson had learned how to perform the nerve block.) The anesthetic provided up to five hours of relief, Jamieson said. Apotemnophiliacs, like transsexuals, anorexics, and others with body-image disorders, often do not seek a "cure" for their condition, and Ramachandran spoke gingerly when he suggested that using both the mirror and the drug could potentially yield powerful results. "It's conceivable—nobody knows—but if you do this repeatedly, and I'm not suggesting that you try this, because I know you don't want to be 'changed,' but if you do it repeatedly, both the injections and the visual amputation, it might actually eliminate this desire," he said.

Ramachandran describes his approach to science as "opportunistic": "You come across something strange—what Thomas Kuhn, the famous historian and philosopher of science, called 'anomalies.' Something seems weird, doesn't fit the big picture of science—people just ignore it, doesn't make any sense. They say, 'The patient is crazy.' A lot of what I've done is to rescue these phenomena from oblivion." Ramachandran is conscious of the fact that this focus might lead some to think that he works on the margins of his field. "Now, you could say that about Oliver," he told me, referring to his friend and colleague Oliver Sacks, the neurologist and author of *The Man Who Mistook His Wife for a Hat.* "'Oh, he

studies spooky things,'" Ramachandran went on. "That's bulls
This man has deep insight into the human condition. He's a poet
of neurology." Ramachandran says that his own interest in oddities
is not for their own sake but for what they can tell us about the normal brain, including, he said, "very enigmatic aspects of the brain
that few people have dared to approach, like what is a metaphor?
How do you construct a body image? Things of that nature."

In 1999 Ramachandran turned his attention to synesthesia, an
intermingling of the senses that causes some people to see each
letter of the alphabet in a particular color. Others identify musical
notes with colors; still others mix touch sensations with strong emotions, so that sandpaper might evoke disgust; velvet, envy; wood
grain, guilt. Vladimir Nabokov described his letter-color synesthesia in *Speak, Memory*: "I see *q* as browner than *k*, while *s* is not the
light blue of *c*, but a curious mixture of azure and mother-of-pearl."
As an artist, Nabokov was, according to Ramachandran's research,
eight times more likely to have synesthesia than someone who is
not an artist; the fact that Nabokov's mother also had the condition suggested a genetic component. (The phenomenon runs in
families.)

The most common synesthesia is number-color. Ramachandran
believed it was not coincidental that the fusiform gyrus, where
number shapes are processed in the brain, lies next to the area
where colors are processed. He suspected that a cross-wiring in the
brain, similar to that in phantom-limb patients, was responsible.
Brain scans confirmed his hunch: in synesthetes, there are excess
neural connections between the two brain centers. This suggested
to Ramachandran that the syndrome arises from a defect in the
gene responsible for pruning away the neural fibers that connect
the various centers of the brain as it develops early in life. "What
do artists, poets, and novelists have in common?" Ramachandran
asked me. "The propensity to link seemingly unrelated things. It's
called metaphor. So what I'm arguing is, if the same gene, instead
of being expressed only in the fusiform gyrus, is expressed diffusely
through the brain, you've got a greater propensity to link seemingly unrelated brain areas in concepts and ideas. So it's a very
phrenological view of creativity."

In the mid-nineties, Ramachandran read a paper by Italian researchers who had discovered that a set of neurons in the frontal

lobes of monkeys fired not only when the monkeys reached for an object but also when they observed another monkey performing the same action. Ramachandran wondered if these so-called "mirror neurons" also exist in humans—a difficult thing to test, since the Italians had inserted electrodes into the brains of living monkeys, a technique that it is impossible to use on people. But Ramachandran knew of experiments from the 1950s in which noninvasive EEG scans were used. These had shown that deliberate movements in humans suppress a kind of brain activity in the motor cortex called mu waves. Ramachandran and a postdoctoral fellow, Eric Altschuler, ran EEGs on volunteers as they observed another person performing an action such as opening and closing a hand. The tests showed that merely witnessing an action in others caused mu-wave suppression in the watcher—evidence that mirror neurons exist in humans, too. Other researchers have since confirmed that people have several systems of mirror neurons, which perform different functions.

"So let's take the broader theoretical implications of this," Ramachandran said one afternoon while we were visiting the San Diego Rehabilitation Institute at Alvarado Hospital, where he had examined a paralyzed stroke patient suffering from limb denial. He was sitting in the hospital cafeteria with the clinic's medical director, Lance Stone. "These mirror-neuron experiments are showing that, through and through, the brain is a dynamic system not only interacting with your skin receptors, up here"—he pointed at his own head—"but with *Lance!*" He pointed across the cafeteria table at Dr. Stone. "Your brain is hooked up to Lance's brain! The only thing separating you from Lance and me is your bloody skin, right? So much for Eastern philosophy." He laughed, but he wasn't kidding. Ramachandran has dubbed mirror neurons "Gandhi neurons"—"because," he said, "they're dissolving the barrier between you and me."

Ramachandran wondered whether mirror neurons were implicated in autism, a condition whose primary characteristic is severe social impairment, including an inability to imitate and a lack of empathy. Ramachandran, Altschuler, and Jaime Pineda, a UCSD colleague, ran EEGs on autistic children. They got normal mu-wave suppression when the subjects moved their own hands. But when the children watched another person move his hand, their brains didn't respond. At a neuroscience conference in 2000, Ramachandran and his coauthors presented their findings and speculated that autism was caused by a deficit in the mirror-neuron system. The idea initially met with resistance from autism researchers, some of whom argue that the disorder is caused primarily by deficits in the cerebellum. Unlike his earlier foray into ichthyology, Ramachandran was entering a sphere of science fraught with politics. "The trouble is, it's a minefield," he told me. "The parents are involved. There's big money involved. Suppose you invested your life in saying that the cerebellum is what's going on, then someone comes along and spends one year on it and says, 'It's the mirror-neuron system'?"

In the past nine years, however, mirror neurons have become a central topic in autism research. Almost at the same time as Ramachandran, a group in Scotland had also suggested the link. Among those who have provided further evidence are researchers at the Helsinki University of Technology, who used MEG scans to show mirror-neuron deficits in autistic teenagers and adults. Lindsay Oberman, a former graduate student of Ramachandran's, who now works as a postdoctoral fellow at Beth Israel Deaconess Medical Center in Boston, has begun using a technology called transcranial magnetic stimulation—a technique that triggers targeted areas of neurons in the brain—to influence brain plasticity in autistics. "So far, we have done some amazing things," Oberman has written. "We have found evidence that we can improve the functioning of the mirror-neuron system and some communication skills following repeated application of TMS."

On the last day of my visit with Ramachandran, I attended the lab discussion that he holds each Monday with his postdoctoral and graduate students at the Center for Brain and Cognition Laboratory, on the second floor of Mandler Hall. The lab, a room of modest size, was dominated by a long central table heaped with the strange tools of Ramachandran's trade: a foam-rubber hand of the type you buy at a horror shop (for a demonstration that Ramachandran likes to do to show visitors how the brain projects touch sensations onto objects that are not part of the body); a mirror ball of the type that M. C. Escher liked to draw; a boxed set of the BBC miniseries of Sherlock Holmes (for inspiration); several plastic minimizing lenses (Ramachandran has found that viewing a painful arm or leg through a lens that makes the limb look smaller dramatically reduces pain); a reflective metal tube that could be twisted into various amoebic shapes (when I asked if this puzzle had "experimental significance," Ramachandran said, "No," then quickly corrected himself: "Well, it's fun"); a series of oddly shaped metal boxes outfitted with slanting mirrors (for inducing perceptual distortions in those who peer through the eyeholes); and a plaster cast of *Minotaurasaurus ramachandrani,* a creature that resembles a medieval gargoyle, with three nasal openings on either side of its ridged and crenellated head. Ramachandran has asked one of his postdocs, Paul McGeoch, to perform a CAT scan of the skull in order to learn about the creature's olfactory lobes, and, in this way, to test Ramachandran's theory that his ankylosaur's heightened sense of smell might allow the beast to sniff out mates or carrion from a great distance (although it was more likely a vegetarian).

Seated around the table were members of Ramachandran's research group. Most were in their middle to late twenties, except for a man in his eighties with a British accent: John Smythies, whom Ramachandran introduced to me as the person who launched the drug revolution in the sixties. Smythies demurred, explaining that as a postdoc at Cambridge in the fifties, while performing psychopharmacology experiments involving mescaline, he had merely introduced Aldous Huxley to a colleague, who then administered to Huxley the hallucinogens that led him to write *The Doors of Perception,* which later became a bible of the Woodstock generation.

Ramachandran, who was dressed in his usual black leather jacket and dark polo shirt, took a seat at the table and fielded questions from his students, helping them to refine their methodologies and using the brisk interchanges to hone ideas for research. At one point, Lisa Williams, a Ph.D. student who specializes in schizophrenia—a disorder that Ramachandran first began exploring about a decade ago—mentioned in passing the difficulty that schizophrenics have in differentiating between phenomena that are internally and externally generated.

"Oh!" Ramachandran cut in. "Speaking of that, I have an idea —I'm sure it's been done—but you know that when people think to themselves you get unconscious movements of the vocal cords?

Now, has anybody done that with schizophrenia to s⌐
hanced?"

"I don't know," Williams said. "I'll look that up."

If such enhanced subvocalization occurs when sch┄┄┄┄┄┄┄
think, that would support Ramachandran's view of the brain as an
organ in dynamic equilibrium—and of mental illnesses as resulting from a neurological disruption that destroys that equilibrium.
In the case of schizophrenia, whose sufferers often complain of
"hearing voices," Ramachandran suspected damage or deficit in
a sensory mechanism in the vocal cords, which, when normal people think, sends a signal to the brain indicating "This is simply a
thought; no one is actually saying this." If this mechanism was dam-
aged, the subconscious movement of the vocal cords could be interpreted as an outside voice speaking in one's head.

"By the way," Ramachandran continued, "I have a theory that if
you take people with carcinoma of the larynx, and you remove the
vocal cords, and they think to themselves, they may actually start
hallucinating. A prediction."

This remark prompted Laura Case, a first-year graduate student
who has focused on autism, to speak. "That could be interesting in
autism, too," Case said. "Because if they lack the robust mirror activation for actions, which they do—"

Ramachandran interjected, "Then they confuse—so they may
confuse their own vocalizations with somebody else's! And people
have linked autism to schizophrenia. The old theory was that it was
early-childhood schizophrenia! Was that a coincidence?"

The discussion proceeded in this freewheeling manner for more
than an hour, with Ramachandran seizing on notions that seemed
to offer fruitful possibilities for further investigation and tactfully
deflecting those which he thought were dead ends. When the discussion ended, at 6 P.M., and Ramachandran's students had departed, I asked him if he thought that his work was aimed at constructing a "grand unified theory" of the brain. He said that
neuroscience was still too young a discipline for such an ambition.
Nevertheless, in recent years he has increasingly focused on the
biggest mystery of the brain: consciousness. Mirror neurons play a
role, he thinks. "One of the theories we put forward," he said, as he
packed up his bag, "is that the mirror-neuron system is used for
modeling someone else's behavior, putting yourself in another per-

son's shoes, looking at the world from another person's point of view. This is called an allocentric view of the world, as opposed to the egocentric view. So I made the suggestion that at some point in evolution this system turned back and allowed you to create an allocentric view of yourself. This is, I claim, the dawn of self-awareness."

Still, Ramachandran said, deciphering how consciousness works will take a supreme creative leap. "It may require a radical revision of the way in which you perceive the universe, the world, the brain," he said, as he stepped into the hallway and locked the lab door behind him. "Just like Einstein had to change your complete perspective in order to really understand time, saying it's part of the whole space-time manifold. Things don't 'pass through' time—that's a human illusion. But if it requires that, some genius is going to have to come along and solve it." He opened the door to the stairwell and started down. "What we're hoping," he went on, "is that we can grope our way toward the answer, finding little bits and pieces, little clues, toward understanding what consciousness is. We've just scratched the surface of the problem. When I say 'we,' not just our lab but the entire world of neuroscience."

By now, we had reached the ground floor of Mandler Hall and were walking outside, past clusters of students. Ramachandran was still speaking excitedly—he had veered into a knotty digression about the brain's role in the evolution of language—when he glanced up and realized that we had reached the parking lot. He stopped talking and looked out over the sea of automobiles.

"Uh-oh," he said.

PART THREE

Natural Beauty

GUSTAVE AXELSON

The Alpha Accipiter

FROM *Minnesota Conservation Volunteer*

ON A PLEASANT SUMMER DAY in Chippewa National Forest, I was strolling down a woodsy trail—until I crossed a boundary where I was unwelcome. A screaming goshawk hurtled out of the forest shadows.

Kee-kee-kee-kee-kee! Its high-pitched, incessant alarm call pulsated like a siren. The hawk fluttered from perch to perch amid the leafy treetops, then settled atop a dead aspen to assume an aggressive posture. Its undertail feathers flared, a snow white, fluffy plume befitting this bird's Latin name: *Accipiter gentilis,* a raptor of gentility.

Kee-kee-kee-kee-kee! The strident refrain continued. The goshawk protested with agitated bobs of its head. Its red eyes burned like hot, glowing coals embedded in its dark gray facial stripes.

Kee-kee-kee-kee-kee! The message was clear: *Leave.*

"She must have nestlings," said the Department of Natural Resources Nongame Wildlife biologist Maya Hamady, indicating that the goshawk's young must have already hatched. "The female doesn't leave the nest when she's sitting on eggs."

Hamady had agreed to guide me to one of the 109 goshawk nest sites in northern Minnesota that the DNR and other agencies have been monitoring since 1991. As she records goshawk nesting activity, Hamady is filling a void in Minnesota's avian annals. No one has ever conducted a comprehensive survey of goshawk populations in Minnesota, so biologists don't really know how many goshawks live in our north woods—or how they are faring.

But this lack of a historical baseline for evaluating the goshawk's conservation status doesn't stop Hamady from looking into the future of Minnesota forests to see if goshawks will have enough closed-canopy habitat for nesting. Habitat fragmentation and declining stands of mature aspens could pose challenges for goshawks as they seek nesting territories over the next two decades and beyond.

A Fierce Reputation

The northern goshawk is the largest Minnesota accipiter, a genus of forest-dwelling, fast-flying hawks that includes Cooper's and sharp-shinned hawks. The goshawk's ferocity is legendary. It has been known to attack other hawks, people, even black bears. Attila the Hun rode into battle wearing a helmet that bore an emblem of this most truculent bird of prey.

Hamady has experienced the accipiter's aggressiveness firsthand on several nest-site checks, when goshawks swooped at her. On one visit, a goshawk flew overhead for most of her quarter-mile retreat to her car.

She also witnessed a prime example of the goshawk's territorial tendencies. "I once got a call from a landowner about a goshawk nest, but when I went out to confirm it I only saw a sharp-shinned hawk. So I thought the landowner was mistaken," Hamady said. Later another raptor researcher visited the nest and found two fledgling goshawks and the skeleton of a sharp-shinned hawk on the ground below.

Goshawks hunt by executing surgical strikes in thick woods — weaving among tree boles, flying at speeds up to fifty-five miles an hour. Relentless in pursuit of prey, a goshawk will crash through brush at top speed and even drop to the ground to run down its quarry on foot.

The goshawk's choice of prey has tarnished its reputation among some hunters, who worry that the "grouse hawk" could depress local game bird populations. But a 2003 research study of northern goshawk food habits in Minnesota revealed a broad diet. Over forty-five days, a pair of breeding goshawks averaged two daily deliveries of prey, including twenty-nine red squirrels, fourteen eastern chipmunks, six crows, five snowshoe hares, five ruffed grouse, two diving ducks, one cottontail rabbit, one blue jay, and thirty-one miscellaneous forest creatures (including pileated woodpeckers, a weasel, and a veery).

Dense Woods Denizen

The prime hunting ground for goshawks is dense forest, where their jet-fighter flight tactics provide an advantage that keeps out other raptors better suited to open spaces. Goshawks prefer habitat with a closed tree canopy, shrubby cover on the forest floor to give prey a false sense of security, and a midway fly zone free of branches. One Minnesota goshawk study showed that they preferred nest trees averaging seventy-two feet tall and forests with 60 percent to 90 percent canopy closure.

This mature woods preference has landed the goshawk in a few spotted owl–type timber harvest controversies. In Arizona and New Mexico, lawsuits were filed in the early 1990s against the U.S. Forest Service to add the goshawk to the endangered species list as a means of halting old-growth logging projects, though the attempts failed due to a lack of evidence that goshawk populations were declining.

In the mid-1990s, the Minnesota offices of the Sierra Club and Audubon filed appeals on timber-harvest proposals in Chippewa National Forest, charging that the USFS hadn't properly considered goshawk habitat needs. Harvest levels in the forest were high in the late 1980s and early 1990s, says the Chippewa wildlife biologist Jim Gallagher, and the harvests were guided by a 1986 forest plan that didn't contain specifics about goshawk management. Gallagher says that the USFS didn't know much about goshawks back then. He says the USFS was aware of only one or two goshawk nests in the entire 1.6-million-acre Chippewa National Forest.

As a result of the Sierra Club and Audubon appeals, the USFS got to work finding out more about goshawks in Chippewa National Forest—and soon discovered the forest's goshawk population was more abundant than previously thought.

"Researchers at other national forests told me they searched all summer and couldn't find any goshawk nests," recalled Gallagher. But he found a couple of nests just by visiting some proposed timber-harvest sites and being attacked by goshawks. Several more nests were reported to him as eagle or osprey nests but turned out to be goshawk nests. And as Gallagher gained experience, he got better at finding goshawks. After a few field seasons, Gallagher had located twenty-one goshawk nests.

Habitat Challenges

Maya Hamady's second goshawk nest visit of the day took us to a different nook of Chippewa National Forest. In her DNR truck, we bumped down an active logging road past fifteen-foot-tall stacks of logs that looked like Paul Bunyan's firewood. We stopped at the edge of a pine plantation, where we immediately saw an ominous sign: a red-tailed hawk perched in a tree.

Red-tailed hawks and great horned owls are the archrivals of goshawks; the presence of redtails and great horned owls often results in nest and territory abandonment by goshawks. Redtails and owls are prime predators of goshawks (which lose their acrobatic flight advantage beyond the forest's edge) and their nestlings. More important, they are harbingers of forest fragmentation, which renders goshawk habitat unsuitable.

We walked a trail until red pines mixed with aspens, where sunlight filtered through an enclosed tree canopy and dappled the forest floor. In the fork of an aspen, about fifty feet high, sat a nest of kindling-sized twigs, patched together with mosses and duff. A goshawk emerged from the nest, circled silently overhead, then flew off into the forest, leaving the echo of a fading call, *KEE-kee-kee*. Hamady said this goshawk might have been a male attempting to draw us away from the nest.

Of the fifteen goshawk territories closely monitored by the DNR Nongame Wildlife Program for habitat requirements, this site has had the most logging—with 35 percent of the upland mature habitat harvested between 1995 and 2005. Yet this nest has also been one of the most productive, with hatchlings most years since 2003.

Gallagher says the USFS and DNR are smarter about balancing timber harvests with goshawk habitat needs now, compared with the 1990s, when logging activity in federal forests used to take all trees except the actual nest tree. In 2003 the DNR issued a set of state forest management considerations that are sensitive to goshawk breeding territories, including nest areas.

In 2004 the revised Chippewa National Forest plan introduced new rules for timber harvests near goshawk nests on federal land: no logging allowed during the nesting season; no logging allowed in an area of fifty acres around the nest; and only selective logging within a five-hundred-acre area around the nest, such that 60 percent of that area remains in suitable habitat condition (tall trees, partially closed canopy). Since the national forest rules were implemented, Gallagher says timber harvests have not harmed monitored goshawk nests in Chippewa National Forest. And there have been no further appeals from the Sierra Club and Audubon.

"Audubon has been watching to see that [the rules] work," said Mark Martell, director of bird conservation for Audubon Minnesota. "We want to give the system a chance to work. It's premature to call it a success after only a few years. But it seems to be working so far."

Changing Forests

Despite the success in Chippewa National Forest, there have still been logging-goshawk conflicts in Minnesota. Federal and state guidelines apply only to national and state forests, which hold 60 percent and 15 percent of known goshawk nests, respectively. About 15 percent of known goshawk nests are on private land, and 10 percent are on county forestland. From 2003 to 2007, DNR goshawk nest surveyors found three instances in which nest stands were logged, twice on private land and once on county land. In two of those cases, the goshawks abandoned their territory.

But individual timber harvest–goshawk nest conflicts don't concern Hamady nearly as much as the forecasts for a changing northern forest composition, which could result in less goshawk nesting and foraging habitat over the next two decades.

Goshawks are very particular in choosing nesting areas, seeking mature forest structure with a closed canopy and trees about fifty to one hundred feet tall. In the Black Hills and Arizona, those nesting areas are in stands of ponderosa pine. In British Columbia, goshawks choose Douglas fir or lodgepole pine. And in Minnesota, that prime goshawk nesting habitat tends to consist of the most dominant trees on the northern forest landscape: aspen.

Aspens aren't inherently suited to goshawk nests, but their abundance here makes aspens the most likely trees to provide suitable nesting structure, at least for now. Red and white pines can also provide that structure, but Minnesota currently has far fewer mature pines on the landscape.

Hamady worries that over the next few decades there will not be enough mature aspen forests, or mature upland forests in general, for goshawks. The Chippewa National Forest plan estimates that mature and old forest will decrease from 49 percent of total upland forest to 43 percent within the next ten to twenty years. In state forests, currently over 30 percent of aspen forests are mature, but long-term goals aim for about 10 to 15 percent of aspen stands in a mature or old forest condition. County forests have more early successional forests and could see a steeper decline in mature forest habitat. All federal, state, and county forestlands have a lack of middle-aged aspens (thirty to fifty years old), which would age over the next two decades into mature forest habitat.

"Whether through harvest or natural processes like wind, decay, or diseases, the old aspen trees that currently exist won't last very long," says the DNR forestry policy and planning supervisor Jon Nelson. He notes that aspen is a short-lived species with a life expectancy of about eighty years. "We can't create middle-aged aspen stands out of thin air to replace them," he says.

Hamady says she sees that conifers will replace old aspens in the fifty- to sixty-year forest plans for the Chippewa and Superior national forests. "But old aspens will [be less abundant] in the next twenty years," Hamady says. "What then?" she asks.

To conserve goshawk habitat, the DNR is identifying contiguous, mature forest complexes in the north woods. DNR biologists are working with foresters to coordinate and maintain that habitat on the landscape through extended-rotation forests — timber stands left to grow longer than their typical harvest age.

"Coordination among land management agencies to conserve goshawk habitat is a crucial part of goshawk conservation," says Hamady. "It's not about stopping logging, but managing forests so there's a dynamic patchwork of mature forest habitat available. That doesn't just benefit goshawks, but fishers, spruce grouse, boreal owls, and many species of warblers, including ovenbirds, black-throated blue warblers, and Blackburnian warblers."

The Superior and Chippewa national forest plans now contain guidance to manage for larger patches of mature forest. "The goshawk was one of the driving species for this large-patch management," says Gallagher.

Likewise, the DNR's forest management planning process is working to maintain and restore larger patches of older forest, such as long-lived conifers, to provide more mature forest habitat.

Gos in the Hand

Months after my summer nest visits, I was still mesmerized by the goshawk—its fearless spirit, its regal form. So in the autumn, I went to Hawk Ridge in Duluth to perchance see a goshawk again. Hundreds of goshawks from Minnesota, Canada, and Alaska migrate by Hawk Ridge every fall.

I was in luck. On the October morning of my visit, the biologists at the bird banding station had captured a goshawk. They handed the bird to me so I could admire it for a few moments before its release. A juvenile with a brown-streaked body, this goshawk shifted within my grip—a bundle of intense power. The eyes were yellow, not yet red, but every bit aflame. The goshawk glared at me like an alpha predator glaring at its prey, like a timber wolf wrapped in feathers.

Upon release, the gos swooped to take a swipe at a bystander —one last act of pugnacity before winging to wintering grounds somewhere farther south.

DON STAP

Flight of the Kuaka

FROM *Living Bird*

IN FEBRUARY, at 37 degrees 12 minutes south latitude, the sun sets late, but night has fallen and the darkness is thick and close. In the hills to the west I see a few dull globes of light from distant houses. Above, the stars glimmer like chipped ice, but they cast little light, and I would like very much to see where I'm stepping. I am in a shallow pond about a hundred yards west of the Firth of Thames on New Zealand's North Island. And I am up to my shins in mud as slick as lard. Having given up all pretense of grace, I wave my arms about with each step, as if I'm on a tightrope. I hope not to fall into this black goop—the manure-enriched runoff from a cow pasture—as two people near me did moments ago.

A few feet away, Nils Warnock, codirector of the Wetlands Ecology Division at California's PRBO Conservation Science, is disentangling a bar-tailed godwit from a mist net. Hundreds of godwits —large, long-legged, cinnamon-breasted sandpipers with up-turned bills—flew to this roosting site when high tide submerged the mud flats of the firth, where they spent the day feeding and preening. Warnock, who speaks softly and slowly, never seems to be in a hurry, but suddenly he is moving quickly in my direction. He's holding three godwits, taking them to a site on dry ground where the birds are being separated into holding crates according to sex. But apparently he has taken one too many birds. "Here," he says, handing me a warm bundle with long, kicking legs. "Don't hold him too tightly." I cradle the bird against my body and stumble through the darkness, the godwit's heart beating like a trapped moth against my chest.

New Zealand is the principal wintering site for the *baueri* race of the bar-tailed godwit, a subspecies that breeds in Alaska. The bird's annual journey to these southern latitudes caught the attention of South Sea Islanders long ago. About 950 AD a group of Polynesians left their homeland, heading out across the South Pacific in seagoing canoes for a land they believed existed somewhere to the south. Their only guide was the flight path of an elegant bird they knew as the *kuaka*—the bar-tailed godwit—which they had observed flying south each year at the same time. Because the kuaka was not a seabird, they reasoned it must go to some land as yet undiscovered.

They were right. The explorers eventually saw a cloud on the horizon that stretched for miles, hanging, they correctly guessed, above a large landmass. So goes a Maori legend describing how their ancestors discovered New Zealand. They were the first humans to set foot on the island they called Aotearoa—"land of the long, white cloud."

Roughly a thousand years later, Nils Warnock and Bob Gill, a wildlife biologist at the Anchorage, Alaska, office of the United States Geological Survey (USGS), followed the flight of a bar-tailed godwit in a more leisurely fashion. From August 29 to September 7, 2007, they sat in their offices and watched their computers produce a line across a map of the Pacific Ocean—the flight path of E7, a *baueri* bar-tailed godwit whose surgically implanted satellite transmitter was sending signals to National Oceanic and Atmospheric Administration satellites 510 miles above Earth.

Prior to her departure, E7 had been feeding voraciously on fingernail-sized clams, worms, and other marine invertebrates on the Yukon-Kuskokwim Delta along with tens of thousands of her kind. On August 29, she probably rose into the air throughout the day with other bar-tails, circling for several minutes, judging the weather, then settling back down. In early evening, a couple of hours before sunset, she rose into the sky and this time kept going, heading southeast. Within a few hours she crossed the Alaska Peninsula and headed out over the North Pacific.

Gill hoped that the battery in her transmitter would last long enough to prove what he had long suspected about *baueri*'s southward journey to New Zealand. Birds that migrate between northern and southern sites in the Pacific Rim—including *baueri*'s close

relative, the Siberian-nesting *menzbieri* race of the bar-tailed godwit—usually follow coastlines as they travel, stopping at shoreline
mud flats along the way to rest and feed. To cross the Pacific Ocean,
essentially a barren desert to land birds not adapted to landing and
feeding on the water, would seem to be a fatal mistake. But rather
than fly toward the Asian coast, E7, as Gill expected, turned south
—toward open ocean. Three days later she was still flying above
blue waters. Around noon, flying at an altitude of perhaps two
miles or more, she must have had a beautiful view of the Hawaiian
archipelago as she passed over it four hundred miles west of Hono-
lulu. Two days later, she crossed the international dateline about
three hundred miles north-northeast of Fiji. And then on the afternoon of September 7, E7 approached North Cape, the northern-
most tip of New Zealand. She adjusted her flight path, and late
that evening touched down at the mouth of the Piako River on
the Firth of Thames, eight miles from where she'd been captured
seven months earlier in the same mud-bottomed ponds where Gill,
Warnock, and crew were now trapping more birds. She had flown
for eight days—nonstop—covering approximately 7,250 miles at
an average speed of nearly 35 miles per hour.

This journey—the longest nonstop migratory flight documented for any bird—seems barely credible. "I just did a talk yesterday for some colleagues at the U.S. Geological Survey," Gill told
me not long after E7 had been tracked to New Zealand. "And I
showed these graphics of E7's flight and said, 'Okay, the flight is
nonstop, no food, no water, no sleep as we know it, flying for eight
days,' and there was just this silence in the room, and I could see
their minds trying to wrap around this—as does mine. I try to be
objective as a scientist, but this just . . ." Gill's sentence trailed off as
he seemed unable to summon up the right word to describe his
reaction.

Although he may be at a loss for words to describe the wonder of
the feat, Gill knows something of what makes such a flight possible.
Like shorebirds in general, the godwit has a sleek, aerodynamic
body and long, tapered wings that reduce drag in flight. Its endur-
ance comes from the enormous reserves of fat the bird builds up
in the weeks preceding migration, when it gorges itself on marine
invertebrates, more than doubling its body weight until, as Gill has
said, it looks like the Concorde when it takes off. Burned off dur-

ing flight, the fat yields more than twice the energy of comparable amounts of carbohydrates or protein. In addition, the godwit's body undergoes a remarkable change: its intestines and gizzard, which the bird makes little use of during migration, shrink, allowing more space to store fat.

In his early sixties, with close-cropped white hair, Gill is the senior member of an international team of scientists analyzing the migration of godwits and curlews as part of a four-year project funded largely by the David and Lucile Packard Foundation. Gill's voice rises with enthusiasm when he speaks of the bar-tails, and here in New Zealand his spirit is all the more buoyant because he's wearing shorts and sandals; when he left Alaska two days ago, it was 10 below zero. Gill, Warnock, and their North American colleagues are working with a number of New Zealand biologists, among them Phil Battley of Massey University, as well as a local shorebird expert, Adrian Riegen. It is Riegen's van that Warnock and I are looking for in the darkness as we cross the cow pasture. There I pass on my godwit to more experienced hands.

At the side of the van, Lee Tibbitts, one of the USGS Alaska crew, holds each godwit up into the light from Riegen's headlamp so he can measure the bird's bill with calipers. Female bar-tails are larger than males, and bill length is an easy way to distinguish the sexes. The team hopes to find several males large enough to carry an implanted transmitter so they can track an equal number of each sex. A transmitter weighs 25 grams, and the rule of thumb is that a bird should not carry anything that is more than 3 percent of its body weight.

According to figures gathered over the past thirty years by the International Shorebird Survey, more than half of all shorebird species show evidence of serious decline. Some populations are dwindling slowly, some plummeting at a rate that, unchecked, will lead to their extinction in the not-too-distant future. Of all the taxonomic groups of shorebirds, the Numeniini (godwits and curlews) are the most threatened. Thirteen species of Numeniini exist worldwide. All but two of them—the black-tailed godwit of Europe and Russia and the Eurasian curlew of Europe and Asia—have received designations ranging from "critically endangered" to "species of high concern" by one or more world conservation organizations. With this tracking project, Gill and Warnock hope to learn more about the timing and routes of migration for several species of Numeniini, valuable information for future conservation efforts.

Of the species that breed in North America—which also include the bristle-thighed curlew, long-billed curlew, Hudsonian godwit, marbled godwit, whimbrel, and upland sandpiper—the bar-tailed godwit is actually the most populous, with an estimated 1.2 million birds worldwide, 120,000 of which are the *baueri* race, which breeds mainly on estuaries in western Alaska. More than a million birds seems like a healthy population, but you don't have to look far to find a cautionary tale. One Numeniini, the Eskimo curlew, very likely exists in name only. The Eskimo curlew's population fell sharply in a few decades in the nineteenth century, dropping from hundreds of thousands of birds—perhaps millions—to a few individuals, the result of indiscriminate hunting and loss of the grasslands that they depended on during migration. Some estimate that anywhere from two dozen to a hundred Eskimo curlews may still be moving between their ancestral wintering sites in the pampas of Argentina and breeding territories in the far north, but this may be wishful thinking. The last confirmed record was a solitary bird shot in Barbados in September 1963.

Godwits and curlews, like most shorebirds, are particularly vulnerable to habitat loss because they congregate in great numbers at a relatively few sites during migration, and many have restricted wintering sites as well. Understanding their migration is vital to any hope of developing effective conservation strategies to halt or reverse population declines. Although studies that began in the 1970s mapped the general migration routes for many shorebirds, it wasn't until birds were captured and outfitted with radio transmitters—and now satellite transmitters—that scientists began to see how individual birds used various sites.

Warnock explains one reason this is important: "If you look at two sites, A and B, and site A has one hundred birds on it on day one, and site B also has a hundred birds on day one, and then you go back ten days later and both sites still have a hundred birds, you might think each site is equally important. But you don't know if it's the same one hundred birds or not. Site A may have a different one hundred birds each day—meaning one thousand birds have used it over a ten-day period—while site B has had the same one hundred birds for ten days. Tracking individual birds can indicate how long a bird typically stays at a site. And that can tell you which areas are really most important."

The Numeniini have significantly different migration strategies. Although the bar-tailed godwit is the ultimate long-distance migrant, the long-billed curlew, which breeds in the western United States, may travel only a few hundred miles from its breeding grounds to wintering sites. The bristle-thighed curlew is an intermediate-distance migrant, moving between Alaska and islands in the Pacific Ocean. Furthermore, the birds do not always follow the same migration paths in spring and fall. The *baueri* bar-tails, for instance, return to Alaska by a different route, and their northward journey is as interesting to Gill and Warnock as their record-setting southward flight.

When they leave New Zealand in the spring, the birds head to the Yellow Sea. The coastline of this large, relatively shallow body of water between mainland China and the Korean Peninsula has 8,000 square miles of intertidal flats that support more than 5 million migratory shorebirds each year. There the *baueri* stop for five weeks or so to rest and feed. Then they launch out across the Pacific again on a beeline to their breeding territories in Alaska. Recent surveys have suggested that a great many of the *baueri* bar-tails stop at one site, the Yalu Jiang National Nature Reserve — 450,000 acres at the mouth of the Yalu Jiang River, which separates North Korea from China.

But fast-growing economies in this heavily populated region of the world are in direct conflict with shorebird habitat. Just down the coastline from Yalu Jiang, in South Korea, lies the Saemangeum estuary, a major stopover site for migrating shorebirds, including 30 percent of the world's population of great knots. In 2006, after years of court battles, the South Korean Supreme Court gave the government permission to complete a twenty-mile-long seawall separating the estuary from the Yellow Sea. The area's extensive tidal flats will thereby be "reclaimed" for agricultural land, and the water that remains will become fresh water from the rivers that feed the estuary. In time, the project will drain an estimated 154 square miles of tidal flats. Some of the new land can already be seen on photographs taken from NASA satellites. Where fertile intertidal flats existed, there is now stark white, barren land.

It may get worse. Mark Barter, an Australian biologist who has worked in mainland China, writes that "80 percent of the significant wetlands in east and southeast Asia [are] classified as threatened in some way; 51 percent of these are under serious threat." Recently he told Gill of two reclamation projects on the Chinese coast of the Yellow Sea, each larger than Saemangeum. "The coastline of the Yellow Sea is just being assaulted," Gill says. It is tricky for Westerners to express outrage over this, considering that the ground we stand on is often the result of dams, dikes, impoundments, and other contrivances that have destroyed our own wetland ecosystems. "It's like San Francisco Bay a century ago," says Gill.

So how safe is Yalu Jiang? How important is it really? How accurate are the surveys? If the need arises to make a case for Yalu Jiang, the bar-tailed godwits carrying satellite transmitters have provided irrefutable evidence: so far, the majority of satellite-tagged birds have stopped there.

Wetlands are not the only threatened habitat. The winds—a "habitat" as tangible to the godwits as mountains, valleys, and wide-open plains—will certainly be affected by global climate change. The godwits "have evolved wind-sensitive migration strategies," Gill says. Presently, the birds must negotiate five different wind systems that rule different regions of the Pacific Ocean. Their departure from Alaska appears to be timed to take advantage of the winds created by storms that the Aleutian Low Pressure System sends to the Alaska coastline on an almost weekly basis during the period when godwits usually begin their journey. The birds ride the tailwinds from the back side of a storm for the first 600 miles or more, the winds boosting their speed up to 80 to 90 miles per hour.

North of the Hawaiian archipelago the Northeasterly Trades blow toward the southwest. These "quartering" tailwinds (midway between a tailwind and a crosswind) push the birds along, but if they do not compensate for the westward wind flow, they will be blown far off course. The equatorial doldrums, a zone of little or no wind that sailors have always feared, neither help nor hinder the birds. At approximately 20 degrees south latitude, they face the Southeasterly Trades, quartering headwinds that they must fly against. And then, in the South Pacific, their fat reserves running low, they face more quartering headwinds as they enter a "convergence zone." In recent years, studies that use computer models to predict how climate change will affect wind systems have come up with different results. But they agree on one thing: the winds will change. Perhaps the godwits will adapt. Or perhaps, over time, the winds will push them toward new, less agreeable wintering sites.

At one in the morning on the second day of capturing godwits, I go with Nils Warnock and Jesse Conklin, a Ph.D. student at Massey University, to release two godwits that now carry satellite transmitters. We drive to a deserted beach north of the shorebird center rather than take them back to the crowded ponds. This will be a kind of post-op recovery room, where they will have some peace and quiet as they adjust to their surroundings. When the birds are released, they do not rush away, but stand motionless for a few minutes. Then, slowly, they walk off toward the sound of water lapping the shoreline.

With the birds out of sight, we stand for a moment enjoying the night air. I look up, once again drawn to the starry sky. A poor student of astronomy, I'm delighted to recognize a constellation — Crux Australis, the Southern Cross. It is, of course, one of the most famous formations in the Southern Hemisphere, remarked upon by virtually every early explorer who sailed south of the equator. In a few weeks the godwits will rise into the air and leave behind Crux Australis for the cold northern skies of Ursa Major. On the tundra of the far north, they will breed and raise young, then move again to their staging grounds to prepare for the long flight southward. The adults will depart first, leaving behind the juveniles. A short while later, the young birds, guided by some deep *baueri* knowledge of the earth and wind and stars, will set off on a 7,000-mile journey to a place they've never seen — the land of the long white cloud.

MATT RIDLEY

Modern Darwins

FROM *National Geographic*

JUST TWO WEEKS BEFORE HE DIED, Charles Darwin wrote a short paper about a tiny clam found clamped to the leg of a water beetle in a pond in the English Midlands. It was his last publication. The man who sent him the beetle was a young shoemaker and amateur naturalist named Walter Drawbridge Crick. The shoemaker eventually married and had a son named Harry, who himself had a son named Francis. In 1953 Francis Crick, together with a young American named James Watson, would make a discovery that has led inexorably to the triumphant vindication of almost everything Darwin deduced about evolution.

The vindication came not from fossils, or from specimens of living creatures, or from dissection of their organs. It came from a book. What Watson and Crick found was that every organism carries a chemical code for its own creation inside its cells, a text written in a language common to all life: the simple, four-letter code of DNA. "All the organic beings which have ever lived on this earth have descended from some one primordial form," wrote Darwin. He was, frankly, guessing. To understand the story of evolution — both its narrative and its mechanism — modern Darwins don't have to guess. They consult genetic scripture.

Consider, for instance, the famous finches of the Galápagos. Darwin could see that their beaks were variously shaped — some broad and deep, others elongated, still others small and short. He surmised (somewhat belatedly) that in spite of these differences, all the Galápagos finches were close cousins. "Seeing this gradation and diversity of structure in one small, intimately related group of birds," he wrote in *The Voyage of the Beagle,* "one might really fancy that from an original paucity of birds in this archipelago, one species had been taken and modified for different ends."

This, too, was inspired guesswork. But by analyzing the close similarity of their genetic codes, scientists today can confirm that the Galápagos finches did indeed descend from a single ancestral species (a bird whose closest living relative is the dull-colored grassquit).

DNA not only confirms the reality of evolution, it also shows, at the most basic level, how it reshapes living things. Recently, Arhat Abzhanov of Harvard University and Cliff Tabin of Harvard Medical School pinned down the very genes responsible for some of those beak shapes. Genes are sequences of DNA letters that when activated by the cell make a particular protein. Abzhanov and Tabin found that when the gene for a protein called BMP4 is activated (scientists use the word "expressed") in the growing jaw of a finch embryo, it makes the beak deeper and wider. This gene is most strongly expressed in the large ground finch *(Geospiza magnirostris),* which uses its robust beak to crack open large seeds and nuts. In other finches, a gene expresses a protein called calmodulin, which makes a beak long and thin. This gene is most active in the large cactus finch, *G. conirostris,* which uses its elongated beak to probe for seeds in cactus fruit.

In another set of islands, off the Gulf Coast of Florida, beach mice have paler coats than mice living on the mainland. This camouflages them better on pale sand: owls, hawks, and herons eat more of the poorly disguised mice, leaving the others to breed. Hopi Hoekstra, also at Harvard, and her colleagues traced the color difference to the change of a single letter in a single gene, which cuts down the production of pigment in the fur. The mutation has occurred since the beach islands formed less than 6,000 years ago.

Darwin's greatest idea was that natural selection is largely responsible for the variety of traits one sees among related species. Now, in the beak of the finch and the fur of the mouse, we can actually see the hand of natural selection at work, molding and modifying the DNA of genes and their expression to adapt the organism to its particular circumstances.

Darwin, who assumed that evolution plodded along at a glacially slow rate, observable only in the fossil record, would be equally delighted by another discovery. In those same Galápagos finches, modern Darwins can watch evolution occur in real time. In 1973 Peter and Rosemary Grant, now of Princeton University, began annual observations of the finch populations on the tiny Galápagos island of Daphne Major. They soon discovered that the finches in fact evolved from one year to the next, as conditions on the island swung from wet to dry and back again. For instance, Daphne Major initially had only two regularly breeding ground finches, one of which was the medium ground finch *(G. fortis),* which fed on small seeds. When severe drought struck the island in 1977 and small seeds became scarce, the medium finches were forced to switch to eating bigger, harder seeds. Those with larger beaks fared better and survived to pass on the trait to their offspring.

Another shift took place after a competitor arrived in 1982: the large ground finch *(G. magnirostris),* which also eats large, tough seeds. For many years the two species coexisted, and in 2002 both became unusually abundant. But then drought struck, and by 2005, only thirteen large and eighty-three medium ground finches remained alive. Remarkably, instead of adjusting to the drought by eating bigger seeds, as they had twenty-eight years before, the surviving medium finches experienced a marked reduction in the size of their beaks, as in competition with their larger cousins they struggled to carve out a niche by surviving on very small seeds. A finch with a smaller beak is not a new species of finch, but Peter Grant reckons it might take only a few such episodes before a new species is established that would not choose to reproduce with its parent species.

The variation seen among the Galápagos finches is a classic example of "adaptive radiation," each species evolving from a common ancestor to exploit a special kind of food. Another famous radiation took place on a different set of islands—islands of water rather than land. The lakes and rivers of Africa's Great Rift Valley contain some 2,000 species of cichlid fish that have evolved from a few ancestors, some in an instant of geologic time. For example, Lake Victoria, the largest of those lakes, was completely dry just 15,000 years ago. Its 500 diverse species of cichlids have all evolved since then from a handful of species of uncertain origin. Like the finches, cichlid fish species have adapted to diets in different habitats, such as rocky or sandy patches of lake beds. Some species eat algae and have densely packed teeth suited to scraping and pulling plant matter, while others feed on snails and have thick, power-ful jaws capable of crushing open their shells. And what gene is responsible for thickening those jaws? The gene for the protein BMP4—the same gene that makes the Galápagos ground finch's beak deep and wide. What better evidence for Darwin's belief in the commonality of all species than to find the same gene doing the same job in birds and fish, continents apart?

In *The Origin of Species,* Darwin tactfully left unspoken how his theory would extend that commonality to include humankind. A decade later he confronted the matter head-on in *The Descent of Man.* He would be delighted to know that a certain gene, called *FOXP2,* is critical for the normal development of both speech in people and song in birds. In 2001 Simon Fisher and his colleagues at the University of Oxford discovered that a mutation in this gene causes language defects in people. He later demonstrated that in mice, the gene is necessary for learning sequences of rapid move-ment; without it, the brain does not form the connections that would normally record the learning. In human beings, presum-ably, *FOXP2* is crucial to learning the sophisticated flickers of lips and tongue with which we express our thoughts.

Constance Scharff of the Free University of Berlin then discov-ered that this very same gene is more active in a part of the brain of a young zebra finch just when the bird learns to sing. With fiendish ingenuity, her group infected finches' brains with a special virus carrying a mirror-image copy of part of the *FOXP2* gene, which stifled the gene's natural expression. The result was that birds not only sang more variably than usual but also inaccurately imitated the songs of adults—in much the same way as children with mutant *FOXP2* genes produce variable and inaccurately copied speech.

Today's Darwins see in detail how pressures such as competition and a changing environment can forge new species. But Darwin also proposed another evolutionary driver: sexual selection. In Lake Victoria, cichlid fish have vision adapted to the light in their surrounding environment—at greater depths, where available light is shifted toward the red end of the spectrum, their visual receptors are biased toward red light, while closer to the surface they see better in blue. Ole Seehausen of the University of Bern and the Swiss Federal Institute for Aquatic Science and Technology has found that male cichlids have evolved conspicuous colors to catch the female eye: typically red nearer the lake bottom and blue at shallower depths. The blue and red populations appear to be genetically diverging—suggesting they represent two separate species in the making.

If natural selection is survival of the fittest (a phrase coined by the philosopher Herbert Spencer, not by Darwin), then sexual selection is reproduction of the sexiest. It has the delightful effect of generating weapons, ornaments, songs, and colors, especially on male animals. Darwin believed that some such ornaments, such as stags' antlers, helped males fight each other for females; others, such as peacocks' tails, helped males "charm" (his word) females into mating. It was, in truth, an idea born of desperation, because useless beauty worried him as an apparent exception to the ruthlessly practical workings of natural selection. He wrote to the American botanist Asa Gray in April 1860 that "the sight of a feather in a peacock's tail, whenever I gaze at it, makes me sick!"

His notion of sexual selection was politely ignored by most Victorian opinion, which was mildly scandalized by the thought of females actively choosing a mate rather than submitting coyly to the advances of males. Even biologists dropped the idea for roughly a century, because they became obsessed with arguing that traits evolve to suit the species rather than to suit the individual. But we now know Darwin was right all along. In all sorts of species, from fish and birds to insects and frogs, females approach the males with the most elaborate displays and invite them to mate.

Darwin did not speculate much on why a female would choose an ornamented male. It is a question that still excites biologists, because they have two equally good answers to it. One is simply fashion: when females are choosing gorgeous males, other females must follow suit or risk having sons that do not attract females. The other is more subtle. The tail of a peacock is an exhausting and dangerous thing for the bird to grow. It can be done well only by the healthiest males: parasites, starvation, and careless preening will result in duller plumage. So bright plumage constitutes what evolutionary biologists call an "honest indicator of fitness." Substandard peacocks cannot fake it. And peahens, by instinctively picking the best males, thereby unknowingly pass on the best genes to their offspring.

In one of his flights of fancy, Darwin argued that sexual selection might account for human racial differences: "We have seen that each race has its own style of beauty . . . The selection of the more attractive women by the more powerful men of each tribe, who would rear on an average a greater number of children, [would] after the lapse of many generations modify to a certain extent the character of the tribe." The jury is still out on that particular idea, but there are hints that Darwin might be at least partly right.

Take blue eyes. Darwin, like many Europeans, had blue eyes. In early 2008, Hans Eiberg and his colleagues at the University of Copenhagen announced that they had found the genetic mutation common to all people having pure blue eyes. The mutation is a single letter change, from A to G, on the long arm of chromosome 15, which dampens the expression of a gene called *OCA2*, involved in the manufacture of the pigment that darkens the eyes. By comparing the DNA of Danes with that of people from Turkey and Jordan, Eiberg calculated that this mutation happened only about 6,000–10,000 years ago, well after the invention of agriculture, in a particular individual somewhere around the Black Sea. So Darwin may have gotten his blue eyes because of a single misspelled letter in the DNA in the baby of a Neolithic farmer.

Why did this genetic change spread so successfully? There is no evidence that blue eyes help people survive. Perhaps the trait was associated with paler skin, which admits more of the sunlight needed for the synthesis of vitamin D. That would be especially important as people in less sunny northern climates became more dependent on grain as a food source, which is deficient in vitamin D. On the other hand, blue-eyed people may have had more descendants chiefly because they happened to be more attractive to the opposite sex in that geographic region. Either way, the explanation leads straight back to Darwin's two theories—natural and sexual selection.

Intriguingly, the spelling change that causes blue eyes is not in the pigment gene itself but in a nearby snippet of DNA scripture that controls the gene's expression. This lends support to an idea that is rushing through genetics and evolutionary biology: evolution works not just by changing genes but by modifying the way those genes are switched on and off. According to Sean Carroll of the University of Wisconsin at Madison, "The primary fuel for the evolution of anatomy turns out not to be gene changes, but changes in the regulation of genes that control development."

The notion of genetic switches explains the humiliating surprise that human beings appear to have very few, if any, special human genes. Over the past decade, as scientists compared the human genome with that of other creatures, it has emerged that we inherit not just the same number of genes as a mouse—fewer than 21,000—but in most cases the very same genes. Just as you don't need different words to write different books, so you don't need new genes to make new species: you just change the order and pattern of their use.

Perhaps more scientists should have realized this sooner than they did. After all, bodies are not assembled like machines in factories; they grow and develop, so evolution was always going to be about changing the process of growth rather than specifying the end product of that growth. In other words, a giraffe doesn't have special genes for a long neck. Its neck-growing genes are the same as a mouse's; they may just be switched on for a longer time, so the giraffe ends up with a longer neck.

Just as Darwin drew lessons from both fossil armadillos and living rheas and finches, his scientific descendants combine insights from genes with insights from fossils to understand the history of life. In 2004 Neil Shubin of the University of Chicago and his colleagues found a 375-million-year-old fossil high in the Canadian Arctic—a creature that fit neatly in the gap between fish and land-living animals. They named it *Tiktaalik*, which means "large freshwater fish" in the local Inuktitut language. Although it was plainly a fish with scales and fins, *Tiktaalik* had a flat, amphibian-style head with a distinct neck and bones inside its fins corresponding to the upper and lower arm bones and even the wrists of land animals: a missing link if ever there was one. It may even have been able to live in the shallows or crawl in the mud when escaping predators.

Equally intriguing, however, is what *Tiktaalik* has taught Shubin and his colleagues in the laboratory. The fossil's genes are lost in the mists of time. But, inspired by the discovery, the researchers studied a living proxy—a primitive bony fish called a paddlefish—and found that the pattern of gene expression that builds the bones in its fins is much the same as the one that assembles the limbs in the embryo of a bird, a mammal, or any other land-living animal. The difference is only that it is switched on for a shorter time in fish. The discovery overturned a long-held notion that the acquisition of limbs required a radical evolutionary event.

"It turns out that the genetic machinery needed to make limbs was already present in fins," says Shubin. "It did not involve the origin of new genes and developmental processes. It involved the redeployment of old genetic recipes in new ways."

Though modern genetics vindicates Darwin in all sorts of ways, it also turns the spotlight on his biggest mistake. Darwin's own ideas on the mechanism of inheritance were a mess—and wrong. He thought that an organism blended together a mixture of its parents' traits, and later in his life he began to believe it also passed on traits acquired during its lifetime. He never understood, as the humble Moravian monk Gregor Mendel did, that an organism isn't a blend of its two parents at all, but the composite result of lots and lots of individual traits passed down by its father and mother from their own parents and their grandparents before them.

Mendel's paper describing the particulate nature of inheritance was published in an obscure Moravian journal in 1866, just seven years after *The Origin of Species*. He sent it hopefully to some leading scientists of the day, but it was largely ignored. The monk's fate was to die years before the significance of his discovery was appreciated. But his legacy, like Darwin's, has never been more alive.

TIM FLANNERY

The Superior Civilization

FROM *The New York Review of Books*

ANTS ARE SO MUCH A PART of our everyday lives that unless we discover them in our sugar bowl we rarely give them a second thought. Yet those minuscule bodies voyaging across the kitchen counter merit a closer look, for as the entomologists Bert Hölldobler and Edward O. Wilson tell us in their latest book, they are part of a superorganism. Superorganisms such as some ant, bee, and termite colonies represent a level of organization intermediate between single organisms and the ecosystem: you can think of them as comprised of individuals whose coordination and integration have reached such a sophisticated level that they function with some of the seamlessness of a human body. The superorganism whose "hand" reaches into your sugar bowl is probably around the size of a large octopus or a garden shrub, and it will have positioned itself so that its vital parts are hidden and sheltered from climatic extremes while it still has easy access to food and water.

The term "superorganism" was first coined in 1928 by the great American ant expert William Morton Wheeler. Over the ensuing eighty years, as debates around sociobiology and genetics have altered our perspectives, the concept has fallen into and out of favor, and Hölldobler and Wilson's book is a self-professed and convincing appeal for its revival. Five years in the making, *The Superorganism* draws on centuries of entomological research, charting much of what we know of the evolution, ecology, and social organization of the ants.

For all its inherent interest to an intelligent lay reader, it's a technical work filled with complex genetics, chemistry, and entomological jargon such as, for example, "gamergate," "eclosed," and "anal trophallaxis." Occasional lapses add to the lay reader's difficulties. The etymology of "gamergate" ("married worker"), for example, which is so useful in understanding the term, is given only many pages after it's first introduced. I fear that *The Superorganism* may reach a smaller audience than it deserves, which is a great pity, for this is a profoundly important book with immediate relevance for anyone interested in the trends now shaping our own societies.

Ants first evolved around 100 million years ago, and they have since diversified enormously. With 14,000 described species, and perhaps as many still awaiting discovery, they have colonized every habitable continent and almost every conceivable ecological niche. They vary enormously in size and shape. The smallest are the leptanilline ants, which are so rarely encountered that few entomologists have ever seen one outside of a museum. They are possibly the most primitive ants in existence, and despite being less than a millimeter in length they are formidable hunters. Packs of these Lilliputian creatures swarm through the gaps between soil particles in search of venomous centipedes much larger than themselves, which form their only prey. The largest ant in existence, in contrast, is the bullet ant, *Dinoponera quadriceps* (of which Hölldobler and Wilson give abundant details, yet frustratingly neglect to inform us precisely how large these formidable-sounding creatures are). Inhabitants of the Neotropics—South and Central America —bullet ants belong to a great group known as the ponerines.

In explaining what a superorganism is, Hölldobler and Wilson draw up a useful set of "functional parallels" between an organism (such as ourselves) and the superorganism that is an ant colony. The individual ants, they say, function like cells in our body, an observation that's given more piquancy when we realize that, like many of our cells, individual ants are extremely short-lived. Depending upon the species, between 1 and 10 percent of the entire worker population of a colony dies each day, and in some species nearly half of the ants that forage outside the nest die daily. The specialized ant castes—such as workers, soldiers, and queens—correspond, they say, to our organs; and the queen ant, which in some instances never moves, but which can lay twenty eggs every minute for all of her decade-long life, is the equivalent of our gonads.

Pursuing the same reasoning, Hölldobler and Wilson argue that the nests of some ants correspond to the skin and skeleton of other creatures. Some ant nests are so enormous that they are akin to the skeletons of whales. Those of one species of leafcutter ant from South America, for example, can contain nearly two thousand individual chambers, some with a capacity of fifty liters, and they can involve the excavation of forty tons of earth and extend over hundreds of square feet. Coordination within such giant colonies, which can house 8 million individual ants, occurs through ant communication systems that are extraordinarily sophisticated and are the equivalent of the human nervous system. Not all ant species have reached this level of organization. Indeed, one of the most successful groups of ants, the ponerines, rarely qualifies for superorganism status.

Parallels between the ants and ourselves are striking for the light they shed on the nature of everyday human experiences. Some ants get forced into low-status jobs and are prevented from becoming upwardly mobile by other members of the colony. Garbage dump workers, for example, are confined to their humble and dangerous task of removing rubbish from the nest by other ants who respond aggressively to the odors that linger on the garbage workers' bodies.

Some of the most fascinating insights into ants have come from researchers who measure the amount of carbon dioxide given off by colonies. This is rather like measuring the respiration rate in humans in that it gives an indication of the amount of work the superorganism is doing. The researchers discovered (perhaps unsurprisingly) that colonies experiencing internal conflict between individuals seeking to become reproductively dominant produce more CO_2 than do tranquil colonies where the social order is long established. But extraordinarily, they also discovered that about three hours after removing a queen ant, the CO_2 emissions from a colony drop. "Removing the queen thus has a clear effect on worker behavior, apparently reducing their inclination to work for the colony," the researchers concluded. While it's dangerous to anthropomorphize, it seems that ants may have their periods of mourning just as we humans do when a great leader passes from us.

However, ants clearly are fundamentally different from us. A whimsical example concerns the work of ant morticians, which recognize ant corpses purely on the basis of the presence of a product

of decomposition called oleic acid. When researchers dau
ants with the acid, the undertakers promptly carry off the
daubed ants to the ant cemetery, despite the fact that they are ꓽ ꓹ ꞓ
and kicking. Indeed, unless they clean themselves very thoroughly
they are repeatedly dragged to the mortuary, despite showing every
other sign of life.

The means that ants use to find their way in the world are fascinating. It has recently been found that ant explorers count their
steps to determine where they are in relation to home. This remarkable ability was discovered by researchers who lengthened the
legs of ants by attaching stilts to them. The stilt-walking ants, they
observed, became lost on their way home to the nest at a distance
proportionate to the length of their stilts.

The principal tools ants use, however, in guiding their movements and actions are potent chemical signals known as pheromones. So pervasive and sophisticated are pheromones in coordinating actions among ants that it's appropriate to think of ants as
"speaking" to each other through pheromones. Around forty different pheromone-producing glands have been discovered in ants,
and, although no single species has all forty glands, enough diversity of signaling is present to allow for the most sophisticated inter-
actions. The fire ant, for example, uses just a few glands to produce
its eighteen pheromone signals, yet this number, along with two visual signals, is sufficient to allow its large and sophisticated colonies to function.

Pheromone trails are laid by ants as they travel, and along well-
used routes these trails take on the characteristics of a superhighway. From an ant's perspective, they are three-dimensional tunnels
perhaps a centimeter wide that lead to food, a garbage dump, or
home. If you wipe your finger across the trail of ants raiding your
sugar bowl, you can demonstrate how important the pheromone
trail is: as the ants reach the spot where your finger erased their
trail they will become confused and turn back or wander. The
chemicals used to mark such trails are extraordinarily potent. Just
one milligram of the trail pheromone used by some species of attine ants to guide workers to leaf-cutting sites is enough to lay an
ant superhighway sixty times around Earth.

Ant sex seems utterly alien. Except for short periods just before the
mating season, when an ant colony is reproducing, it is composed

entirely of females, and among some primitive species virgin births are common. All the offspring of such virgin mothers, however, are winged males that almost invariably leave the nest. If a female ant mates, however, all of her fertilized eggs become females. In many ant societies, reproduction is the prerogative of a single individual — the queen. She mates soon after leaving her natal colony and stores the sperm from that mating (or from multiple matings) all of her life, using it to fertilize (in some cases) millions of eggs over ten or more years.

Some ant species do not have queen ants in the strict sense. Instead, worker ants (which are all female) that have mated with a male ant become the dominant reproductive individuals. These are the gamergates, or "married workers," and their sex life can be brutal. In one species the gamergates venture outside of the nest to attract a male, engage him in copulation, then carry him into the nest before snipping off his genitals and throwing away the rest of his body. The severed genitals continue to inseminate the gamergate for up to an hour, after which they too are discarded. The fertilized gamergates then vie for dominance, causing disruptive conflict in the nest. Sometimes an oligarchy of gamergates is established, but in other instances a single gamergate triumphs.

You might think that such an established gamergate would watch the colony carefully for signs of emerging rivals, but this is not the case. Instead it's the worker ants that do so by taking a keen interest in the sexual status of their sisters. If they sense that one is becoming a sexually active gamergate, they will turn on her, either assaulting her or watching carefully until she produces eggs, which they promptly consume. It's intriguing that the sterile workers play the role of monitoring and regulating the sexual life of the colony. In a stretch of the imagination, I can see parallels between this behavior and the role of policing and censuring the sex lives of the rich and famous that gossip magazines play in our own society.

The ponerines are the most diverse of all the ant groups and are global in distribution. They cannot really be thought of as sophisticated superorganisms, however, for they tend to live in small colonies of a few tens to a few thousand individuals, with one Australian species living in colonies of just a dozen. Like Stone Age human hunters who specialized in killing woolly mammoths, the ponerines tend to specialize in hunting one or a few kinds of prey. That the great success of the ponerines is achieved despite their primitive social organization presents entomologists with what is known as the ponerine paradox. It lacks a widely accepted solution, but researchers suspect that it's the ponerines' predilection to seek specialized types of prey that limits their colony size (for such specialized hunters cannot gather enough food to develop large and sophisticated colonies). If this is the case, then the very characteristic that helps the ponerines to diversify and survive in a wide variety of environments also prevents them from attaining superorganism status.

The progress of ants from this relatively primitive state to the complexity of the most finely tuned superorganisms leaves no doubt that the progress of human evolution has largely followed a path taken by the ants tens of millions of years earlier. Beginning as simple hunter-gatherers, some ants have learned to herd and milk bugs, just as we milk cattle and sheep. There are ants that take slaves, ants that lay their eggs in the nests of foreign ants (much as cuckoos do among birds), leaving the upbringing of their young to others, and there are even ants that have discovered agriculture. These agricultural ants represent the highest level of ant civilization, yet it is not plants that they cultivate but mushrooms. These mushroom farmers are known as attines, and they are found only in the New World. Widely known as leafcutter ants, they are doubtless familiar from wildlife documentaries.

The attines, say Hölldobler and Wilson, are "Earth's ultimate superorganisms," and there is no doubt that their status is due to their agricultural economy, which they developed 50 to 60 million years before humans sowed the first seed. Indeed, it is in the changes wrought in attine societies by agriculture that the principal interest for the student of human societies lies. The most sophisticated of attine ant species has a single queen in a colony of millions of sterile workers that vary greatly in size and shape, the largest being two hundred times heavier than the smallest. Their system of worker specialization is so intricate that it recalls Swift's ditty on fleas:

> So, naturalists observe, a flea
> Has smaller fleas that on him prey;
> And these have smaller still to bite 'em;
> And so proceed *ad infinitum*.

In the case of the attines, however, the varying size classes have specific jobs to do. Some cut a piece from a leaf and drop it to the ground, while others carry the leaf fragment to a depot. From there others carry it to the nest, where smaller ants cut it into fragments. Then ants that are smaller still take these pieces and crush and mold them into pellets, which even smaller ants plant out with strands of fungus. Finally, the very smallest ants, known as minims, weed and tend the growing fungus bed. These minute and dedicated gardeners do get an occasional outing, however, for they are known to walk to where the leaves are being cut and hitch a ride back to the nest on a leaf fragment. Their purpose in doing this is to protect the carrier ants from parasitic flies that would otherwise attack them. Clearly, not only did the attines beat us to agriculture, but they exemplified the concept of the division of labor long before Adam Smith stated it.

You may not believe it, but, like the sailors of old, the leafcutter ants "sing" as they work. Leaf-cutting is every bit as strenuous for the ants as hauling an anchor is for human beings, and their singing, which takes the form of stridulation (a sound created by the rubbing together of body parts), assists the ants in their work by imparting vibrations to the mandible that is cutting the leaf, enhancing its action in a manner akin to the way an electric knife helps us cut roasts. The leafcutters also use stridulation to cry for help, for example when workers are trapped in an underground cave-in. These cries for help soon prompt other ants to rush in and begin digging until they've reached their trapped sisters.

The fungus farmed by the leafcutter ants grows in underground chambers whose temperature, humidity, and acidity are precisely regulated to optimize its growth. The fungus, which produces a tiny mushroom, grows nowhere else, and genetic studies reveal that various attine ant species have been cultivating the same fungus strain for millions of years. In truth, after tens of millions of years of coevolution, such is their interdependence that the ants cannot live without the fungus nor the fungus without the ants. The system is not perfect, however, for the ants' fungal gardens are occasionally devastated by pests. One of the worst is an invasive fungus known as *Escovopsis,* whose depredations can become so severe that the leafcutters must desert their hard-won gardens and start elsewhere anew. Often a colony so beset evicts a smaller attine colony, taking over the premises and enlarging them to suit.

Fortunately, the ants possess a potent defense against this fungal weed that usually prevents its proliferation. Their fungicide is produced by a bacterium that is found only in pits located on specific parts of the ants' bodies and is known to exist nowhere else. These bacteria produce secretions that not only destroy the *Escovopsis* pest but promote the growth of the fungus the ants wish to cultivate. Thus these special bacteria must be considered as comprising the third element in a triumvirate of coevolved organisms, whose fate is now so closely interwoven that they are utterly interdependent and form a single, functional whole. Humanity's dependence upon a few grains—principally wheat and rice—and the complete dependence on cultivated varieties of these plants by human farmers presents a similar symbiosis.

One curious aspect of the agricultural enterprise of the attines is that the worker ants rarely eat the fungus they cultivate. Studies show that the adults gain most of their nutrition from plant sap, deriving a mere 5 percent from fungus. The balance of nutrients in the fungus, as it happens, is poorly suited to the needs of adult ants but is perfect for their growing young. The mushroom gardens are thus cultivated principally for the delectation of the ant larvae. Indeed it forms their only source of food.

When growing fungus on such a large scale, waste management becomes a crucial issue, and the attines have developed a finely tuned solution. Their sanitation teams comprise one group of workers that gather the refuse from inside the colony and dump it at depots outside. From there dump managers that work exclusively outside the nest carry the waste to great disposal sites far from the colony. The dump managers that work outside are mostly older ants that have only a short time to live in any case, which is a good thing, for the great refuse dumps they toil at teem with pathogens and toxins. This system effectively quarantines the colony from a dangerous threat and at the same time minimizes loss of worker life. Curiously, humans have found a use for the ant refuse. So strong is the ants' aversion to it that South American farmers gather it and sprinkle it around young plants they wish to protect from attacks by leafcutters.

One can hardly help but admire the intelligence of the ant colony, yet theirs is an intelligence of a very particular kind. "Nothing in the brain of a worker ant represents a blueprint of the social order," Hölldobler and Wilson tell us, and there is no overseer or "brain caste" that carries such a master plan in its head. Instead, the ants have discovered how to create strength from weakness by pooling their individually limited capacities into a collective decision-making system that bears an uncanny resemblance to our own democratic processes.

This capacity is perhaps most clearly illustrated when an ant colony finds reason to move. Many ants live in cavities in trees or rocks, and the size, temperature, humidity, and precise form and location of the chamber are all critically important to the success of the superorganism. Individual ants appear to size up the suitability of a new cavity using a rule of thumb called Buffon's needle algorithm. Each one does this by laying a pheromone trail across the cavity that is unique to that individual ant, then walking about the space for a given period of time. The more often they cross their own trail, the smaller the cavity is.

This yields only a rough measure of the cavity's size, for some ants using it may choose cavities that are too large, and others will choose cavities that are too small. The cavity deemed most suitable by the majority, however, is likely to be the best. The means employed by the ants to "count votes" for and against a new cavity is the essence of elegance and simplicity, for the cavity visited by the most ants has the strongest pheromone trail leading to it, and it is in following this trail that the superorganism makes its collective decision. The band of sisters thus sets off with a unity of purpose, dragging their gargantuan queen and all their eggs and young to a new home that gives them the greatest chance of a comfortable and successful life.

Reflecting on our own societies when armed with knowledge of the ants as provided in *The Superorganism,* it's hard to avoid the conclusion that we are in the process of metamorphosing into the largest, most formidable superorganism of all time. Yet even the creation of a superorganism on this colossal scale is not entirely new, for just thirty years ago another gargantuan superorganism came into existence, and it was the ants that created it. This superorganism is composed of fire ants, and already it covers most of the southern United States. It consists of billions of individuals whose ancestors were accidentally imported from South America to Mobile, Alabama, in the 1930s.

In their native land fire ants form discrete colonies, with just one or a few queen ants at the center of each. This is how most ants live, but something very strange happened to the fire ants soon after they reached the United States. They gave up founding colonies by the traditional method of sending off flights of virgin queens, and instead began producing many small queens, which spread the colony rather in the way an amoeba spreads, by establishing extensions of the original body. Astonishingly, at the same time the ants ceased to defend colony boundaries against other fire ants. As Hölldobler and Wilson put it, "With territorial boundaries erased, local populations now coalesce into a single sheet of intercompatible ants spread across the inhabited landscape." This remarkable shift was caused by a change in the frequency of a single gene.

Is it possible, *The Superorganism* left me wondering, that the invention of the Internet is leading to a similar social evolution of our own species? The proliferation of conflict, much of it prompted by defense of national boundaries, may make us doubt it, but other trends are occurring that give pause for thought. As we strive to avert a global economic disaster or agree on a global treaty to prevent catastrophic climate change, we inevitably build structures that, as with the ants, allow the superorganism to function more efficiently. But of course it's possible that we'll fail to make the grade—that our destructive path will catch up with us before we can make the transition to a seamlessly working superorganism.

When conferring an honorary degree upon the man who invented the term "superorganism," President Lowell of Harvard University said of William Morton Wheeler that he had demonstrated how ants "like human beings can create civilizations without the use of reason." Create perhaps, but there is no question of maintaining this first global civilization without resort to humanity's defining faculty. As the twenty-first century progresses we'll doubtless find ourselves trying to shape our planet-sized nest as carefully as an ant colony does, but the great difference is this: in the case of the human superorganism it will be our intelligence that will guide us. We have to hope that we shall find ourselves living sustainably in a global superorganism whose own self-created intelligence has been bent to the management and the maintenance of its life systems for the greater good of life as a whole.

KENNETH BROWER

Still Blue

FROM *National Geographic*

IN ACAPULCO HARBOR, amid the white yachts, R.V. *Pacific Storm* stood out: a working boat, black hulled, a West Coast trawler in a previous life, reborn now as a research vessel. There were bigger, more opulent boats in the harbor—fortunes are invested in the white yachts of Acapulco—but this eighty-five-foot trawler, with its grim mien and high black bow, was the ship for me. Asked to choose, from all this fleet, the vessel to carry me on a month-long cruise in pursuit of blue whales, I would not have hesitated. As Flip Nicklin and I passed our gear up the trawler's ladder and stowed it in our cabin, I felt an almost savage contentment.

Call me Ishmael, if you like, but whenever I find myself growing grim about the mouth; whenever it is a damp, drizzly November in my soul; whenever I have spent too many consecutive months at the computer keyboard, in artificial light, like some sort of troglo-dyte, self-imprisoned, pecking out my living, I account it high time to get to sea as soon as I can. I jumped at the assignment on *Pacific Storm*. As the voyage was to depart on the third of January, I made three New Year's resolutions: I would try to be an affable shipmate. I would strip all the blubber from my prose. I would refrain from making a single allusion to Herman Melville.

Did I mention we were after a white whale?

It's true. In the eastern North Pacific population of blue whales —the group that summers mostly off California and whose migra-tion we were following south—there is a white blue whale, maybe an albino. An inflatable skiff from *Pacific Storm* had satellite-tagged this whale off Santa Barbara four months before, but his tag, number 4172, had ceased transmitting a few weeks after implantation, and now his whereabouts were a mystery. The sun-synchronous, polar-orbiting TIROS N satellites could no longer track him, but he was one of the animals we hoped to see off Central America.

When we had settled in on *Pacific Storm,* Nicklin, cross-legged on his bunk, set up his Nikon D200 with its Sea & Sea underwater dome. He squeezed a dab of silicone grease from a small tube onto his fingertip and ran it around the rim of the dome's blue O-ring. He opened the back of the camera and gave a similar treatment to the O-ring at the stern. Nicklin is a new kind of whaler. His job is not to render the oil but to capture the essence of cetaceans, and the Nikon is his favorite harpoon.

Pacific Storm put to sea. We sailed a leg due south to avoid the Tehuantepec winds along the eastward bend of Central America, then turned southwest toward the temperature anomaly that was our destination.

The Costa Rica Dome is an upwelling of cold, nutrient-rich water generated by a meeting of winds and currents west of Central America. The location is not fixed; it meanders a bit, but the dome is reliably encountered somewhere between three hundred and five hundred miles offshore. The upwelling brings the thermocline—the boundary layer between deep, cold water and the warm water of the surface—up as high as thirty feet from the top. Upwelling with the cold, oxygen-poor water from the depths come nitrate, phosphate, silicate, and other nutrients. This manna, or antimanna—a gift not from heaven but from the deep—makes for an oasis in the sea. The upwelling nutrients of the dome fertilize the tiny plants of the phytoplankton, which feed the tiny animals of the zooplankton, which bring bigger animals, some of which are very big indeed.

The blue whale, *Balaenoptera musculus,* is the largest creature ever to live. Linnaeus derived the genus name from the Latin *balaena,* "whale," and the Greek *pteron,* "fin" or "wing." His species name, *musculus,* is the diminutive of the Latin *mus,* "mouse"—apparently a Linnaean joke. The "little mouse whale" can grow to 200 tons and 100 feet long. A single little mouse whale weighs as much as the entire National Football League. Just as an elephant might pick up a little mouse in its trunk, so the elephant, in its turn, might be taken up by a blue whale and carried along on the colossal tongue.

Had Jonah been injected intravenously, instead of swallowed, he could have swum the arterial vessels of this whale, boosted along every ten seconds or so by the slow, godlike pulse.

The great swimming speed of the blue whale, together with the remoteness of its stronghold—where three of Earth's oceans merge in the ice-cold waters around Antarctica—protected most of the species until early in the twentieth century. With the invention of explosive harpoons and fast, steam-powered catcher boats, the stronghold was breached. Through the first six decades of the twentieth century, 360,000 blue whales were killed. The population around South Georgia Island was extirpated, along with those that once fed in the coastal waters of Japan. Some blue whale populations were reduced by ninety-nine one-hundredths, and the species tipped at the very brink of extinction.

For Bruce Mate and John Calambokidis, the head scientists aboard *Pacific Storm,* the irony is deep and poignant. The blue whales they study, the two thousand animals that summer off western North America, once just a splinter group, now make up a significant population.

Mate, director of the Marine Mammal Institute at Oregon State University, is the world's most inventive and prolific satellite-tagger of whales. The dome first caught his attention in 1995, when a blue whale he had tagged off California in summer began transmitting off Costa Rica in winter. Calambokidis, a cofounder of Cascadia Research, in Olympia, Washington, is the West Coast's most prolific photo-identifier of whales. A tall, lean biologist with a Quaker seaman's beard and a monomaniacal dedication to bringing back diagnostic images, Calambokidis was tantalized by the reports from the satellite. In 1999 he made a reconnaissance of the dome by sailboat. The voyage was plagued by bad weather, and the sailboat was too small for its mission, yet at the dome Calambokidis managed to photo-identify ten whales that he had photographed off California.

Why would a blue whale depart from its feeding grounds at the end of summer and migrate thousands of miles to spend winter in this tropical zone of upwelling? Mate and Calambokidis thought they knew. The satellite data showed that some of the tagged whales lingered five months or more at the dome, arriving early in the southern migration and departing late—a pattern that, in other species of baleen whales, is seen in pregnant females and new mothers. It had never been noted in blue whales, for the best of reasons: no one has ever witnessed the birth of a blue whale.

Gray, humpback, and right whales — the baleen species that have been studied at their calving grounds — seem to feed little, if at all, at those grounds. But there is evidence that the blue whale might be different. Given its great size and enormous energy requirements, the blue whale may be forced to find winter grounds where it can do more than snack. The oasis of the Costa Rica Dome would satisfy this requirement. Plus, the productivity of the upwelling would help nursing mothers convert schools of krill into the barrels of milk required by the calves to put on their two hundred pounds a day.

Balaenoptera musculus received international protection in the mid-1960s, yet, for reasons not fully understood, it has scarcely rebounded. If the greatest of creatures is to come back, Mate and Calambokidis believe, its demographics and its movements need to be charted. The largest remaining population of the species is most vulnerable in tropical waters where it gives birth to dainty, twenty-five-foot-long, three-ton calves.

As we followed the corridor of the blue whale migration southward, we took turns standing whale watch on the bridge, searching the horizon for blows. Whales 5801 and 23043 had already arrived at the dome, according to the satellite, and number 5670 was nearing it. The scientists were particularly interested in 23043, because they knew the sex, female, and because she had arrived at the dome early, as one might expect of a mother-to-be. The white blue whale, 4172, if he was migrating to the dome this year, was out there somewhere in the host moving south. The Pacific is a big ocean, however, and we saw not a single spout.

Now and again, day and night, the ship shifted to neutral, and the researchers put gear overboard: a CTD sensor, an echo sounder, and a hydrophone. The CTD sensor recorded conductivity (a measure of salinity), temperature, and depth. The echo sounder searched for concentrations of krill, upon which the blue whale subsists almost entirely. "We're doing some control observation on the way down," Mate explained. "If there's no krill, will the whales pass through? If there are big concentrations of krill, will they hang around? We're looking for poop. We'll try to scoop it up, see if they're feeding. And checking their breath, which is fouler when they've eaten. I don't find blue whale breath offensive — certainly not in comparison to gray whale breath, which is really foul — but blue whale breath can be strong."

The hydrophone was to detect blue whale voices. The simple song of the blue whale bull — the thumping, stentorian, basso profundo pulse of the A call, followed by the continuous tone of the B call — is the mightiest song in the sea, theoretically capable of propagating halfway across an ocean basin. But big baleen whales often run silent. Except for a few dubious snatches of song, we heard nothing at all.

When we reached the Costa Rica Dome, three days out of Acapulco, the ocean looked no different, just blue horizon and marching swells. It took a sounding by the CTD sensor to detect the thermocline lying just sixty feet under the surface. We had arrived. "Blow at eleven o'clock!" Calambokidis called down the next morning from the crosstrees, our crow's-nest, over his walkie-talkie. We saw two more blows side by side in quick succession — our first blue whales — and we launched the tagging boats, beginning the repetitive ritual that would occupy us for the next three weeks.

The boats were Coast Guard surplus, a pair of diesel-powered RHIBs, or rigid-hull inflatable boats. Sticking with meteorological nomenclature, we called the big one *Hurricane* and the small one *Squall.* I generally went out on *Hurricane.* Its commander was Bruce Mate. The second mate, and also the second Mate, was Mary Lou, the expedition videographer and the professor's wife of forty years. I was the biopsy guy. My first job was to cock my crossbow, take a biopsy bolt from the cooler that served as ammunition box, nock the bolt, and then remove the sheath of aluminum foil protecting the tip from contamination by extraneous DNA. The bolt, when shot into the whale, would excise a plug of skin and blubber. About three inches back from its tip, the bolt was blocked by an oblong ball of yellow rubber that prevented the projectile from going in too deep and also served to bounce it off the whale.

Mounted on the rubber bow of *Hurricane* was a metal bowsprit, the "pulpit," custom-made for this work. Each time we closed on whales, I would follow Professor Mate up onto the narrow grate of the pulpit deck. From its holster, which was a transparent plastic tube lashed to the pulpit rail, Mate withdrew the satellite-tag "applicator," a long-barreled, red-metal blunderbuss with a wooden rifle stock. This device, originally a Norwegian invention for shooting line between ships, is powered by compressed air from a scuba tank. The pop is adjustable. For blue whales, Mate sets the dial at 85 pounds per square inch of pressure. For sperm whales, which have very tough skin, he sets the pressure at 120 pounds. Both Mate and I wore waist harnesses, which we clipped into slings on the pulpit rail, freeing up our hands for the shooting.

The first we saw of a whale was almost always its blow.

When the sun was behind us, we sometimes saw a prismatic scatter of color in the explosive expansion of spray and vapor — a few milliseconds of rainbow — before the color shimmered out and the spout faded to white.

Whenever a blue whale surfaced to blow nearby, I was struck by the blowhole — a pair of nostrils countersunk atop the tapering mound of the splash guard, built up almost into a kind of nose on the back of the head. Other baleen whales have splash guards too, but not like this. This nose was almost Roman. It seemed disproportionately large, even for the biggest of whales. Its size explained that loud, concussive exhalation — less a breath than a detonation — and its size explained the thirty-foot spout. It was a mighty blow, followed quickly by a mighty inhalation.

The second thing we saw of the whale was its back.

The blue whale is "a light bluish gray overall, mottled with gray or grayish white," as one field guide describes it, and the back is often, indeed, this advertised color, but just as often, depending on the light, the back shows as silvery gray or pale tan. Whichever the color, the back always has a glassy shine. When you are close, you see the water sluicing off the vast back, first in rivulets and sheets, and then in a film that flows in lovely, pulsed patterns downhill to the sea.

If blue whales above water are only putatively blue, then below the surface they go indisputably turquoise. *Balaenoptera musculus* is a pale whale, and when seen through the blue filter of the ocean, its pallor goes turquoise or aquamarine. This view of the whale, downward through twenty to fifty feet of water, is for me the most haunting and evocative.

If the most beautiful hue of the blue whale is turquoise, then the most beautiful form, the finest sculpture, is in the flukes. In the first week of our tagging efforts, the tail always seemed to be waving goodbye. "Ta-ta," it signaled. "Nice try. Better luck next time." When a whale showed its flukes—when the two palmate blades poised high in the air—we would break off the chase, because elevated flukes meant a deep dive.

But sometimes we saw the flukes close under the surface. They were huge, wider than the boat, and in motion they were hypnotically lovely. "In no living thing are the lines of beauty more exquisitely defined than in the crescentic borders of these flukes," Melville writes in *Moby Dick*.

The last thing we saw of the whale was its "flukeprint."

When a whale or dolphin swims at shallow depths, turbulence from its flukes rises to form a circular slick on the surface: the footprint or flukeprint. The flukeprints of blue whales are large and surprisingly persistent. The smooth patch lingers long after the whale is gone. "It's a measure of how much energy is in the stroke," Mate told me one afternoon when he caught me staring at one of these slicks. The circle of the flukeprint is perfectly smooth, except for a few faint curves that mark the continued upwelling of energy. Eventually the chop of the ocean begins to erode the slick from the outside inward, but only slowly.

The emphatic flukeprint was another of those discouraging signs that caused us to call off a chase. "Holy smokes!" Mate said one afternoon, as we motored into the middle of a huge one. Ladd Irvine, a research assistant who served as helmsman, laughed in admiration: "We're not going to see him again for a while."

Out on the pulpit, the professor spread his feet for balance, rested the butt of his applicator on the grating of the pulpit deck, and gripped the barrel just below the muzzle-loaded, chiseled tip of his satellite tag. His quick-dry khaki pants luffed and billowed in the sea wind, and now and again the breeze brought a powerful smell of staleness and mold, mixed sometimes with an alarming flatulence. Whew, Bruce! I thought on more than one occasion. Then one day, as the wind rippled in his khakis and we closed in on the spout ahead, the professor emitted a blast so powerful, inhuman, and malodorous that I realized he had to be completely innocent. What I had been smelling, all along, was not our leader. I had been smelling the bad breath of blue whales.

For almost a week at the dome, every whale slipped away from us. On our sixth day our luck changed. We saw three spouts to the southeast that morning and launched *Hurricane*.

The first two whales toyed with us, as usual, allowing us close, then pulling away. The third allowed us to get in perfect position. We paced the great turquoise shape, keeping abreast of the flukes as the whale coursed along underwater to starboard. As the animal surfaced to blow, it angled up from turquoise abstraction into photo-realism. Irvine gunned the engine. Up in the pulpit I clicked off my crossbow's safety. Mate tucked the rifle stock of the tag applicator into his shoulder, leaned outward over the pulpit rail, and aimed the long, red barrel almost straight downward at the rising whale, now just ten feet underwater. The whale blew, and the glistening wall of its flank erupted in a steep curve above the sea.

My instructions as biopsy guy were to wait for the bang of the tag applicator before firing my crossbow. The smooth flank of the whale filled my whole field of view; there was no way I could miss. At the bang of the applicator, I pulled my trigger. The bolt left the crossbow, and a black hole, small but inky, appeared where I had been aiming. It took a millisecond for me to understand that I was responsible for it, and I felt a pang of regret and guilt. *I did that?* I thought, like a boy whose pop fly has gone through a stained-glass window.

Then my sense of proportion returned. In relation to the vastness of this whale, my hole was just a mosquito bite. This was not a crime; it was a blow for science. On the pulpit, Mate and I unclipped our harnesses and shook hands.

The blue whale writes a kind of longhand on the surface of the sea. There is the ovoid slick that forms above the head the moment before emergence, the long, narrow slick left by the arching back, and the circular slick of the flukeprint. There are the sputtering white fountains that a blue whale raises by blowing early, still gliding under the surface—a sequence of premature spouts. There are bubble blasts. I saw my first of these just ahead of the bowsprit, about twelve feet deep, as the blowhole of a whale erupted a big bolus of bubbles. It expanded toward the surface, vitreous and glittery, like a crystal chandelier falling upward. "Bubble blast," observed Mate.

This particular bubble blast seemed to be commentary directed at our persistent and irritating little boat—some kind of whale expletive, probably. It rose above the whale's head like a speech balloon in a Gary Larson cartoon. Its message was something like "@*#&%$!?!"

Of all the marks of blue whale cursive, the most colorful was the defecation trail. The first defecation we saw was in a yearling, a little fifty-footer. This whale blew forty yards away, and behind it the ocean brightened in a long, red-orange contrail. "We have a defecation," Irvine announced. This contrail, a brick red streak of processed krill, more watery than particulate, was our first direct evidence that blue whales were feeding in winter at the Costa Rica Dome. As this was one of the hypotheses this expedition had been launched to test, Mate scrambled to find a Ziploc bag to collect a sample.

The evidence for feeding that we observed firsthand in the defecation trails was corroborated in the ship's laboratory. On her computer screen, Robyn Matteson, Mate's graduate student, monitored the echo sounder and the concentrations of krill it detected at the dome. Krill distribution was patchier than anyone had imagined, but dense schools of the small crustaceans were plainly here. Across the lab table, at their own computers, Calambokidis and Erin Oleson of Scripps Institution of Oceanography studied the dive profiles recorded by acoustic tags they had succeeded in applying to several whales. The acoustic tags, deployed by pole and attached by suction cups, stay on the whale for hours, not months, like the more invasive satellite tags. Here at the dome, the depth recorders on the tags showed dives to eight hundred feet and deeper. The vertical line marking each dive, on reaching its greatest depth, began to zigzag in the sawtooth pattern characteristic of blue whales when lunge-feeding on krill.

The evidence for calving at the Costa Rica Dome proved more elusive, but after many fruitless days, it arrived finally, to starboard, by way of a mother and her calf.

The pair were moving slowly, spending a lot of time at the surface. The mother surprised us by allowing her calf to turn toward *Pacific Storm*. A mother whale often interposes herself between her calf and potential danger, but this mother was an easygoing, Montessori sort of parent, and she let her baby explore.

John Calambokidis drove *Squall* out to snap surface pictures for photo identification. Nicklin and cameraman Ernie Kovacs grabbed their gear and went along. On nearing the whales, they pulled on their fins and slipped overboard. At first they saw nothing through their dive masks but blue. Then Kovacs, looking for the youngster, was startled to see it pass, maybe five feet below his fins. This whale was just a baby, yet its blue back seemed to pass under him endlessly. The calf, gliding by Nicklin, rolled slightly to bring an eye to bear on him. It peered into the glass orb of the camera housing, and Nicklin's shutter winked back.

After twenty-one days at the Costa Rica Dome, we could stay no longer and turned north for Acapulco.

On the voyage home, we took stock. There had been disappointments: we wished we had satellite-tagged more whales, had seen more calves, had experienced more underwater encounters with blue whales. We were sorry not to have glimpsed whale 4172, the white bull. But for the most part we were satisfied.

In three weeks spent crisscrossing the dome, we had succeeded in finding three whales satellite-tagged in California and tracked down here. Each time we homed in on the transmissions of one of these telemetric whales, we had found it in the company of "clean" whales. Satellite-tagging had proved itself an efficient method for locating concentrations of the untagged. We had satellite-tagged three new blue whales (but one tag failed to transmit), affixed acoustic tags to six more, and photo-identified about seventy. Thirteen of those seventy were from California. The voyage proved that the dome is visited by large numbers of blue whales. We saw many threesomes, the romantic triangles of the blue whale, and we witnessed much boisterous courtship behavior, all suggesting that the dome is a mating ground. We demonstrated beyond a doubt that blue whales do feed here in the winter. With sonobuoys and acoustic tags, we eavesdropped on A and B calls of the blue whale song and on the D calls whales make between bouts of feeding, and thus began notation of the winter music in this patch of ocean.

The news from the dome is good.

The grandest creature in all creation has been hunted by our kind, the thinking ape, to near extinction. Its numbers still are low, but it was hard not to feel optimistic. In my bunk with Nicklin's laptop, lingering over his digital portraits of the curious calf, I thought I could read, in its strange visage, a gargantuan impishness. I found this cheering. The young do give us hope.

On the voyage home, we found time for reflection, and I understood why the blue whale's flukeprint so mesmerized me each time I saw it at the dome. That big, circular slick is the signature of the species, the John Hancock of flukeprints, outsize and insistent. It jumps out boldly from the parchment. Its uncanny persistence on the sea's surface, defying the choppiness, is a good omen. Appearing at the dome, this winter haven, it suggests that the blue whale might after all defy the chop of history.

"Still here!" the flukeprint says.

JANE GOODALL

The Lazarus Effect

FROM *Discover*

IN 2008, DURING MY LECTURE TOUR in Australia, a very large, very black, very friendly Lord Howe Island stick insect (*Dryococelus australis*) crawled across my hands, my face, and my head. The encounter sent shivers up my spine—knowing, as I did, the incredible story of how it came to be there.

The forests of Lord Howe Island, about three hundred miles off the coast of New South Wales, Australia, were the only known home of the Lord Howe Island phasmid, also called a stick insect or walking stick—a creature about the size of a large cigar, four or five inches long and half an inch wide. In 1918 black rats arrived on the island after a shipwreck, relentlessly adapting to their new environment and probably finding easy and delicious prey in the giant phasmid, which lacked wings. At some point in the 1920s, the Lord Howe Island phasmid was presumed extinct.

Then, in 1964, rock climbers found the dried-out remains of a giant stick insect on Ball's Pyramid, an 1,800-foot-tall spire of volcanic rock fourteen miles from Lord Howe Island. Five years later, other rock climbers found two other dried bodies incorporated into a bird's nest on a remote pinnacle of the spire, a place almost entirely without vegetation. It seemed impossible that a large, forest-loving vegetarian insect could be surviving in such a bleak environment. And so biologists ignored these reports until, in February 2001, a small group of people—David Priddel, the senior research scientist of the Department of Environment and Climate Change in New South Wales, his colleague Nicholas Carlile, and two other intrepid souls—decided to settle the matter once and for all.

The seas around Ball's Pyramid are rough, and the team of three men and one woman had to leap from their small boat onto the rocks. ("Swimming would have been much easier, but there are too many sharks," Carlile said.) They put up a small camp and set off to climb about 500 feet up the spire of rock where the main vegetative patches clung to life. They searched thoroughly but found nothing other than some big crickets, and eventually the heat and lack of water drove them back down. Then, in a crevice 225 feet above the sea, they came upon another tiny patch of comparatively lush vegetation, dominated by a single melaleuca bush. Here they found the fresh droppings of some large insect.

Back in camp, over supper, they discussed the situation. Priddel knew that stick insects were nocturnal and that the group would have a better chance of seeing them if they went back to that bush at night. Carlile and team member Dean Hiscox—a local ranger and expert rock climber—volunteered to make the almost suicidal climb in the dark. Finally they reached the vegetated area and saw one and then two enormous, shining, black-looking bodies spread out on the bush. "It felt like stepping back into the Jurassic age, when insects ruled the world," Carlile said.

Early the next morning, the whole team climbed back up and made a thorough search. They found some frass (the proper terminology for insect poo) and about thirty eggs in the soil. They were all convinced that the only population in the world of Lord Howe Island's giant phasmid lived on that one melaleuca shrub.

How did the little colony get to that isolated pillar of rock? Perhaps a female, full of eggs, had made the fourteen-mile journey from Lord Howe Island clinging to the leg of some seabird or floating on some vegetation after a storm. And once there, she had found the one and only suitable habitat on the entire pyramid, that little bush. The point is, she got there somehow. How her descendants survived for eighty years in that desolate environment we shall never know.

As soon as they returned, the biologists got to work on a recovery plan for the stick insect. They faced many battles with bureaucracy, and two years elapsed before they had permission to return—and they were allowed to catch only four individuals. When they arrived, they found that there had been a big rockslide on Ball's Pyramid. How easily the entire population could have been wiped out during those two frustrating years. However, on Valen-

tine's Day in 2003, they found the colony still thriving on its one
bush. To transport the incredibly rare insects, a special container
had been prepared, and this presented a problem when they arrived in Australia. It was not long after 9/11, and security was very
tight, yet the scientists had to convince officials not to open the
precious box! One pair of insects went to a private breeder in Sydney, and the other two, Adam and Eve, went to the Melbourne Zoo.
To everyone's delight and relief, Eve soon began laying pea-size
eggs.

Within two weeks of arriving in Australia, the pair in Sydney died
and Eve became very, very sick. Patrick Honan, a member of the
Invertebrate Conservation Breeding Group, worked every night
for a month desperately trying to cure her. He scoured the Internet for help, but no one knew anything about the veterinary care
of giant stick insects! Eventually, based on gut instinct, Patrick con-
cocted a mixture that included calcium and nectar and fed it to his
patient, drop by drop, as she lay curled up in his hand. To his joy,
she seemed to get better, and she laid eggs for a further eighteen
months. But the only ones that hatched were the thirty or so that
she had laid before she fell sick.

In 2008, when I visited the Melbourne Zoo, Patrick showed me
his rows of incubating eggs: 11,376 at the last count, with about
700 adults in the captive population. He showed me a photo of
how they sleep at night, in pairs, the male with three of his legs
protectively over the female beside him. As further insurance for
the survival of the species, eggs are now being sent to other zoos
and private breeders in Australia and overseas. The 200 eggs that
were sent to the San Antonio Zoo in Texas have already begun to
hatch.

My second story is about a very small and very beautiful breed of
horse and an American woman, Louise Firouz, who "discovered"
and rescued the animals from obscurity in Iran. Louise had married a young man from the Iranian royal family, Narcy Firouz, and
had become a princess. In 1957 the young couple established the
Norouzabad Equestrian Center, where the wealthier Iranian families sent their children to learn to ride. But the horses were typically too big for the smaller children, including their own three.
And so when, in 1965, Louise heard rumors of a small pony in the

Elburz Mountains near the Caspian Sea, she determined to investigate. She set out on horseback with a few women friends, and she found the "ponies." They were being used as work animals, pulling carts, malnourished and covered with ticks.

Almost at once Louise realized that these were not ponies at all—they had the distinctive gait, temperament, and facial bone structure of horses. Very small, narrow horses to be sure, standing just over forty inches, but horses for all that.

As she pondered the nature of this little horse, Louise suddenly remembered seeing, on the walls of the ancient palace in Persepolis, relief carvings of a horse that looked very much like the one she had just found. The Lydian horse depicted in those carvings had the same small, prominent skull formation. With a sense of excitement, Louise began to wonder whether, hidden beneath the matted coats of these work animals, there was a true representative of the ancient lost breed of the royals, considered extinct for a thousand years. She found that there were still five purebred horses in the village, and she bought three of them. After extensive DNA testing, archaeozoologists and genetic specialists agreed with Louise that these animals were indeed Caspian horses, the ancestral form of the Arabian horse. What an incredible find!

At first Louise and Narcy financed the breeding themselves, but then in 1970 a Royal Horse Society was formed in Iran. The society's mission was to protect Iran's native breeds, and it bought all of Louise's Caspian horses, which by then numbered twenty-three. Louise and Narcy then started a second, private herd near the Turkmenistan border. When two mares and a foal were killed by wolves, Louise, wanting to ensure that some of the horses would be kept safe, arranged for eight of them to be exported to Britain in 1977. The Royal Horse Society was angered, presumably because it had not been consulted. The society immediately banned all further exports of Caspian horses and began collecting all of the animals that remained in Iran, including all but one of the Firouzes' second herd.

Then came the 1979 Islamic revolution. Because of their connections with the royal family, the Firouzes were imprisoned. Narcy was jailed for six months, but Louise for only a few weeks, for she remembered advice given to her by a friend: that if she went to prison, she should go on a hunger strike. This worked, but during that time most of the Caspian horses were auctioned for use as beasts of burden or slaughtered for meat.

Still passionate about saving her beloved Caspian horses, Louise managed to rescue some of those that remained from starvation and slaughter and established, for the third time, a small herd in Iran. And once again she managed to export some of them to safety.

The last such effort was in the early 1990s, when Louise sent seven horses on a dangerous journey to the United Kingdom. They had to pass through the Belarus war zone, where bandits attacked and robbed the convoy. The horses arrived safely, but it had been a costly business. Soon after, in 1994, Louise's husband died, and she could no longer afford her breeding program.

With Iran's many political upheavals—the overthrow of the shah, the Iran-Iraq war, the very real threat of famine—as well as the Caspian's former association with royalty—the fate of these horses was ever in the balance. One moment they were considered a national treasure, the next they were seized as wartime food. But thanks to Louise Firouz, who had exported a total of nine stallions and seventeen mares, the future of this ancient line has been ensured. Today they can be found in England, France, Australia, Scandinavia, New Zealand, and the United States.

DAVID QUAMMEN

Darwin's First Clues

FROM *National Geographic*

THE JOURNEY OF YOUNG CHARLES DARWIN aboard His Majesty's Ship *Beagle,* during the years 1831–36, is one of the best known and most neatly mythologized episodes in the history of science. As the legend goes, Darwin sailed as ship's naturalist on the *Beagle,* visited the Galápagos archipelago in the eastern Pacific Ocean, and there beheld giant tortoises and finches. The finches, many species of them, were distinguishable by differently shaped beaks, suggesting adaptations to particular diets. The tortoises, island by island, carried differently shaped shells.

These clues from the Galápagos led Darwin (immediately? long afterward? here the mythic story is vague) to conclude that Earth's living diversity has arisen by an organic process of descent with modification—evolution, as it's now known—and that natural selection is the mechanism. He wrote a book called *The Origin of Species* and persuaded everyone, except the Anglican Church establishment, that it was so.

Well, yes and no. This cartoonish account of the *Beagle* voyage and its consequences contains a fair bit of truth, but it also confuses, distorts, and omits much. For instance, the finches weren't as illuminating as the diversity of the islands' mockingbirds, at least initially, and Darwin couldn't make sense of them until a bird expert back in England helped. The Galápagos stopover was a brief anomaly near the end of an expedition devoted mostly to surveying the South American coastline. Darwin hadn't signed on to the *Beagle* as its official naturalist; he was a twenty-two-year-old Cambridge graduate pointed rather indifferently toward a career as a country clergyman, invited on the voyage as a dining companion for the captain, a mercurial young aristocrat named Robert Fitzroy. Darwin did assume the role of naturalist, and thought of himself that way, as time went on. But his theory developed slowly, secretively, and *The Origin of Species* (full title: *On the Origin of Species by Means of Natural Selection, or the Preservation of Favoured Races in the Struggle for Life*) didn't appear until 1859. Many scientists, along with some Victorian clergymen, resisted its evidence and arguments for decades afterward. The reality of evolution became widely accepted during Darwin's lifetime, but his particular theory—with natural selection as prime cause—didn't triumph until about 1940, after it had been successfully integrated with genetics.

Apart from those clarifications, the most interesting point missed by the simplified tale is this: Darwin's first real clue toward evolution came not in the Galápagos but three years before, on a blustery beach along the north coast of Argentina. And it didn't take the form of a bird's beak. It wasn't even a living creature. It was a trove of fossils. Never mind the notion of Darwin's finches. For a fresh view of the *Beagle* voyage, start with Darwin's armadillos and giant sloths.

In September 1832, during the first year of its mission, the *Beagle* anchored near Bahía Blanca, a settlement at the head of a bay about four hundred miles southwest of Buenos Aires. A certain General Rosas was waging a genocidal war against the Indians, and Bahía Blanca stood as a fortified outpost, occupied mostly by soldiers. For more than a month the *Beagle* remained in that area, some of its crew occupied with surveying, others assigned to shore duties—digging a well, gathering firewood, hunting for meat. The landscape round about was classic Argentine Pampas, fertile grassland, giving way to grass-anchored sand dunes along the coast. The hunters brought back deer, agoutis, and other game, including several armadillos and a large flightless bird Darwin loosely called an "ostrich." Of course it wasn't an ostrich (which is native to Africa and, formerly, the Middle East); it was a rhea, specifically *Rhea americana,* ostrichlike in appearance but endemic to South America and the heaviest bird on the continent.

"What we had for dinner to day would sound very odd in England," Darwin wrote in his diary on September 18, reveling in the exoticism of his new regimen: "Ostrich dumpling & Armadilloes."

He was out for a romping adventure, not just a natural history field trip, and his shipboard diary (later transformed into a travel book that came to be known as *The Voyage of the Beagle*) reflects his attention to cultures, peoples, and politics as well as to science. The red meat of the big bird resembled beef, he recorded. The armadillos, peeled out of their shells, tasted and looked like ducks. His culinary experiences here on the Pampas, and later in Patagonia, besides being part of his voracious tour of discovery, would eventually play a role in his evolutionary thinking.

A few days afterward, on September 22, 1832, Darwin and Fitzroy took a small boat to visit a site called Punta Alta, ten miles from their anchorage, where they found some rocky outcrops overlooking the water. "These are the first I have seen," Darwin wrote, "& are very interesting from containing numerous shells & the bones of large animals."

Despite the name, Punta Alta ("high point") was not very high, its reddish mudstone cliff rising only about twenty feet. But if the headland wasn't dramatic, the exposed fossils were: big shapes, unusual shapes, and abundant. Darwin and a helper went to work on the soft rock with pickaxes. Between that session and later efforts, he harvested from Punta Alta the remains of nine great mammals, all unknown or barely known to science. They were extinct Pleistocene giants, unique to the Americas in an age sometime before 12,000 years ago.

The most famous of them was *Megatherium,* an elephant-size ground sloth that had already been named and described by the French anatomist Georges Cuvier on the basis of one set of fossils found in Paraguay. Living sloths are native to Central and South America, and only there; *Megatherium* shared many of their anatomical traits but was far too large for climbing trees. Darwin's finds also included at least three other giant ground sloths, an extinct form of horse, and a protective carapace of small bony scutes fitted closely together, remnant from some big beast that must have strongly resembled an armadillo. He was already familiar with flesh-and-blood armadillos, having eaten those shucked, ducky ones with his ostrich dumplings. He had also watched local gauchos kill armadillos and roast them in the shells. Of the twenty species of living armadillo, all are confined to the Americas and several are common on the Pampas; the roasted animals may have been six-banded armadillos *(Euphractus sexcinctus),* plentiful there-

abouts and reputed to taste terrible, which might not have dis-
suaded those unfussy gauchos, who sometimes lived off the land
for weeks. "Like to snails, all their property is on their backs & their
food around them," Darwin wrote, referring to the cowboys, not
the armadillos.

A month later, thirty miles up the coast from Punta Alta, Darwin discovered another fossil-rich sea cliff, this one rising 120 feet
and marking a place called Monte Hermoso. There he unearthed
the stony remains of several gnawing creatures, which variously put
him in mind of an agouti, a capybara, and a smaller South American rodent, the tuco-tuco, except that again, in each case, the
match between fossil and living species was close but not identical.
Still later and farther south on the Argentine coast, he excavated a
third set of mammal bones, which, to an anatomist who eventually
examined them, suggested an extinct form of camel. That creature
became known as *Macrauchenia*. The camel family includes two
wild South American species, the guanaco and the vicuña, as well
as their domesticated forms, the llama and the alpaca. Darwin was
well aware that living guanacos inhabited that area, having shot
one himself just days earlier.

These discoveries, analogies, and juxtapositions went into his
memory and imagination, to ferment there as the voyage continued and for years afterward. Meanwhile the fossils themselves were
crated up for shipping back to England, mostly to the care of John
Stevens Henslow, the gentle botanist who had been Darwin's mentor at Cambridge.

"I have been lucky with fossil bones," he told Henslow in a letter.
He mentioned the giant rodent, the ground sloths, and the section
of bony polygonal scutes, commenting on the last: "Immediately I
saw them I thought they must belong to an enormous Armadillo,
living species of which genus are so abundant here." And he added:
"If it interests you sufficiently to unpack them, I shall be very curious to hear something about them."

It's important not to overstate how clearly Darwin could even
identify, let alone interpret, what he had found. Most of his fossils,
apart from the *Megatherium*, represented species not yet familiar to
experts, and he was no expert. He wasn't a comparative anatomist,
like the great Cuvier; he wasn't especially knowledgeable about
mammals; and the very word "paleontologist" hadn't yet come into
use. Darwin entrusted the description and identification of his fos-

sils to a brilliant young anatomist back in London named Richard
Owen, an up-and-coming authority on extinct mammals. It was
Owen who gave names to the unknown sloths, and Owen who sug-
gested (mistakenly, later correcting himself) the affinity between
Macrauchenia and a camel.

Darwin himself was no Owen. He was just a highly attentive field
man, greedy for specimens, learning as he went. The *Beagle* invitation had rescued him from an unsuitable future as a country pas-
tor, and since his first days aboard ship he had applied himself dili-
gently, maturing fast to assume (and then transcend) the role of
ship's naturalist. His best qualifications for interpreting the fossils
were his intense curiosity, his talent for close observation, and his
instinctive sense that everything in the natural world is somehow
connected with everything else. Also, he wasn't afraid to speculate
boldly—so long as he could do it in private.

Another small but suggestive datum reached him months later,
while the *Beagle* lingered off northern Patagonia and Darwin spent
time ashore among another congenial group of gauchos. First it
was hearsay: the gauchos mentioned a rare form of ostrich, smaller
than the common one, with shorter legs, and more easily killed
using their bolas, but otherwise similar. The possibility of finding
that bird slipped Darwin's mind until one of his shipmates shot
such a smaller "ostrich" (another rhea) for its meat. Darwin paid
little attention, assuming it was a juvenile. "The bird was skinned
and cooked before my memory returned," he wrote, in a passage
so candid you can almost see him smacking his forehead with a
palm. "But the head, neck, legs, wings, many of the larger feathers,
and a large part of the skin, had been preserved." He rescued those
scraps and sent them to England, where they were stitched into a
presentable specimen for the museum of the Zoological Society.
The ornithologist John Gould, to whom Darwin would consign his
Galápagos finches and mockingbirds for identification, also got a
first look at this creature. Gould confirmed that it was a distinct
species and called it *Rhea darwinii* (a name later changed because
of taxonomic technicalities) for the man who had rescued it from
the midden.

What intrigued Darwin most about the two rhea species was that,
similar as they were, they overlapped very little in geographic distribution. The greater rhea inhabited the Pampas and northern

Patagonia as far south as Argentina's Río Negro, which drained to
the coast at about 41 degrees south latitude; the lesser rhea replaced it beyond the Río Negro and occupied southern Patagonia.
Together with the evidence of extinct South American mammals,
the implications of rhea diversity and distribution would prove almost as suggestive to Darwin as the patterns he would later find
among the finches and mockingbirds of the Galápagos.

How do species originate, and how do they come to be where they
are? The orthodox story, still firmly embraced by European science at the time of the *Beagle* voyage, was that God had created
species independently, in sequential batches (to compensate for
extinctions), and had chosen to place them, almost arbitrarily, in
their particular locales — kangaroos in Australia, giraffes and zebras in Africa, rheas and sloths and armadillos in South America,
extinct and living forms clustered closely in space and time. But to
Darwin, both the extinct mammals (along with their living coun-
terparts among sloths and armadillos) and the two rheas (occupy-
ing adjacent regions of habitat) suggested something more rational: the ideas of relatedness and succession among closely allied
species. The living tree sloths and armadillos seemed to have succeeded earlier such forms in time, inhabiting roughly the same
terrain during different epochs of Earth's history. (Those earlier
forms of sloth were true sloths; the earlier armored creatures are
now known as glyptodonts, a family distinct from but closely related to living armadillos.) The two rheas, similar but not identical, likewise seemed to succeed each other — but in space, across
the horizontal dimension of landscape. The clustering in time and
in space thus hinted that each group had descended, with modification, from common ancestors: rheas from rheas, sloths from earlier sloths, armadillos from an armadilloish or glyptodontish precursor, possibly far larger than armadillos living today. That's the
explanation to which Darwin felt drawn, because it seemed more
economical, more inductive, and more persuasive than the cre-
ationist scenario.

How important were the South American data in shaking his
faith in the orthodox view — persuading him that evolution was a
reality for which he should seek a material explanation? Darwin
himself would give several answers to that question over the length

of his lifetime. His answers ranged, in essence, from very important, but less so than the Galápagos birds, to crucially important,
period.

He hinted at the subject in 1845, in the second edition of his
Beagle narrative, revised by him to include coy hints about the theory he was still unprepared to publish. The relationships between
fossil and living forms among the rodents, the sloths, the camels,
and the armadillos were "most interesting facts," he noted. Further
work by other investigators had meantime revealed the same kind
of pattern in Brazil—fossil and living forms of anteater, of tapir,
of monkey and peccary and possum. "This wonderful relationship
in the same continent between the dead and the living," Darwin
wrote, would "throw more light on the appearance of organic beings on our earth, and their disappearance from it, than any other
class of facts." But what sort of light? What would that light reveal?
Throwing light was one of his favorite metaphors, and it would return, but not for a decade and a half—not until he was ready to
shine the blinding beam of his theory in public.

There's another intriguing question about the South American
fossils and rheas: When did this evidence register on Darwin, tip-
ping him toward the idea of evolution? The widely accepted view
is that he returned from the *Beagle* voyage not yet an evolutionist,
merely puzzled by what he had seen, and that he made the big leap
to evolutionary thinking after his consultations in London with
John Gould and Richard Owen about the bird and fossil specimens
he had consigned to them. (Soon after that he began using a new
term for the process: "transmutation.") But not everyone agrees.

"I think he was personally converted much earlier," a historian
of paleontology named Paul D. Brinkman told me. We were sitting
in his office at the North Carolina Museum of Natural Sciences in
Raleigh, amid a portrait of young Darwin, a *Jurassic Park* poster,
and photos of old ground sloth and glyptodont specimens. "Why
would there be this resemblance between the fossil fauna and the
extant fauna of this area? Why would they be so similar?" he asked,
repostulating questions that Darwin must have framed. The ancient rodents and the living agoutis, the glyptodonts and the armadillos—why? "I think one of the possible explanations he was
mulling over, even as early as 1832, was that one begat the other.
Transmutation." But even Brinkman admits that there is only tenu-

ous evidence, "no smoking gun," for his hypothesis about Darwin having converted to evolutionism long before ever striding ashore in the Galápagos.

One cryptic piece of testimony came from Darwin himself, near the end of his life, in the private autobiography he wrote for his family. "During the voyage of the *Beagle*," he reminisced, "I had been deeply impressed by discovering in the Pampean formation great fossil animals covered with armour like that on the existing armadillos." He alluded also to the rheas and to the Galápagos species, differing island by island. "It was evident," Darwin wrote, "that such facts as these, as well as many others, could be explained on the supposition that species gradually become modified; and the subject haunted me." In years since, it has also haunted scholars.

The *Beagle*, having completed its South American survey work, then spent a year circumnavigating the world; it reached England in October 1836. Darwin, then twenty-seven and a seasoned naturalist, weary of travel, eager for home, was a changed man in other ways too. He no longer saw himself serving time in a country parsonage; he was committed to a life of science. And he had at least started to lose his belief in the immutability of species. It's not possible to know with certainty, but he seems by then to have identified the great question, though not yet the great answer, that would dominate the rest of his working life.

With his specimens outsourced for expert identification — the birds to Gould, the fossil mammals to Owen, the reptiles to a zoologist named Thomas Bell — he set about putting his thoughts in order and following out his suspicions. He brainstormed in his most private notebook about ostriches, guanacos, and whether "one species does change into another." If so, how might such transmutation occur? About a year and a half later, after adding one crucial piece to his thinking (the idea of excess reproduction and struggle for existence, adopted from an essay on human population by Thomas Malthus), Darwin hit upon his theory: natural selection, whereby the best adapted individuals of each population survive to leave offspring and others don't. Then he nurtured, refined, developed, and concealed that theory for twenty years, until a younger man named Alfred Russel Wallace struck upon the same idea, forcing Darwin to rush to get his own ready for print.

That was in 1858. By then Darwin had begun writing a long, detailed, heavily footnoted treatise on natural selection, but it was only half finished. Panicked, feeling proprietary, yet also reawakened to the wondrous immediacy of the story he had to tell, he shoved the big book aside and quickly composed a more streamlined account. This shorter, slapdash version would be merely an "abstract" of the theory and its supporting data, he claimed. He called it "my abominable volume" because, after decades of cogitation and delay, the writing process was so hurried and painful. He wanted to title it *An Abstract of an Essay on the Origin of Species and Varieties Through Natural Selection,* but his publisher persuaded him to accept something at least marginally more snappy. It appeared in November 1859, titled *On the Origin of Species by Means of Natural Selection* et cetera, and was a sellout success immediately.

Five more editions went to print during Darwin's lifetime. Almost inarguably, it's the most significant single scientific book ever published. After 150 years, people still venerate it, people still deplore it, and *The Origin of Species* continues to exert an extraordinary influence — though, unfortunately, not many people actually read it.

And the forgotten clues that led him to his theory are still largely forgotten. Anyway, they're omitted from the mythic account. Scholars still dispute the significance of those extinct and living Argentine creatures, especially the ground sloths and glyptodonts, the tree sloths and armadillos and rheas. Evidence is mixed, even among the various comments on the matter left behind by Darwin himself. The most telling of those comments, in my view, is one so conspicuously placed that it tends to get overlooked. It comprises the first two sentences of *The Origin of Species,* beginning the book on a nostalgic note. It says:

"When on board H.M.S. 'Beagle,' as naturalist, I was much struck with certain facts in the distribution of the inhabitants of South America, and in the geological relations of the present to the past inhabitants of that continent. These facts seemed to me to throw some light on the origin of species . . ."

The finches of the Galápagos make their appearance about four hundred pages later.

PART FOUR

The Environment:
Gloom and Doom

JIM CARRIER

All You Can Eat

FROM *Orion*

THE GREEN DUMPSTER behind Red Lobster was nearly empty when I lifted the lid. Through the effluvium of yesterday's supper, way down, sat a couple of pretty blue boxes. I hitched myself over the rim, leaned in, and took one.

I am not a regular dumpster diver. I was driven by a hunger for knowledge. Inside the restaurant, where the décor, ambience, soundtrack—all but the smell—reeked of the sea, I asked the server who laid before me the first plate of Red Lobster's "endless shrimp" where they came from.

"Farms," she said.

"Where are these farms?" I asked.

"Different places." She gave a shrug. "Do you want another beer?"

I ate only eight grilled shrimp from Red Lobster's "endless" supply. Something was stuck in my craw. An hour before, I had been in a community hall in Brownsville, Texas, with forty-three angry, tearful American shrimpers. In a country awash in shrimp, they were going bankrupt. They had gathered to hear more bad news: severe new rules limiting what they could catch.

"What about Red Lobster?" I asked the group.

"Red Lobster!" one man shouted. "They're our enemy. They haven't bought a shrimp since the 1980s."

The restaurant walls were covered with shrimp boats—striking photos of trawlers at docks, at sea, in sunset silhouettes. The Gulf of Mexico was a mile away. Yet while I sat eating, real shrimp boats sat rusting, their outriggers raised as if surrendering.

The box from the dumpster gave me a clue: "Product of Ecuador. Farm Raised."

I am farm raised. I nurse a nostalgia for what those words used to mean. Holding that fetid box, I began to question my own clueless consumption. From a springboard both pure and naive, I dove into all-you-can-eat shrimp.

Shrimp, in my youth in upstate New York, were rare and pricey. I remember a 1960s shrimp cocktail at the Rainbow Room atop Rockefeller Center. I don't remember my date's face, but I do recall a scent of privilege. Thirty years would pass before another shrimp scene would be as sharply etched on my mind. It was late October in the Carolinas. I had tied my sailboat to an old wood dock in front of a village on the Intracoastal Waterway. Around me marsh grass tinted gold by the sunset was slowly emerging from ebbing waters. Barely a month into living aboard, I'd opened a beer to toast my good fortune when a man from the village walked onto the dock with a bucket and a ball of netting.

With a practiced arabesque he threw the net. It blossomed into a ten-foot parachute that dappled a circle and sank. In a few seconds, he pulled the line, and wet flopping creatures spilled onto the dock. He sorted through them, discarded several, and repeated the motion. Half a dozen throws later the bucket held two handfuls of shrimp.

"Supper," he said, and walked off.

The scene was magical, almost biblical: its grace and bounty, its sense of proportion—one man, one meal—evoked a sustaining ocean. As I sailed farther south I began to see shrimp everywhere. Shrimp boats seining night and day. Roadside stands selling shrimp from coolers. All-you-can-eat shrimp buffets for a few dollars. These waters, a federal survey reported in 1884, contained "immense schools" of shrimp, so many that a man could catch bushels on a "pleasant evening." It appeared that nothing had changed. In fact, everything had changed during a century in which shrimp had gone from lowly regional fare, caught by hand, to America's favorite seafood.

In 1913, one hundred miles down the coast from where I had watched the man cast his net, Billy Corkum, a Massachusetts fishing captain, introduced the otter trawl to Amelia Island, Florida. An ungainly contraption of ropes, cables, wooden doors, and nets, the trawl was dragged through the water just above the ocean floor, its mouth open like a whale's. Modified with a drooping chain to "tickle" mud-dwelling shrimp into jumping into the maw, diesel-pulled trawls scooped shrimp by the billions.

"We never had so darned many shrimp," old-timer Anthony Taranto told the Southern Foodways Alliance, a University of Mississippi institute that studies southern food culture. "You couldn't hardly sell them and couldn't hardly do nothing with them."

Shrimp are a perfect protein delivery system, built with a head and carapace that twist off easily, revealing a muscle that can be cooked in three minutes. The Chinese and Greeks loved them. Apicius included shrimp in his Roman cookbook. But it took decades for shrimp to whet appetites outside the American South. Packed in barrels of ice and shipped by rail, shrimp were served in tulip glasses as "cocktails" in upstate New York in 1914. As cookbooks added Low Country recipes, canning and, in 1943, a shrimp-peeling machine—invented by a teenager, J. M. Lapeyre, in Houma, Louisiana, who noticed how easily shrimp meat could be squished out of its shell by his rubber boot—made shrimp available nationwide.

Trawls soon emptied the shallows of southern waters and moved deeper. For seventy years growing fleets of bigger boats galloped from one gold strike to another as veins of shrimp were discovered off Louisiana (white, 1933), Mexico (brown, 1940), Dry Tortugas (pink, 1949), and Key West, where in 1957 huge, royal red shrimp were discovered a thousand feet down.

"Greater riches are being brought up than all the gold ever sunk off the Spanish Main," gushed *National Geographic* in 1957. Many shrimpers became millionaires.

"We were outlaws," Wallace Beaudreaux, of Brownsville, Texas, eighty-one, told me, describing raids into Mexican waters. It was not unusual for boats to gross $10,000 to $25,000 on a single trip.

I felt rich in 1998, buying a pound of shrimp for a mere three dollars right off the boats near where I anchored in Key West. I had only one question: with thousands of boats endlessly trawling and millions like me endlessly gobbling, how could there be any shrimp left in the sea?

"Shrimp are a crop, like wheat," shrimpers replied. "You can't overfish them."

I was asking the wrong question. I should have wondered where all these shrimp were coming from and how they could cost three dollars a pound. I happened to sail into the Deep South in time to witness the crash of a culture bound to, and blinded by, endless shrimp dreams.

Shrimp have been around since Gondwana. Their tracks are found alongside dinosaurs', which explains their astounding diversity—more than two thousand species in every body of water in the world. They are a major food source for Salt Lake gulls, ocean whales, gulf red snapper—virtually every marine critter, which makes them ideal bait.

But the shrimp's life cycle was understood only in the 1960s. Shrimp don't ascend rivers to spawn, as once thought, but reverse the process in a complicated and delicate cycle. Adult shrimp mate in deep water, holding each other feet-to-feet. He inserts a capsule of sperm, and she spews half a million microscopic eggs that resemble milk spilled in water. These babies molt through a dozen stages as tiny, spiderlike creatures, finally emerging shrimplike in a month.

With mysterious instinct they move up and down in the water column, catching waves, currents, and winds that sweep them into shallow bays. In the gulf this cycle coincides with a shift from northerly to southerly winds, a warming of bay waters, and an increase in freshwater runoff from rivers, which reduces salinity. In these brackish, rich estuaries, protected by reeds and organic muck, they begin devouring one-celled algae called diatoms and growing at the rate of one inch a month. In two or three months, triggered apparently by increased salinity, they begin to walk—literally—and flick their tails back to the sea, traveling as far as two hundred miles. Left alone, a shrimp grows to a length of six to eight inches, developing a tail as big around as a man's thumb. At this stage they are in deep water, ready to spawn before dying or being eaten by a predator.

I learned all this aboard Leslie Hartman's runabout one May day in Mobile Bay. She was out there, as she is every week of the year, her long brown ponytail swinging like a pendulum as she heaved a miniature trawl off the stern. As Alabama's shrimp biologist, Hartman's job is to constantly sample the size of shrimp returning to the sea and determine when they are large enough to open the state's shrimp season.

After fifteen minutes, she stopped the boat, hauled in the net, and dumped the catch into a white bucket. She knelt and fingered through glistening life. Little rays, horned blowfish, baby snapper, and a bunch of crabs were thrown back. Left in the bottom were a set of creatures that ranged from transparent globules half an inch long to juvenile shrimp up to two inches. She counted, measured, and logged the sample and sped off for another drag elsewhere.

The threshold for legal shrimp in Alabama is 68 shrimp per pound. A "68" shrimp is pretty small, often canned, tossed into macaroni salad, or breaded and fried as "popcorn" shrimp. Shrimp cocktails use a minimum size of 40 to 50 per pound. When I look at shrimp in a grocer's case I usually choose "20–25," the size of my little finger. Hartman's task was to calculate when the average of her samples reached 68. She was always anxious to reach that point, for she considered herself a friend of the industry.

"We want our great-grandchildren to be shrimpers a hundred and fifty years from now," she said.

"Go," shouted Joe Skinner, releasing brakes that governed two winches. Squeals, grinds, the sounds of cable and rope under stress on the throbbing bed of a major diesel smothered the splash of green nets on the water. As cables let out, the nets disappeared behind the boat. At six A.M. at the start of the 2005 Alabama shrimp season, the *A. S. Skinner* was trawling.

We were in Mobile Bay. A rising sun, barely burning through haze, added a band of pink to a formless horizon. Around us a circus of boats—trawls, skiffs, outboards—were out for opening day. "It's a madhouse," said Mike Skinner, Joe's brother at the helm. A black radar screen set at one-mile resolution was dotted with forty or fifty green moving spots. "I'll be glad when this day is over."

A. S. Skinner, for whom the boat was named, had been a jeweler, as was his son. But grandson Gary left gold in the showcase to seek his fortune with shrimp. Great-grandsons Mike and Joe joined him at age five. By high school, the last formal education they sought, they were taking boats out by themselves. "When I came out of high school, we done good," said Mike, tall and angular, dressed in blue-jean shorts and a white T-shirt. The first year they grossed $200,000. "We didn't work that hard. Dad had two good years and then it started dropping."

The Skinners, aged thirty-two and thirty, each with a one-year-old son, reminded me of cowboys I'd known out West, still pining for pastures before barbed wire. Their dreams of a commons, free to exploit, had once been our dream, so woven into our national DNA that we, like they, mourned its passing. Each spring they rode out after a myth, only to find that the world had changed.

The sea stopped giving in the 1980s. Catches flattened worldwide. There were, in fact, only so many shrimp in the sea. And because of overfishing for half a century, the average shrimp size caught in the gulf had shrunk from "50" to "75."

There was also growing dismay that shrimpers wasted more than they caught. Down below, in the channel made famous by the Union admiral David Farragut's cry, "Damn the torpedoes. Full speed ahead," was a kind of "fishing" that was nothing short of marine clear-cutting.

In the gold rush days, before Joe and Mike were born, shrimpers killed ten pounds of sea life for every pound of harvested shrimp —waste that reached one billion pounds a year in the gulf. Once called "trash," now called "by-catch," this sea life included sea turtles driven to the brink of extinction and juvenile red snapper, a good eating fish. Under environmental regulations requiring escape hatches in nets, the ratio of by-catch to shrimp has been reduced to four to one, still a startling sight when the Skinners dumped their twin nets on deck. Using grain shovels, they transferred this squirming pile into a large wooden box of seawater mixed with Cargil Boat and Boil salt. The shrimp sank to the bottom, and the by-catch, mostly dead, floated to the surface. This they skimmed and threw overboard.

Gulf shrimpers, the last cowboys of the sea, were corralled in 2006 when the U.S. government, trying to balance the gulf's ecosystem with a sustainable supply of shrimp for a viable commercial fishery, capped the federal-waters shrimp fleet at 2,700 boats, down from a gold rush high of 7,500, and ordered federal clerks to be randomly stationed aboard to record by-catch. The goal was a "maximum sustainable yield," roughly 110 million pounds a year, which left 22 billion shrimp to reproduce, according to modeling by Dr. Jim Nance, head of the NOAA Fisheries Service Galveston Laboratory. This figure was half the natural shrimp population before the arrival of the trawl, estimated Bill Hogarth, the former head of the agency.

The Skinners grossed $1,000 on opening day—not a bad haul, I thought, until I learned that it was half the price they got when they were teenagers. They made a living but not a killing selling their shrimp to their father, who ran a roadside stand on Dauphin Island. "The last few years, we're just paying for fuel," said Joe, sitting below their federal license framed on the Masonite wall of their boat's dinette. "If it weren't for the shop . . ." His voice trailed off.

What really ended the Skinners' dreams, what really brought shrimpers to their knees and tears in Mobile Bay, Brownsville, New Orleans, Biloxi, and Bayou la Batre — all along the Gulf Coast — was not regulation or lack of shrimp but good old global supply and demand. "Because of imported, farmed shrimp from the Far East," said Joe Skinner, "wholesale shrimp prices in the U.S. are the same as when Dad started thirty years ago."

The story of farmed shrimp begins with a Japanese dish called "dancing shrimp," a casserole that arrives at your table with the unmistakable sound of something inside striking the cover. Jumping about on a bed of hot rice are Kuruma prawns — live. The object is to grab one between chopsticks and pop it wiggling into your mouth. Kuruma, large, meaty shrimp found in limited quantities in the Sea of Japan, sell for a hundred dollars a pound. Seventy-five years ago this rarity prompted an ichthyology student at Tokyo University to try growing Kuruma in captivity.

Until 1933, when Motosaku Fujinaga first spawned and hatched shrimp in a lab, aquaculture had been an ancient artisanal practice. Tides swept fish and shrimp into estuaries, and weirs were built to prevent their escape. The shrimp grew to eating size in naturally replenished waters.

Out of their element, though, shrimp proved to be finicky eaters, fragile and prone to diseases. It took Fujinaga twenty-five years of trying, interrupted by World War II, to be able to grow ten kilograms of shrimp to adulthood. In 1967, when he spoke to the United Nations Food and Agriculture Organization's first world conference on shrimp culture in Mexico City, Fujinaga envisioned a world where capitalism and altruism could coexist in the "vast and boundless marshes, swamps, or jungles in the tropics." Shrimp farms, he predicted, "will greatly contribute toward the increased supply of animal protein to the human race."

It was a lovely thought. A Blue Revolution. But his success fueled a global grab in which protein and profits flowed one way — north toward the moneyed. One year after his speech, a group of Japanese businessmen bought Fujinaga's technology, won a U.S. patent, and approached DuPont for money. DuPont declined, but two officials who heard the pitch, Paul Bente and John Rutledge Cheshire, were so excited they quit their jobs, put up $200,000 of Cheshire's family money, and opened Marifarms in a bay near Panama City, Florida.

Aided by research at the U.S. lab in Galveston, Marifarms harvested a disappointing 6,000 pounds in 1970, according to Cheshire's book, *Memoir of a Shrimp Farmer.* The same year another venture, Sea Farms, was digging canals in a Florida key to grow shrimp.

Because of environmental issues—Marifarms scooped up pregnant white shrimp and confined them in a public bay, while Sea Farms flew in nonindigenous shrimp from Central America, a practice Florida soon prohibited—shrimp farming moved south. Supported by USAID, World Bank loans, and willing developing-world officials, the corporate giants United Fruit, Armour, Conagra, and Ralston Purina launched shrimp farms in Honduras, Brazil, Panama, and Ecuador, according to oral histories collected by Bob Rosenberry of *Shrimp News International.* Learning as they went, the farmed-shrimp industry laid waste to mangroves, fishing communities, and ecosystems. The word "plundering" comes to mind.

A shrimp farm is a saltwater feedlot. There can be as many as 170,000 shrimp larvae in a one-acre pond that is one to two meters deep. So-called intensive ponds can yield 6,000 to 18,000 pounds of shrimp in that acre in three to six months. (A good wheat yield is 3,600 pounds per acre.) Because of this density, the waste they swim in, and their susceptibility to disease, most farmed shrimp are treated with antibiotics, only some of them legal in the United States. A wide array of poisons is used to kill unwanted sea life and cleanse ponds for reuse, creating what Public Citizen calls a "chemical cocktail." In random sampling of imported shrimp, health officials in the United States, Japan, and the European Union have found chloramphenicol, a dangerous antibiotic banned in food.

The industry acknowledges that 5 percent of the world's mangroves, hundreds of thousands of acres, have been destroyed creating shrimp ponds. In some estuaries 80 percent of the mangroves are gone. A commons was privatized, ruining artisanal fishing and driving indigenous fishermen to work raising shrimp. By removing the thick coastal barrier of trees, shrimp farms have undoubtedly aggravated damage from hurricanes and tsunamis. And salt intrusion has sterilized once fertile estuaries.

Even in the best-run farms, two to four pounds of sea life is caught and ground up as feed for every pound of shrimp raised. Mortality rates of 30 percent are common. The dead shrimp, shrimp excrement, and chemical additives are often flushed into coastal waters.

By the mid-1970s, farmed shrimp from South and Central America, at less than half the cost of gulf shrimp, began arriving at Red Lobster restaurants — and everywhere else. All-you-can-eat shrimp dinners became a standard, filling both waistlines and Red Lobster's coffers. That box of shrimp I retrieved from the dumpster cost $2.50 a pound and sold, in my case, for $25 a pound, a markup that bettered the beer's.

Quietly, farmed shrimp took over the market, its source hidden behind the motif of a picturesque but actually sinking shrimp fleet. By 1980 half of America's shrimp consumption came from foreign farms. By 2001 shrimp passed canned tuna as America's favorite seafood. Today, 90 percent of our shrimp — more than 1 billion pounds a year — come from foreign farms. Virtually every restaurant chain, from Captain D's to Red Lobster, serves farmed shrimp. Foreign farmed shrimp were peddled for years by vendors at the National Shrimp Festival in Alabama — until they were caught — and at happy hour for the Gulf of Mexico Fishery Management Council meeting in Birmingham, Alabama, in March 2005, where government officials finalized a ten-year freeze on 2,700 shrimp boat licenses. The sight of government biologists slurping Vietnamese shrimp after reining in American shrimpers was an irony sharper than cocktail sauce. Even in New Orleans, where a handful of high-end chefs brag about their Louisiana shrimp, imported shrimp are the norm in most restaurants. A new Louisiana law requires restaurateurs to tell the truth — if asked.

To get a sense of the pink tsunami on U.S. shores, I flew to Long Beach, California, the single largest shrimp port, where among the 5 million containers arriving each year are several thousand filled with shrimp, 265 million pounds of it in a year.

On the day I visited, five ships were docking with nine containers—412,000 pounds—of shrimp from Peru, Ecuador, Venezuela, and China. One container, a semitrailer load, holds an astounding amount. Laid out in a customs warehouse, boxes holding 30,000 pounds of shrimp covered a 12-by-100-foot area chest high. Based on our average consumption, this one container held a year's supply of shrimp for 12,000 Americans.

The container in question had been seized and opened because of suspicions that the beautiful bags of store-ready "26/30" frozen raw shrimp, labeled "farm raised in Indonesia," may, in fact, have come from China and been relabeled in Singapore, a common cat-and-mouse game that customs officials call "transshipment." A bag was dispatched to a government lab in Savannah, Georgia, to try a new sniffing tool that might determine its source. Transshipping is used to evade special import taxes or restrictions, such as one imposed on Chinese shrimp and four other species in 2007 after malachite green, gentian violet, and other carcinogens were found in farmed fish.

"It's very, very difficult to prove a transshipment issue," said Jeff DeHaven, the deputy director of fines, penalties, and forfeitures. So great is their volume of business that importers just walk away from seized containers, he said. Moreover, U.S. customs is concerned primarily with duty issues, not food safety. "We don't look at that much shrimp," admitted an enforcement chief.

The Food and Drug Administration, responsible for imported food safety, samples less than 1 percent of the 1 billion pounds, a "sorry" record, according to U.S. Representative John Dingell, who in 2007 chaired food safety hearings before the House Energy and Commerce Committee. Mindful of consumer fears fanned by poisoned seafood arriving from China, the Global Aquaculture Alliance—an industry group underwritten by Wal-Mart, Red Lobster, and multinational seafood importers—has written standards that, if enforced, could produce clean, safe shrimp without damaging people or the environment. But that will take years, admitted GAA president George Chamberlain. Only 45 shrimp farms are certified by the alliance—out of more than 100,000 worldwide.

Today, if you live more than a hundred miles from the Gulf Coast, the shrimp you eat most likely come from a foreign farm. You can tour these farms while standing at your supermarket seafood freezer and reading labels. The top ten countries exporting to the United States are Thailand, Indonesia, Ecuador, China, Vietnam, Malaysia, Mexico, India, Bangladesh, and Guyana. The wholesale value of their shrimp is $4 billion a year.

Despite that income, citizens in the developing world have protested shrimp farms—and have been killed for doing so. *The Blues of a Revolution,* a book published in 2003 by a consortium of environmental and indigenous groups, described Honduran shrimp farms ringed by barbed wire and watchtowers and armed guards. Between 1992 and 1998, in the Bay of Fonseca near large shrimp farms, "11 fishermen have been found dead by shooting or by machete injuries . . . no one has been brought to justice."

One story from the book I cannot shake involved Korunamoyee Sardar, a Bangladeshi woman who, on November 7, 1990, joined a protest against a new shrimp farm near Harin Khola. She was shot in the head, cut into pieces, and thrown into a Bangladesh river. A monument stands where she was murdered. It reads: "Life is struggle, struggle is life."

Red Lobster, which buys 5 percent of the world's shrimp, is Bangladesh's biggest U.S. customer. The restaurant did not respond to repeated requests for an interview.

FELIX SALMON

A Formula for Disaster

FROM *Wired*

A YEAR AGO, it was hardly unthinkable that a math wizard like David X. Li might someday earn a Nobel Prize. After all, financial economists—even Wall Street quants—have received the Nobel in economics before, and Li's work on measuring risk has had more impact, more quickly, than previous Nobel Prize–winning contributions to the field. Today, though, as dazed bankers, politicians, regulators, and investors survey the wreckage of the biggest financial meltdown since the Great Depression, Li is probably thankful he still has a job in finance at all. Not that his achievement should be dismissed. He took a notoriously tough nut—determining correlation, or how seemingly disparate events are related—and cracked it wide open with a simple and elegant mathematical formula, one that would become ubiquitous in finance worldwide.

For five years, Li's formula, known as a Gaussian copula function, looked like an unambiguously positive breakthrough, a piece of financial technology that allowed hugely complex risks to be modeled with more ease and accuracy than ever before. With his brilliant spark of mathematical legerdemain, Li made it possible for traders to sell vast quantities of new securities, expanding financial markets to unimaginable levels.

His method was adopted by everybody from bond investors and Wall Street banks to ratings agencies and regulators. And it became so deeply entrenched—and was making people so much money—that warnings about its limitations were largely ignored.

Then the model fell apart. Cracks started appearing early on, when financial markets began behaving in ways that users of Li's formula hadn't expected. The cracks became full-fledged canyons in 2008—when ruptures in the financial system's foundation swallowed up trillions of dollars and put the survival of the global banking system in serious peril.

David X. Li, it's safe to say, won't be getting that Nobel anytime soon. One result of the collapse has been the end of financial economics as something to be celebrated rather than feared. And Li's Gaussian copula formula will go down in history as instrumental in causing the unfathomable losses that brought the world financial system to its knees.

How could one formula pack such a devastating punch? The answer lies in the bond market, the multitrillion-dollar system that allows pension funds, insurance companies, and hedge funds to lend trillions of dollars to companies, countries, and home buyers.

A bond, of course, is just an IOU, a promise to pay back money with interest by certain dates. If a company—say, IBM—borrows money by issuing a bond, investors will look over its accounts very closely to make sure it has the wherewithal to repay them. The higher the perceived risk—and there's always *some* risk—the higher the interest rate the bond must carry.

Bond investors are very comfortable with the concept of probability. If there's a 1 percent chance of default but they get an extra 2 percentage points in interest, they're ahead of the game overall—like a casino, which is happy to lose big sums every so often in return for profits most of the time.

Bond investors also invest in pools of hundreds or even thousands of mortgages. The potential sums involved are staggering: Americans now owe more than $11 trillion on their homes. But mortgage pools are messier than most bonds. There's no guaranteed interest rate, since the amount of money homeowners collectively pay back every month is a function of how many have refinanced and how many have defaulted. There's certainly no fixed maturity date: money shows up in irregular chunks as people pay down their mortgages at unpredictable times—for instance, when they decide to sell their house. And most problematically, there's no easy way to assign a single probability to the chance of default.

Wall Street solved many of these problems through a process called tranching, which divides a pool and allows for the creation of safe bonds with a risk-free triple-A credit rating. Investors in the first tranche, or slice, are first in line to be paid off. Those next in line might get only a double-A credit rating on their tranche of bonds but will be able to charge a higher interest rate for bearing the slightly higher chance of default. And so on.

The reason that ratings agencies and investors felt so safe with the triple-A tranches was that they believed there was no way hundreds of homeowners would all default on their loans at the same time. One person might lose his job, another might fall ill. But those are individual calamities that don't affect the mortgage pool much as a whole: everybody else is still making their payments on time.

But not all calamities are individual, and tranching still hadn't solved all the problems of mortgage-pool risk. Some things, like falling house prices, affect a large number of people at once. If home values in your neighborhood decline and you lose some of your equity, there's a good chance your neighbors will lose theirs as well. If, as a result, you default on your mortgage, there's a higher probability they will default, too. That's called correlation—the degree to which one variable moves in line with another—and measuring it is an important part of determining how risky mortgage bonds are.

Investors *like* risk, as long as they can price it. What they hate is uncertainty—not knowing how big the risk is. As a result, bond investors and mortgage lenders desperately want to be able to measure, model, and price correlation. Before quantitative models came along, the only time investors were comfortable putting their money in mortgage pools was when there was no risk whatsoever—in other words, when the bonds were guaranteed implicitly by the federal government through Fannie Mae or Freddie Mac.

Yet during the nineties, as global markets expanded, there were trillions of new dollars waiting to be put to use lending to borrowers around the world—not just mortgage seekers but also corporations and car buyers and anybody running a balance on their credit card—if only investors could put a number on the correlations between them. The problem is excruciatingly hard, especially when you're talking about thousands of moving parts. Whoever solved it would earn the eternal gratitude of Wall Street and quite possibly the attention of the Nobel committee as well.

To understand the mathematics of correlations better, consider something simple, like a kid in an elementary school—let's call her Alice. The probability that her parents will get divorced this year is about 5 percent, the risk of her getting head lice is about 5 percent, the chance of her seeing a teacher slip on a banana peel is about 5 percent, and the likelihood of her winning the class spelling bee is about 5 percent. If investors were trading securities based on the chances of those things happening only to Alice, they would all trade at more or less the same price.

But something important happens when we start looking at two kids rather than one—not just Alice but also the girl she sits next to, Britney. If Britney's parents get divorced, what are the chances that Alice's parents will get divorced, too? Still about 5 percent: the correlation there is close to zero. But if Britney gets head lice, the chance that Alice will get head lice is much higher, about 50 percent—which means the correlation is probably up in the 0.5 range. If Britney sees a teacher slip on a banana peel, what is the chance that Alice will see it, too? Very high indeed, since they sit next to each other: it could be as much as 95 percent, which means the correlation is close to 1. And if Britney wins the class spelling bee, the chance of Alice winning it is zero, which means the correlation is negative: −1.

If investors were trading securities based on the chances of these things happening to *both* Alice *and* Britney, the prices would be all over the place, because the correlations vary so much.

But it's a very inexact science. Just measuring those initial 5 percent probabilities involves collecting lots of disparate data points and subjecting them to all manner of statistical and error analysis. Trying to assess the conditional probabilities—the chance that Alice will get head lice *if* Britney gets head lice—is an order of magnitude harder, since those data points are much rarer. As a result of the scarcity of historical data, the errors there are likely to be much greater.

In the world of mortgages, it's harder still. What is the chance that any given home will decline in value? You can look at the past history of housing prices to give you an idea, but surely the nation's macroeconomic situation also plays an important role. And what is the chance that if a home in one state falls in value, a similar home in another state will fall in value as well?

Enter Li, a star mathematician who grew up in rural China in the mid-1960s. He excelled in school and eventually got a master's degree in economics from Nankai University before leaving the country to get an MBA from Laval University in Quebec. That was followed by two more degrees: a master's in actuarial science and a Ph.D. in statistics, both from Ontario's University of Waterloo. In 1997 he landed at Canadian Imperial Bank of Commerce, where his financial career began in earnest; he later moved to Barclays Capital and by 2004 was charged with rebuilding its quantitative analytics team.

Li's trajectory is typical of the quant era, which began in the mid-1980s. Academia could never compete with the enormous salaries that banks and hedge funds were offering. At the same time, legions of math and physics Ph.D.'s were required to create, price, and arbitrage Wall Street's ever more complex investment structures.

In 2000, while working at JPMorgan Chase, Li published a paper in the *Journal of Fixed Income* titled "On Default Correlation: A Copula Function Approach." (In statistics, a copula is used to couple the behavior of two or more variables.) Using some relatively simple math—by Wall Street standards, anyway—Li came up with an elegant way to model default correlation without even looking at historical default data. Instead, he used market data about the prices of instruments known as credit default swaps.

If you're an investor, you have a choice these days: you can either lend directly to borrowers or sell investors credit default swaps, insurance against those same borrowers defaulting. Either way, you get a regular income stream—interest payments or insurance payments—and either way, if the borrower defaults, you lose a lot of money. The returns on either strategy are nearly identical, but because an unlimited number of credit default swaps can be sold against each borrower, the supply of swaps isn't constrained the way the supply of bonds is, so the CDS market managed to grow extremely rapidly. Though credit default swaps were relatively new when Li's paper came out, they soon became a bigger and more liquid market than the bonds on which they were based.

When the price of a credit default swap goes up, that indicates that default risk has risen. Li's breakthrough was that instead of waiting to assemble enough historical data about actual defaults, which are rare in the real world, he used historical prices from the CDS market. It's hard to build a historical model to predict Alice's or Britney's behavior, but anybody could see whether the price of credit default swaps on Britney tended to move in the same direction as that on Alice. If it did, then there was a strong correlation between Alice's and Britney's default risks, as priced by the market. Li wrote a model that used price rather than real-world default data as a shortcut (making an implicit assumption that financial markets in general, and CDS markets in particular, can price default risk correctly).

It was a brilliant simplification of an intractable problem. And Li didn't just radically dumb down the difficulty of working out correlations; he decided not to even bother trying to map and calculate all of the nearly infinite relationships between the various loans that made up a pool. What happens when the number of pool members increases or when you mix negative correlations with positive ones? Never mind all that, he said. The only thing that matters is the final correlation number—one clean, simple, all-sufficient figure that sums up everything else.

The effect on the securitization market was electric. Armed with Li's formula, Wall Street's quants saw a new world of possibilities. And the first thing they did was start creating a huge number of brand-new triple-A securities. Using Li's copula approach meant that ratings agencies like Moody's—or anybody wanting to model the risk of a tranche—no longer needed to puzzle over the underlying securities. All they needed was that correlation number, and out would come a rating telling them how safe or risky the tranche was.

As a result, just about anything could be bundled and turned into a triple-A bond—corporate bonds, bank loans, mortgage-backed securities, whatever you liked. The resulting pools were often known as collateralized debt obligations, or CDOs. You could tranche that pool and create a triple-A security even if none of the components were themselves triple-A. You could even take lower-rated tranches of *other* CDOs, put them in a pool, and tranche them—an instrument known as a CDO-squared, which at that point was so far removed from any actual underlying bond or loan or mortgage that no one really had a clue what it included. But it didn't matter. All you needed was Li's copula function.

The CDS and CDO markets grew together, feeding on each other. At the end of 2001, there was $920 billion in credit default swaps outstanding. By the end of 2007, that number had skyrocketed to more than $62 *trillion.* The CDO market, which stood at $275 billion in 2000, grew to $4.7 trillion by 2006.

At the heart of it all was Li's formula. When you talk to market participants, they use words like *beautiful, simple,* and, most commonly, *tractable.* It could be applied anywhere, for anything, and was quickly adopted not only by banks originating new bonds but also by traders and hedge funds dreaming up complex trades between those bonds.

"The corporate CDO world relied almost exclusively on this copula-based correlation model," says Darrell Duffie, a Stanford University finance professor who served on Moody's Academic Advisory Research Committee. The Gaussian copula soon became such a universally accepted part of the world's financial vocabulary that brokers started quoting prices based on their correlations. "Correlation trading has spread through the psyche of the financial markets like a highly infectious thought virus," wrote the derivatives guru Janet Tavakoli in 2006.

The damage was foreseeable and, in fact, foreseen. In 1998, before Li had even invented the copula function, Paul Wilmott wrote that "the correlations between financial quantities are notoriously unstable." Wilmott, a quantitative-finance consultant and lecturer, argued that no theory should be built on such unpredictable parameters. And he wasn't alone. During the boom years, everybody could reel off reasons why the Gaussian copula function wasn't perfect. Li's approach made no allowance for unpredictability: it assumed that correlation was a constant rather than something mercurial. Investment banks would regularly phone Stanford's Duffie and ask him to come in and talk to them about exactly what Li's copula was. Every time he would warn them that it was not suitable for use in risk management or valuation.

In hindsight, ignoring those warnings looks foolhardy. But at the time, it was easy. Banks dismissed them, partly because the managers empowered to apply the brakes didn't understand the arguments between various arms of the quant universe. Besides, they were making too much money to stop.

In finance, you can never reduce risk outright; you can only try to set up a market in which people who don't want risk sell it to those who do. But in the CDO market, people used the Gaussian copula model to convince themselves they didn't have any risk at all, when in fact they just didn't have any risk 99 percent of the time. The other 1 percent of the time they blew up. Those explosions may have been rare, but they could destroy all previous gains, and then some.

Li's copula function was used to price hundreds of billions of dollars' worth of CDOs filled with mortgages. And because the copula function used CDS prices to calculate correlation, it was forced to confine itself to looking at the period of time when those credit default swaps had been in existence: less than a decade, a period when house prices soared. Naturally, default correlations were very low in those years. But when the mortgage boom ended abruptly and home values started falling across the country, correlations soared.

Bankers securitizing mortgages knew that their models were highly sensitive to house-price appreciation. If it ever turned negative on a national scale, a lot of bonds that had been rated triple-A, or risk-free, by copula-powered computer models would blow up. But no one was willing to stop the creation of CDOs, and the big investment banks happily kept on building more, drawing their correlation data from a period when real estate only went up.

"Everyone was pinning their hopes on house prices continuing to rise," says Kai Gilkes of the credit research firm CreditSights, who spent ten years working at ratings agencies. "When they stopped rising, pretty much everyone was caught on the wrong side, because the sensitivity to house prices was huge. And there was just no getting around it. Why didn't rating agencies build in some cushion for this sensitivity to a house-price-depreciation scenario? Because if they had, they would have never rated a single mortgage-backed CDO."

Bankers should have noted that very small changes in their underlying assumptions could result in very large changes in the correlation number. They also should have noticed that the results they were seeing were much less volatile than they should have been—which implied that the risk was being moved elsewhere. Where had the risk gone?

They didn't know, or didn't ask. One reason was that the outputs came from "black box" computer models and were hard to subject to a commonsense smell test. Another was that the quants, who should have been more aware of the copula's weaknesses, weren't the ones making the big asset-allocation decisions. Their managers, who made the actual calls, lacked the math skills to understand what the models were doing or how they worked. They could, however, understand something as simple as a single correlation number. That was the problem.

"The relationship between two assets can never be captured by a single scalar quantity," Wilmott says. For instance, consider the share prices of two sneaker manufacturers: when the market for sneakers is growing, both companies do well and the correlation between them is high. But when one company gets a lot of celebrity endorsements and starts stealing market share from the other, the stock prices diverge and the correlation between them turns negative. And when the nation morphs into a land of flip-flop-wearing couch potatoes, both companies decline and the correlation becomes positive again. It's impossible to sum up such a history in one correlation number, but CDOs were invariably sold on the premise that correlation was more of a constant than a variable.

No one knew all of this better than David X. Li: "Very few people understand the essence of the model," he told the *Wall Street Journal* way back in fall 2005.

"Li can't be blamed," says Gilkes of CreditSights. After all, he just invented the model. Instead, we should blame the bankers who misinterpreted it. And even then, the real danger was created not because any given trader adopted it but because *every* trader did. In financial markets, everybody doing the same thing is the classic recipe for a bubble and inevitable bust.

Nassim Nicholas Taleb, hedge fund manager and author of *The Black Swan,* is particularly harsh when it comes to the copula. "People got very excited about the Gaussian copula because of its mathematical elegance, but the thing never worked," he says. "Co-association between securities is not measurable using correlation," because past history can never prepare you for that one day when everything goes south. "Anything that relies on correlation is charlatanism."

Li has been notably absent from the current debate over the causes of the crash. In fact, he is no longer even in the United

States. Several months ago, he moved to Beijing to head up the risk-management department of China International Capital Corporation. In a recent conversation, he seemed reluctant to discuss his paper and said he couldn't talk without permission from the PR department. In response to a request, CICC's press office sent an e-mail saying that Li was no longer doing the kind of work he did in his previous job and therefore would not be speaking to the press.

In the world of finance, too many quants see only the numbers before them and forget about the concrete reality the figures are supposed to represent. They think they can take a few years' worth of data and come up with probabilities for things that models say should happen only once every 10,000 years. Then people invest on the basis of those probabilities, without stopping to wonder whether the numbers make any sense.

As Li himself said of his own model: "The most dangerous part is when people believe everything coming out of it."

DAWN STOVER

Not So Silent Spring

FROM *Conservation Magazine*

A MALE EUROPEAN BLACKBIRD was terrorizing the neighborhood. For several months, he started singing at around 5 A.M. each day, but this was no ordinary song. The bird imitated the sounds of ambulance sirens and car alarms at a jarringly lifelike volume. It even produced cell-phone ring tones that went unanswered for hours.

The tale of the annoying blackbird in Somerset, United Kingdom, was not unique. Hans Slabbekoorn, an assistant professor of behavioral biology at Leiden University in the Netherlands, had heard similar stories—but he was skeptical that such bizarre reports could be true. So he started asking people to send him recordings of the off-kilter blackbirds. Sure enough, what he got back was pitch-perfect imitations of urban noises, including not just sirens and car alarms but even the distinctive sound of a golf cart backing up—mimicked by blackbirds living near a golf course.

While the sounds seemed artificial, the reason birds were making them was surprisingly natural. Living amid a growing cacophony of man-made noises, the blackbirds started incorporating human sounds into their repertoire. And Slabbekoorn says the unusual strategy might actually help the birds: song variety indicates maturity in male blackbirds, and female blackbirds prefer older guys.

Blackbirds aren't the only animals changing their tunes. As human noise intrudes on nature—from freeway traffic noise to jets screaming over the rainforest—scientists are starting to believe that the acoustic environment is far more intricate and fragile than they ever imagined. Long regarded as a random collection of bird songs and animal cries, the natural soundscape might actually be a coordinated symphony, with animal calls spread carefully across the acoustic spectrum. Now researchers are getting the first glimpses of what happens when humanity's choir drowns out whole sections of that spectrum. Animals ranging from blackbirds to beluga whales are changing their calls or switching them to new frequencies. Others are adapting in ways so powerful that they may be triggering the first steps in an evolutionary shakeup. And some animals are disappearing altogether.

Scientists have traditionally studied animal sounds by focusing on individual species and their vocalizations. Bernie Krause, a bioacoustician who has spent forty years recording the calls of the wild, has hatched a radically different approach and, in turn, a revolutionary vision of the relationship between animals, their environment, and the sounds they emit.

It all started in 1968 when Krause, then a musician, was having lunch with Van Dyke Parks of the Beach Boys. Parks suggested that Krause "do an album on ecology." Krause and his musical partner had introduced the Moog synthesizer to pop music and had contributed to hundreds of albums and movie soundtracks, but Krause knew little about ecology beyond his recent reading of Rachel Carson's *Silent Spring*. Intrigued by Parks's suggestion, Krause went alone with his equipment to Muir Woods, north of San Francisco. "I was so intrigued by what I heard that I made a decision that this was what I wanted to do for the rest of my life," he recalls.

Overtaken by his newfound passion for ecology, Krause eventually sold his music company, enrolled in graduate school, and got his doctorate in bioacoustics. But his musical training never left him. In fact, it helped spawn a startling notion that came to him early one morning while camping in Kenya's Masai Mara reserve. Krause had been up for thirty hours, recording the sounds of insects, owls, hyenas, bats, and elephants. Exhausted and "completely out of it," Krause was suddenly struck by the idea that the animal sounds around him were . . . orchestral. "These critters are vocalizing in relationship to one another," he thought to himself.

Back in his studio, Krause examined sonograms of the recording session. It was clear to him that what he had heard was a sequence of sounds so carefully partitioned that they read like a musical score. Different species vocalize at specific frequencies or times so they can be heard above the other animals—in the same way you can make out the individual sounds of the trumpets, violins, and clarinets as Beethoven's Fifth builds to a crescendo. Krause dubbed the spectrum of animal sounds "biophony" and distinguishes them from human sounds, which he calls "anthrophony."

Krause wasn't alone in his conclusions. More than a decade later, in a laboratory far away from Krause's California headquarters, Hans Slabbekoorn picked up on the same distinction. And, taking Krause's work a step further, he started piecing together its startling implications.

Before studying blackbirds, Slabbekoorn worked in the tropics of Cameroon, testing a theory that habitat constraints can drive birds to sing at different frequencies. Sounds that mask birdcalls may cause difficulty or ambiguity in communication, and Slabbekoorn thought he could get to the bottom of this by comparing the songs of birds living in dense rainforest with those in more open habitats.

His test subject was a bird called the little greenbul. He discovered that little greenbuls in the dense forest sang more often at lower frequencies than their relatives living in open areas, probably because lower frequencies transmit more effectively through thick vegetation—and also because the dense forest is filled with the sounds of high-pitched insects. "It was often so noisy in these habitats that I could hardly make good recordings," Slabbekoorn recalls.

After returning to the Netherlands, Slabbekoorn suspected a similar dynamic was taking place in the city. Most urban noise comes from cars, trucks, buses, and trains, whose sounds are concentrated at low frequencies. If the rainforest's high-frequency noise drives birds to use lower frequencies, Slabbekoorn reasoned, then the low-frequency sounds of the city should pressure birds to use higher frequencies. As his research progressed, Slabbekoorn found that great tits and European blackbirds are indeed switching to higher frequencies to be heard.

That's not the only way birds cope with human noise. In 2007 Richard Fuller and two other scientists at the University of Sheffield in the United Kingdom discovered that European robins living in noisy urban areas have radically departed from their normal behavior of singing during the day. Now they sing almost exclusively at night, presumably to evade interference from the human din. If you hear a bird singing on your way home from a night at the pub, it's probably a robin, Fuller says.

In Berlin, nightingales have taken a different approach, raising their singing volume in response to traffic noise, according to a study by Henrik Brumm of that city's Free University, published in the *Journal of Animal Ecology*. In what is known as the Lombard effect, where a singer or speaker raises his voice to be heard, the nightingales try to counteract rush-hour clamor by singing louder on weekday mornings than on weekend mornings.

Most research on the impact of human sounds has focused on birds, but there is growing evidence that what's happening in the avian world is playing out across the animal kingdom, even in remote places that might seem impervious to human sounds—places like the deep ocean.

Cities are getting louder, but underwater noise is increasing even faster. There are about twice as many ships plying the world's oceans now as in the 1960s, and these ships are faster, more powerful, and collectively far noisier than their predecessors. When scientists at the Scripps Institution of Oceanography and the Colorado-based company Whale Acoustics compared sound levels west of California's San Nicolas Island in 2003 and 2004 with measurements made during the 1960s at the same site, they found that ambient noise levels had increased about tenfold. And this area may be representative of the entire northeast Pacific, the scientists say in a report published in the *Journal of the Acoustical Society of America*.

Instead of containing this noise within shipping lanes or coastal areas, the ocean's unique dynamics actually help the sounds travel hundreds or even thousands of kilometers. The so-called "deep sound channel" is a layer of water where sound travels slowly but encounters little resistance or interference. Some scientists suspect that humpback whales dive down to this channel and then sing into it, communicating with other humpbacks hundreds of kilometers away. And when noises from commercial shipping, offshore drilling, and other human activities get caught in the channel, they too are carried far from the original source.

Researchers have only just begun investigating these sounds' exact impacts, but a few studies suggest man-made noise is forcing marine mammals to respond in much the same way as birds do. Scientists at the Institute of Ocean Sciences in Sidney, Canada, have shown that beluga whales change their vocal patterns in response to the presence of icebreakers, whose systems interfere with the belugas' preferred frequencies. Belugas also switch the frequencies of their echolocation clicks when background noise increases. Elsewhere, orcas in the Pacific Northwest have changed their calls, perhaps in response to increased traffic by commercial ships and whale-watching boats.

In more extreme cases, human sounds have forced whales to abandon their preferred habitat. For instance, gray whales have long used Baja California's Guerrero Negro lagoon as a calving ground. But when construction at a nearby salt factory spurred increases in ship traffic and dredging activities, the whales stayed away from the lagoon for several years, returning only after construction ebbed.

Some researchers interpret these adaptations as a heartening sign, pointing out that some animals will simply change along with the soundscape. But Slabbekoorn cautions that some species could be wiped out by the human din.

Because low-frequency traffic noise accounts for most of humans' clamor, animals that use low-frequency calls and can't switch to higher frequencies are threatened most. Slabbekoorn says birds such as orioles, great reed warblers, and house sparrows fit this category. Populations of house sparrows are declining throughout Europe; researchers haven't pinpointed the cause, but Slabbekoorn suspects human noise is a factor.

Bernie Krause has witnessed a similar phenomenon among spadefoot toads in the Mono Lake basin east of Yosemite National Park. Using its big front claws, the toad buries itself one meter below the desert floor and can survive there for up to six years. When rain finally comes, the toad emerges and joins others to sing in chorus, which makes it harder for predators such as owls and coyotes to get a bead on where the sound is coming from.

The problem is that during nighttime periods when the toads do their singing, military jet planes often use the basin for training.

Flying only one hundred meters above the ground, the planes are so loud that the toads can't hear each other. Even after the planes leave, it takes twenty to forty-five minutes for the toads to resume their synchronized chorus, and in the meantime they're vulnerable to predators. Krause believes the noise is partly responsible for a precipitous decline in spadefoot populations, which he has studied since 1984.

Even adaptable species may be altered in fundamental ways. For instance, if changing calls or switching frequencies helps male birds be heard, they could earn an advantage when it comes to attracting female mates. Over time, this dynamic could force evolutionary changes, splitting populations of birds into localized species with specialized reactions to the sounds in their vicinity.

Slabbekoorn and his colleague Erwin Ripmeester think these noise-driven evolutionary forces may already be separating European blackbirds into urban and rural subspecies. The two researchers have even begun testing whether rural birds can recognize their urban brethren's hip new calls. If Slabbekoorn and Ripmeester's hunch is correct, it could mean that humans, already powerful conductors of the material world, may be extending their fierce control to the audible one.

ELIZABETH KOLBERT

The Catastrophist

FROM *The New Yorker*

A FEW MONTHS AGO, James Hansen, the director of NASA's Goddard Institute for Space Studies, in Manhattan, took a day off from work to join a protest in Washington, D.C. The immediate target of the protest was the Capitol Power Plant, which supplies steam and chilled water to congressional offices, but more generally its object was coal, which is the world's leading source of greenhouse-gas emissions. As it happened, on the day of the protest it snowed. Hansen was wearing a trench coat and a wide-brimmed canvas boater. He had forgotten to bring gloves. His sister, who lives in D.C. and had come along to watch over him, told him that he looked like Indiana Jones.

The march to the power plant was to begin on Capitol Hill, at the Spirit of Justice Park. By the time Hansen arrived, thousands of protesters were already milling around, wearing green hard hats and carrying posters with messages like POWER PAST COAL and CLEAN COAL IS LIKE DRY WATER. Hansen was immediately surrounded by TV cameras.

"You are one of the preeminent climatologists in the world," one television reporter said. "How does this square with your science?"

"I'm trying to make clear what the connection is between the science and the policy," Hansen responded. "Somebody has to do it."

The reporter wasn't satisfied. "Civil disobedience?" he asked, in a tone of mock incredulity. Hansen said that he couldn't let young people put themselves on the line "and then I stand back behind them."

The reporter still hadn't got what he wanted: "We've heard that you all are planning, even hoping, to get arrested today. Is that true?"

"I wouldn't hope," Hansen said. "But I do want to draw attention to the issue, whatever is necessary to do that."

Hansen, who is sixty-eight, has greenish eyes, sparse brown hair, and the distracted manner of a man who's just lost his wallet. (In fact, he frequently misplaces things, including, on occasion, his car.) Thirty years ago, he created one of the world's first climate models, nicknamed Model Zero, which he used to predict most of what has happened to the climate since. Sometimes he is referred to as the "father of global warming," and sometimes as the grandfather.

Hansen has now concluded, partly on the basis of his latest modeling efforts and partly on the basis of observations made by other scientists, that the threat of global warming is far greater than even he had suspected. Carbon dioxide isn't just approaching dangerous levels; it is already there. Unless immediate action is taken — including the shutdown of all the world's coal plants within the next two decades — the planet will be committed to change on a scale society won't be able to cope with. "This particular problem has become an emergency," Hansen said.

Hansen's revised calculations have prompted him to engage in activities — like marching on Washington — that aging government scientists don't usually go in for. Last September, he traveled to England to testify on behalf of anticoal activists who were arrested while climbing the smokestack of a power station to spray-paint a message to the prime minister. (They were acquitted.) Speaking before a congressional special committee last year, Hansen asserted that fossil fuel companies were knowingly spreading misinformation about global warming and that their chairmen "should be tried for high crimes against humanity and nature." He has compared freight trains carrying coal to "death trains," and he wrote to the head of the National Mining Association, who sent him a letter of complaint, that if the comparison "makes you uncomfortable, well, perhaps it should."

Hansen insists that his intent is not to be provocative but conservative: his only aim is to preserve the world as we know it. "The science is clear," he said, when it was his turn to address the protesters blocking the entrance to the Capitol Power Plant. "This is our one chance."

The fifth of seven children, Hansen grew up in Denison, Iowa, a small, sleepy town close to the western edge of the state. His father was a tenant farmer who, after World War II, went to work as a bartender. All the kids slept in two rooms. As soon as he was old enough, Hansen went to work, too, delivering the Omaha *World-Herald*. When he was eighteen, he received a scholarship to attend the University of Iowa. It didn't cover housing, so he rented a room for twenty-five dollars a month and ate mostly cereal. He stayed on at the university to get a Ph.D. in physics, writing his dissertation on the atmosphere of Venus. From there he went directly to the Goddard Institute for Space Studies—GISS, for short—where he took up the study of Venusian clouds.

By all accounts, including his own, Hansen was preoccupied by his research and not much interested in anything else. GISS's offices are a few blocks south of Columbia University; when riots shut down the campus, in 1968, he barely noticed. At that point, GISS's computer was the fastest in the world, but it still had to be fed punch cards. "I was staying here late every night, reading in my decks of cards," Hansen recalled. In 1969 he left GISS for six months to study in the Netherlands. There he met his wife, Anniek, who is Dutch; the couple honeymooned in Florida, near Cape Canaveral, so they could watch an Apollo launch.

In 1973 the first Pioneer Venus mission was announced, and Hansen began designing an instrument—a polarimeter—to be carried on the orbiter. But soon his research interests began to shift earthward. A trio of chemists—they would later share a Nobel Prize—had discovered that chlorofluorocarbons and other manmade chemicals could break down the ozone layer. It had also become clear that greenhouse gases were rapidly building up in the atmosphere.

"We realized that we had a planet that was changing before our eyes, and that's more interesting," Hansen told me. The topic attracted him for much the same reason Venus's clouds had: there were new research questions to be answered. He decided to try to adapt a computer program that had been designed to forecast the weather to see if it could be used to look further into the future.

What would happen to Earth if, for example, greenhouse-gas levels were to double?

"He never worked on any topic thinking it might be any use for the world," Anniek told me. "He just wanted to figure out the scientific meaning of it."

When Hansen began his modeling work, there were good theoretical reasons for believing that increasing CO_2 levels would cause the world to warm, but little empirical evidence. Average global temperatures had risen in the 1930s and '40s; then they had declined, in some regions, in the 1950s and '60s. A few years into his project, Hansen concluded that a new pattern was about to emerge. In 1981 he became the director of GISS. In a paper published that year in *Science,* he forecast that the following decade would be unusually warm. (That turned out to be the case.) In the same paper, he predicted that the 1990s would be warmer still. (That also turned out to be true.) Finally, he forecast that by the end of the twentieth century a global-warming signal would emerge from the "noise" of natural climate variability. (This, too, proved to be correct.)

Later, Hansen became even more specific. In 1990 he bet a roomful of scientists that that year or one of the following two would be the warmest on record. (Within nine months, he had won the bet.) In 1991 he predicted that owing to the eruption of Mt. Pinatubo, in the Philippines, average global temperatures would drop and then, a few years later, recommence their upward climb, which was precisely what happened.

From early on, the significance of Hansen's insights was recognized by the scientific community. "The work that he did in the seventies, eighties, and nineties was absolutely groundbreaking," Spencer Weart, a physicist turned historian who has studied the efforts to understand climate change, told me. He added, "It does help to be right."

"I have a whole folder in my drawer labeled 'Canonical Papers,'" Michael Oppenheimer, a climate scientist at Princeton, said. "About half of them are Jim's."

Because of its implications for humanity, Hansen's work also attracted considerable attention from the world at large. His 1981 paper prompted the first front-page article on climate change that ran in the *Times*—STUDY FINDS WARMING TREND THAT COULD RAISE SEA LEVELS, the headline read—and within a few years he was regularly being invited to testify before Congress. Still, Hansen says, he didn't imagine himself playing any role besides that of a research scientist. He is, he has written, "a poor communicator" and "not tactful."

"He's very shy," Ralph Cicerone, the president of the National Academy of Sciences, who has known Hansen for nearly forty years, told me. "And, as far as I can tell, he does not enjoy a lot of his public work."

"Jim doesn't really like to look at anyone," Anniek Hansen told me. "I say, 'Just look at them!'"

Throughout the 1980s and '90s, the evidence of climate change —and its potential hazards—continued to grow. Hansen kept expecting the political system to respond. This, after all, was what had happened with the ozone problem. Proof that chlorofluorocarbons were destroying the ozone layer came in 1985, when British scientists discovered that an ozone "hole" had opened up over Antarctica. The crisis was resolved—or, at least, prevented from growing worse—by an international treaty phasing out chlorofluorocarbons, which was ratified in 1987.

"At first, Jim's work didn't take an activist bent at all," the writer Bill McKibben, who has followed Hansen's career for more than twenty years and who helped organize the anticoal protest in D.C., told me. "I think he thought, as did I, If we get this set of facts out in front of everybody, they're so powerful—overwhelming—that people will do what needs to be done. Of course, that was naive on both our parts."

As recently as the George W. Bush administration, Hansen was still operating as if getting the right facts in front of the right people would be enough. In 2001 he was invited to speak to Vice President Dick Cheney and other high-level administration officials. For the meeting, he prepared a detailed presentation titled "The Forcings Underlying Climate Change." In 2003 he was invited to Washington again, to meet with the head of the Council on Environmental Quality at the White House. This time he offered a presentation on what ice-core records show about the sensitivity of the climate to changes in greenhouse-gas concentrations. But by 2004 the administration had dropped any pretense that it was interested in the facts about climate change. That year NASA, reportedly at the behest of the White House, insisted that all communications between GISS scientists and the outside world be routed through political appointees at the agency. The following year, the administration prevented GISS from posting its monthly temperature data on its website, ostensibly on the ground that proper protocols had not been followed. (The data showed that 2005 was likely to be the warmest year on record.) Hansen was also told that he couldn't grant a routine interview to National Public Radio. When he spoke out about the restrictions, scientists at other federal agencies complained that they were being similarly treated, and a new term was invented: government scientists, it was said, were being "Hansenized."

"He had been waiting all this time for global warming to become the issue that ozone was," Anniek Hansen told me. "And he's very patient. And he just kept on working and publishing, thinking that someone would do something." She went on, "He started speaking out, not because he thinks he's good at it, not because he enjoys it, but because of necessity."

"When Jim makes up his mind, he pursues whatever conclusion he has to the end point," Michael Oppenheimer said. "And he's made up his mind that you have to pull out all the stops at this point, and that all his scientific efforts would come to naught if he didn't also involve himself in political action." Starting in 2007, Hansen began writing to world leaders, including Prime Minister Gordon Brown, of Britain, and Yasuo Fukuda, then the prime minister of Japan. In December 2008, he composed a personal appeal to Barack and Michelle Obama.

"A stark scientific conclusion, that we must reduce greenhouse gases below present amounts to preserve nature and humanity, has become clear," Hansen wrote. "It is still feasible to avert climate disasters, but only if policies are consistent with what science indicates to be required." Hansen gave the letter to Obama's chief science adviser, John Holdren, with whom he is friendly, and Holdren, he says, promised to deliver it. But Hansen never heard back, and by the spring he had begun to lose faith in the new administration. (In an e-mail, Holdren said that he could not discuss "what I have or haven't given or said to the President.")

"I had had hopes that Obama understood the reality of the issue and would seize the opportunity to marry the energy and climate and national-security issues and make a very strong program,"

Hansen told me. "Maybe he still will, but I'm getting bad feelings about it."

There are lots of ways to lose an audience with a discussion of global warming, and new ones, it seems, are being discovered all the time. As well as anyone, Hansen ought to know this; still, he persists in trying to make contact. He frequently gives public lectures; just in the past few months, he has spoken to Native Americans in Washington, D.C.; college students at Dartmouth; high school students in Copenhagen; concerned citizens, including King Harald, in Oslo; renewable-energy enthusiasts in Milwaukee; folk music fans in Beacon, New York; and public health professionals in Manhattan.

In April I met up with Hansen at the state capitol in Concord, New Hampshire, where he had been invited to speak by local anti-coal activists. There had been only a couple of days to publicize the event; nevertheless, more than 250 people showed up. I asked a woman from the town of Ossipee why she had come. "It's a once-in-a-lifetime opportunity to hear bad news straight from the horse's mouth," she said. For the event, Hansen had, as usual, prepared a PowerPoint presentation. It was projected onto a screen beside a faded portrait of George Washington. The first slide gave the title of the talk, "The Climate Threat to the Planet," along with the disclaimer "Any statements relating to policy are personal opinion."

Hansen likes to begin his talk with a highly compressed but still perilously long discussion of climate history, beginning in the early Eocene, some 50 million years ago. At that point, CO_2 levels were high and, as Hansen noted, the world was very warm: there was practically no ice on the planet, and palm trees grew in the Arctic. Then CO_2 levels began to fall. No one is entirely sure why, but one possible cause has to do with weathering processes that, over many millennia, allow carbon dioxide from the air to get bound up in limestone. As CO_2 levels declined, the planet grew cooler; Hansen flashed some slides on the screen, which showed that between 50 million and 35 million years ago, deep ocean temperatures dropped by more than 10 degrees. Eventually, around 34 million years ago, temperatures sank low enough that glaciers began to form on Antarctica. By around 3 million years ago—perhaps earlier—permanent ice sheets had begun to form in the Northern

Hemisphere as well. Then, about 2 million years ago, the world entered a period of recurring glaciations. During the ice ages — the most recent one ended about 12,000 years ago — CO_2 levels dropped even further.

What is now happening, Hansen explained to the group in New Hampshire, is that climate history is being run in reverse and at high speed, like a cassette tape on rewind. Carbon dioxide is being pumped into the air some ten thousand times faster than natural weathering processes can remove it.

"So humans now are in charge of atmospheric composition," Hansen said. Then he corrected himself: "Well, we're determining it, whether we're in charge or not."

Among the many risks of running the system backward is that the ice sheets formed on the way forward will start to disintegrate. Once it begins, this process is likely to be self-reinforcing. "If we burn all the fossil fuels and put all that CO_2 into the atmosphere, we will be sending the planet back to the ice-free state," Hansen said. "It will take a while to get there — ice sheets don't melt instantaneously — but that's what we will be doing. And if you melt all the ice, sea levels will go up two hundred and fifty feet. So you can't do that without producing a different planet."

There's no precise term for the level of CO_2 that will assure a climate disaster; the best that scientists and policymakers have been able to come up with is the phrase "dangerous anthropogenic interference," or DAI. Most official discussions have been premised on the notion that DAI will not be reached until CO_2 levels hit 450 parts per million. Hansen, however, has concluded that the threshold for DAI is much lower.

"The bad news is that it's become clear that the dangerous amount of carbon dioxide is no more than three hundred and fifty parts per million," he told the crowd in Concord. The *really* bad news is that CO_2 levels have already reached 385 parts per million. (For the ten thousand years prior to the industrial revolution, carbon dioxide levels were about 280 parts per million, and if current emissions trends continue they will reach 450 parts by around 2035.)

Once you accept that CO_2 levels are already too high, it's obvious, Hansen argues, what needs to be done. He displayed a chart of known fossil fuel reserves represented in terms of their carbon content. There was a short bar for oil, a shorter bar for natural gas, and a tall bar for coal.

"We've already used about half of the oil," he observed. "And we're going to use all of the oil and natural gas that's easily available. It's owned by Russia and Saudi Arabia, and we can't tell them not to sell it. So, if you look at the size of these fossil fuel reservoirs, it becomes very clear. The only way we can constrain the amount of carbon dioxide in the atmosphere is to cut off the coal source, by saying either we will leave the coal in the ground or we will burn it only at power plants that actually capture the CO_2." Such power plants are often referred to as "clean-coal plants." Although there has been a great deal of talk about them lately, at this point there are no clean-coal plants in commercial operation, and, for a combination of technological and economic reasons, it's not clear that there ever will be.

Hansen continued, "If we had a moratorium on any new coal plants and phased out existing ones over the next twenty years, we could get back to three hundred and fifty parts per million within several decades." Reforestation, for example, if practiced on a massive scale, could begin to draw global CO_2 levels down, Hansen says, "so it's technically feasible." But "it requires us to take action promptly."

Coincidentally, that afternoon a vote was scheduled in the New Hampshire state legislature on a proposal involving the state's largest coal-fired power plant, the Merrimack Station, in the town of Bow. The station's owner was planning to spend several hundred million dollars to reduce mercury emissions from the plant—a cost that it planned to pass on to ratepayers. Hansen, who said he thought the plant should simply be shut, called the plan a "terrible waste of money." A lawmaker sympathetic to this view had introduced a bill calling for more study of the project, but, as several people who came up to speak to Hansen after his talk explained, it was opposed by the state's construction unions and seemed headed for defeat. (Less than an hour later, the bill was rejected in committee by a unanimous vote.)

"I assume you're used to telling policymakers the truth and then having them ignore you," one man said to Hansen.

Hansen smiled ruefully. "You're right."

*

In scientific circles, worries about DAI are widespread. During the past few years, researchers around the world have noticed a disturbing trend: the planet is changing faster than had been anticipated. Antarctica, for example, had not been expected to show a net loss of ice for another century, but recent studies indicate that the continent's massive ice sheets are already shrinking. At the other end of the globe, the Arctic ice cap has been melting at a shocking rate; the extent of the summer ice is now only a little more than half of what it was just forty years ago. Meanwhile, scientists have found that the arid zones that circle the globe north and south of the tropics have been expanding more rapidly than computer models had predicted. This expansion of the subtropics means that highly populated areas, including the American Southwest and the Mediterranean basin, are likely to suffer more and more frequent droughts.

"Certainly, I think the shrinking of the Arctic ice cap made a very strong impression on a lot of scientists," Spencer Weart, the physicist, told me. "And these things keep popping up. You think, What, another one? Another one? They're almost all in the wrong direction, in the direction of making the change worse and faster."

"In nearly all areas, the developments are occurring more quickly than had been assumed," Hans Joachim Schellnhuber, the head of Germany's Potsdam Institute for Climate Impact Research, recently observed. "We are on our way to a destabilization of the world climate that has advanced much further than most people or their governments realize."

Obama's science adviser, John Holdren, a physicist on leave from Harvard, has said that he believes "any reasonably comprehensive and up-to-date look at the evidence makes clear that civilization has already generated dangerous anthropogenic interference in the climate system."

There is also broad agreement among scientists that coal represents the most serious threat to the climate. Coal now provides half the electricity in the United States. In China, that figure is closer to 80 percent, and a new coal-fired power plant comes online every week or two. As oil supplies dwindle, there will still be plenty of coal, which could be—and in some places already is being—converted into a very dirty liquid fuel. Before Steven Chu, a Nobel

Prize–winning physicist, was appointed to his current post as energy secretary, he said in a speech, "There's enough carbon in the ground to really cook us. Coal is my worst nightmare." (These are lines that Hansen is fond of invoking.) A couple of months ago, seven prominent climate scientists from Australia wrote an open letter to the owners of that country's major utility companies urging that "no new coal-fired power stations, except ones that have ZERO emissions," be built. They also recommended an "urgent program" to phase out old plants.

"The unfortunate reality is that genuine action on climate change will require that existing coal-fired power stations cease to operate in the near future," the group wrote.

But if Hansen's anxieties about DAI and coal are broadly shared, he is still, among climate scientists, an outlier. "Almost everyone in the scientific community is prepared to say that if we don't do something now to reverse the direction we're going in we either already are or will very, very soon be in the danger zone," Naomi Oreskes, a historian of science and a provost at the University of California at San Diego, told me. "But Hansen talks in stronger terms. He's using adjectives. He has started to speak in moral terms, and that always makes scientists uncomfortable."

Hansen is also increasingly isolated among climate activists. "I view Jim Hansen as heroic as a scientist," Eileen Claussen, the president of the Pew Center on Global Climate Change, said. "He was there at the beginning, he's faced all kinds of pressures politically, and he's done a terrific job, I think, of keeping focused. But I wish he would stick to what he really knows. Because I don't think he has a realistic view of what is politically possible, or what the best policies would be to deal with this problem."

In Washington, the only approach to limiting emissions that is seen as having any chance of being enacted is a so-called cap and trade system. Under such a system, the government would set an overall cap for CO_2 emissions, then allocate allowances to major emitters, like power plants and oil refineries, which could be traded on a carbon market. In theory, at least, the system would discourage fossil fuel use by making emitters pay for what they are putting out. But to the extent that such a system has been tried, by the members of the European Union, its results so far are inconclusive, and Hansen argues that it is essentially a sham. (He recently referred to it as "the Temple of Doom.") What is required, he insists, is a direct tax on carbon emissions. The tax should be significant at the start—equivalent to roughly a dollar per gallon for gasoline—and then grow steeper over time. The revenues from the tax, he believes, ought to be distributed back to Americans on a per capita basis, so that households that use less energy would actually make money, even as those that use more would find it increasingly expensive to do so.

"The only defense of this monstrous absurdity that I have heard," Hansen wrote a few weeks ago, referring to a cap and trade system, "is 'Well, you are right, it's no good, but the train has left the station.' If the train has left, it had better be derailed soon or the planet, and all of us, will be in deep doo-doo."

GISS's headquarters, at 112th Street and Broadway, sits above Tom's Restaurant, the diner made famous by *Seinfeld* and Suzanne Vega. Hansen has occupied the same office, on the seventh floor, since he became the director of the institute, almost three decades ago. One day last month, I went to visit him there. Hansen told me that he had been trying to computerize his old files; still, the most striking thing about the spacious office, which is largely taken up by three wooden tables, is that every available surface is covered with stacks of paper.

During the week, Hansen lives in an apartment just a few blocks from his office, but on weekends he and Anniek frequently go to an eighteenth-century house that they own in Bucks County, Pennsylvania, and their son and daughter, who have children of their own, come to visit. Hansen dotes on his grandchildren—in many hours of conversation with me, just about the only time that he spoke with unalloyed enthusiasm was when he discussed planting trees with them this spring—and he claims they are the major reason for his activism. "I decided that I didn't want my grandchildren to say 'Opa understood what was happening, but he didn't make it clear,'" he explained.

The day that I visited Hansen's office, the House Energy and Commerce Committee was beginning its markup of a cap and trade bill cosponsored by the committee's chairman, Henry Waxman, of California. The bill—the American Clean Energy and Security Act—has the stated goal of cutting the country's carbon emissions by 17 percent by 2020. It is the most significant piece of climate legislation to make it this far in the House. Hansen pointed out that the bill explicitly allows for the construction of new coal plants and predicted that it would, if passed, prove close to meaningless. He said that he thought it would probably be best if the bill failed so that Congress could "come back and do it more sensibly."

I said that if the bill failed I thought it was more likely that Congress would let the issue drop, and that was one reason most of the country's major environmental groups were backing it.

"This is just stupidity on the part of environmental organizations in Washington," Hansen said. "The fact that some of these organizations have become part of the Washington 'go along, get along' establishment is very unfortunate."

Hansen argues that politicians willfully misunderstand climate science; it could be argued that Hansen just as willfully misunderstands politics. In order to stabilize carbon dioxide levels in the atmosphere, annual global emissions would have to be cut by something on the order of three-quarters. In order to draw them down, agricultural and forestry practices would have to change dramatically as well. So far, at least, there is no evidence that any nation is willing to take anything approaching the necessary steps. On the contrary, almost all the trend lines point in the opposite direction. Just because the world desperately needs a solution that satisfies both the scientific and the political constraints doesn't mean one necessarily exists.

For his part, Hansen argues that while the laws of geophysics are immutable, those of society are ours to determine. When I said that it didn't seem feasible to expect the United States to give up its coal plants, he responded, "We can point to other countries being fifty percent more energy-efficient than we are. We're getting fifty percent of our electricity from coal. That alone should provide a pretty strong argument."

Then what about China and India?

Both countries are likely to suffer very severely from dramatic climate change, he said. "They're going to recognize that. In fact, they already are beginning to recognize that.

"It's not unrealistic," he went on. "But the policies have to push us in that direction. And, as long as we let the politicians and the people who are supporting them continue to set the rules, such that 'business as usual' continues, or small tweaks to 'business as usual,' then it is unrealistic. So we have to change the rules." He said that he was thinking of attending another demonstration soon, in West Virginia coal country.

ELIZABETH KOLBERT

The Sixth Extinction?

FROM *The New Yorker*

THE TOWN OF El Valle de Antón, in central Panama, sits in the middle of a volcanic crater formed about a million years ago. The crater is almost four miles across, but when the weather is clear you can see the jagged hills that surround the town, like the walls of a ruined tower. El Valle has one main street, a police station, and an open-air market that offers, in addition to the usual hats and embroidery, what must be the world's largest selection of golden-frog figurines. There are golden frogs sitting on leaves and—more difficult to understand—golden frogs holding cell phones. There are golden frogs wearing frilly skirts, and golden frogs striking dance poses, and ashtrays featuring golden frogs smoking cigarettes through a holder, after the fashion of FDR. The golden frog, which is bright yellow with dark brown splotches, is endemic to the area around El Valle. It is considered a lucky symbol in Panama—its image is often printed on lottery tickets—though it could just as easily serve as an emblem of disaster.

In the early 1990s, an American graduate student named Karen Lips established a research site about two hundred miles west of El Valle, in the Talamanca Mountains, just over the border in Costa Rica. Lips was planning to study the local frogs, some of which, she later discovered, had never been identified. In order to get to the site, she had to drive two hours from the nearest town—the last part of the trip required tire chains—and then hike for an hour through the rainforest.

Lips spent two years living in the mountains. "It was a wonderland," she recalled recently. Once she had collected enough data, she left to work on her dissertation. She returned a few months later, and, though nothing seemed to have changed, she could hardly find any frogs. Lips couldn't figure out what was happening. She collected all the dead frogs that she came across—there were only a half dozen or so—and sent their bodies to a veterinary pathologist in the United States. The pathologist was also baffled: the specimens, she told Lips, showed no signs of any known disease.

A few years went by. Lips finished her dissertation and got a teaching job. Since the frogs at her old site had pretty much disappeared, she decided that she needed to find a new location to do research. She picked another isolated spot in the rainforest, this time in western Panama. Initially, the frogs there seemed healthy. But before long, Lips began to find corpses lying in the streams and moribund animals sitting on the banks. Sometimes she would pick up a frog and it would die in her hands. She sent some specimens to a second pathologist in the United States, and, once again, the pathologist had no idea what was wrong.

Whatever was killing Lips's frogs continued to move, like a wave, east across Panama. By 2002 most frogs in the streams around Santa Fé, a town in the province of Veraguas, had been wiped out. By 2004 the frogs in the national park of El Copé, in the province of Coclé, had all but disappeared. At that point, golden frogs were still relatively common around El Valle; a creek not far from the town was nicknamed Thousand Frog Stream. Then, in 2006, the wave hit.

Of the many species that have existed on Earth—estimates run as high as 50 billion—more than 99 percent have disappeared. In light of this, it is sometimes joked that all of life today amounts to little more than a rounding error.

Records of the missing can be found everywhere in the world, often in forms that are difficult to overlook. And yet extinction has been a much contested concept. Throughout the eighteenth century, even as extraordinary fossils were being unearthed and put on exhibit, the prevailing view was that species were fixed, created by God for all eternity. If the bones of a strange creature were found, it must mean that that creature was out there somewhere.

"Such is the economy of nature," Thomas Jefferson wrote, "that no instance can be produced, of her having permitted any one race of her animals to become extinct; of her having formed any link in her great work so weak as to be broken." When, as president, he dispatched Meriwether Lewis and William Clark to the Northwest, Jefferson hoped that they would come upon live mastodons roaming the region.

The French naturalist Georges Cuvier was more skeptical. In 1812 he published an essay on the "Revolutions on the Surface of the Globe," in which he asked, "How can we believe that the immense mastodons, the gigantic megatheriums, whose bones have been found in the earth in the two Americas, still live on this continent?" Cuvier had conducted studies of the fossils found in gypsum mines in Paris and was convinced that many organisms once common to the area no longer existed. These he referred to as *espèces perdues,* or lost species. Cuvier had no way of knowing how much time had elapsed in forming the fossil record. But, as the record indicated that Paris had, at various points, been under water, he concluded that the *espèces perdues* had been swept away by sudden cataclysms.

"Life on this earth has often been disturbed by dreadful events," he wrote. "Innumerable living creatures have been victims of these catastrophes." Cuvier's essay was translated into English in 1813 and published with an introduction by the Scottish naturalist Robert Jameson, who interpreted it as proof of Noah's flood. It went through five editions in English and six in French before Cuvier's death in 1832.

Charles Darwin was well acquainted with Cuvier's ideas and the theological spin they had been given. (He had studied natural history with Jameson at the University of Edinburgh.) In his theory of natural selection, Darwin embraced extinction; it was, he realized, essential that some species should die out as new ones were created. But he believed that this happened only slowly. Indeed, he claimed that it took place more gradually even than speciation: "The complete extinction of the species of a group is generally a slower process than their production." In *On the Origin of Species,* published in the fall of 1859, Darwin heaped scorn on the catastrophist approach: "So profound is our ignorance, and so high our presumption, that we marvel when we hear of the extinction of an organic being; and as we do not see the cause, we invoke cataclysms to desolate the world."

By the start of the twentieth century, this view had become d nant, and to be a scientist meant seeing extinction as Darwin But Darwin, it turns out, was wrong.

Over the past half-billion years, there have been at least twenty mass extinctions, when the diversity of life on Earth has suddenly and dramatically contracted. Five of these — the so-called Big Five — were so devastating that they are usually put in their own category. The first took place during the late Ordovician period, nearly 450 million years ago, when life was still confined mainly to water. Geological records indicate that more than 80 percent of marine species died out. The fifth occurred at the end of the Cretaceous period, 65 million years ago. The end-Cretaceous event exterminated not just the dinosaurs but 75 percent of all species on Earth.

The significance of mass extinctions goes beyond the sheer number of organisms involved. In contrast to ordinary, or so-called background, extinctions, which claim species that, for one reason or another, have become unfit, mass extinctions strike down the fit and the unfit at once. For example, brachiopods, which look like clams but have an entirely different anatomy, dominated the ocean floor for hundreds of millions of years. In the third of the Big Five extinctions — the end-Permian — the hugely successful brachiopods were nearly wiped out, along with trilobites, blastoids, and eurypterids. (In the end-Permian event, more than 90 percent of marine species and 70 percent of terrestrial species vanished; the event is sometimes referred to as "the mother of mass extinctions" or "the great dying.")

Once a mass extinction occurs, it takes millions of years for life to recover, and when it does it generally has a new cast of characters; following the end-Cretaceous event, mammals rose up (or crept out) to replace the departed dinosaurs. In this way, mass extinctions, though missing from the original theory of evolution, have played a determining role in evolution's course; as Richard Leakey has put it, such events "restructure the biosphere" and so "create the pattern of life." It is now generally agreed among biologists that another mass extinction is underway. Though it's difficult to put a precise figure on the losses, it is estimated that if current trends continue, by the end of this century as many as half of Earth's species will be gone.

*

The El Valle Amphibian Conservation Center, known by the acronym EVACC (pronounced "e-vac"), is a short walk from the market where the golden-frog figurines are sold. It consists of a single building about the size of an average suburban house. The place is filled, floor to ceiling, with tanks. There are tall tanks for species that, like the Rabb's fringe-limbed tree frog, live in the forest canopy, and short tanks for species that, like the big-headed robber frog, live on the forest floor. Tanks of horned marsupial frogs, which carry their eggs in a pouch, sit next to tanks of casque-headed frogs, which carry their eggs on their backs.

The director of EVACC is a herpetologist named Edgardo Griffith. Griffith is tall and broad-shouldered, with a round face and a wide smile. He wears a silver ring in each ear and has a large tattoo of a toad's skeleton on his left shin. Griffith grew up in Panama City and fell in love with amphibians one day in college when a friend invited him to go frog hunting. He collected most of the frogs at EVACC—there are nearly six hundred—in a rush, just as corpses were beginning to show up around El Valle. At that point, the center was little more than a hole in the ground, and so the frogs had to spend several months in temporary tanks at a local hotel. "We got a very good rate," Griffith assured me. While the amphibians were living in rented rooms, Griffith and his wife, a former Peace Corps volunteer, would go out into a nearby field to catch crickets for the frogs' dinner. Now EVACC raises bugs for the frogs in what looks like an oversized rabbit hutch.

EVACC is financed largely by the Houston Zoo, which initially pledged $20,000 to the project and has ended up spending ten times that amount. The tiny center, though, is not an outpost of the zoo. It might be thought of as a preserve, except that instead of protecting the amphibians in their natural habitat, the center's aim is to isolate them from it. In this way, EVACC represents an ark built for a modern-day deluge. Its goal is to maintain twenty-five males and twenty-five females of each species—just enough for a breeding population.

The first time I visited, Griffith pointed out various tanks containing frogs that have essentially disappeared from the wild. These include the Panamanian golden frog, which, in addition to its extraordinary coloring, is known for its unusual method of communication; the frogs signal to one another using a kind of semaphore. Griffith said that he expected between a third and a half of all Panama's amphibians to be gone within the next five years. Some species, he said, will probably vanish without anyone's realizing it: "Unfortunately, we are losing all these amphibians before we even know that they exist."

Griffith still goes out collecting for EVACC. Since there are hardly any frogs to be found around El Valle, he has to travel farther afield, across the Panama Canal, to the eastern half of the country.

One day this winter, I set out with him on one of his expeditions, along with two American zookeepers who were also visiting EVACC. The four of us spent a night in a town called Cerro Azul and, at dawn the next morning, drove in a truck to the ranger station at the entrance to Chagres National Park. Griffith was hoping to find females of two species that EVACC is short of. He pulled out his collecting permit and presented it to the sleepy officials manning the station. Some underfed dogs came out to sniff around.

Beyond the ranger station, the road turned into a series of craters connected by ruts. Griffith put Jimi Hendrix on the truck's CD player, and we bounced along to the throbbing beat. (When the driving got particularly gruesome, he would turn down the volume.) Frog collecting requires a lot of supplies, so Griffith had hired two men to help with the carrying. At the very last cluster of houses, in the village of Los Ángeles, they materialized out of the mist. We bounced on until the truck couldn't go any farther; then we all got out and started walking.

The trail wound its way through the rainforest in a slather of red mud. Every few hundred yards, the main path was crossed by a narrower one; these paths had been made by leaf-cutter ants, making millions—perhaps billions—of trips to bring bits of greenery back to their colonies. (The colonies, which look like mounds of sawdust, can cover an area the size of a suburban back yard.) One of the Americans, Chris Bednarski, from the Houston Zoo, warned me to avoid the soldier ants, which will leave their jaws in your shin even after they're dead. "Those'll really mess you up," he observed. The other American, John Chastain, from the Toledo Zoo, was carrying a long hook, for use against venomous snakes. "Fortunately, the ones that can really mess you up are pretty rare," Bednarski said. Howler monkeys screamed in the distance. Someone pointed out jaguar prints in the soft ground.

After about five hours, we emerged into a small clearing. While we were setting up camp, a blue morpho butterfly flitted by, its wings the color of the sky.

That evening, after the sun set, we strapped on headlamps and clambered down to a nearby stream. Many amphibians are nocturnal, and the only way to see them is to go looking in the dark, an exercise that's as tricky as it sounds. I kept slipping and violating Rule No. 1 of rainforest safety: never grab onto something if you don't know what it is. After one of my falls, Bednarski showed me a tarantula the size of my fist that he had found on a nearby tree.

One technique for finding amphibians at night is to shine a light into the forest and look for the reflecting glow of their eyes. The first amphibian sighted this way was a San José Cochran frog, perched on top of a leaf. San José Cochran frogs are part of a larger family known as "glass frogs," so named because their translucent skin reveals the outlines of their internal organs. This particular glass frog was green with tiny yellow dots. Griffith pulled a pair of surgical gloves out of his pack. He stood entirely still and then, with a heronlike gesture, darted to scoop up the frog. With his free hand, he took what looked like the end of a Q-Tip and swabbed the frog's belly. Finally, he put the Q-Tip in a little plastic vial, placed the frog back on the leaf, and pulled out his camera. The frog stared into the lens impassively.

We continued to grope through the blackness. Someone spotted a La Loma robber frog, which is an orangey red, like the forest floor; someone else spotted a Warzewitsch frog, which is bright green and shaped like a leaf. With every frog, Griffith went through the same routine — snatching it up, swabbing its belly, photographing it. Finally, we came upon a pair of Panamanian robber frogs locked in amplexus — the amphibian version of sex. Griffith left these two alone.

One of the frogs that Griffith was hoping to catch, the horned marsupial frog, has a distinctive call that's been likened to the sound of a champagne bottle being uncorked. As we sloshed along, the call seemed to be emanating from several directions at once. Sometimes it sounded as if it were right nearby, but then, as we approached, it would fall silent. Griffith began imitating the call,

making a cork-popping sound with his lips. Eventually, he decided
that the rest of us were scaring the frogs with our splashing. He
waded ahead, while we stood in the middle of the stream, trying
not to move. When Griffith gestured us over, we found him standing in front of a large yellow frog with long toes and an owlish
face. It was sitting on a tree limb, just above eye level. Griffith
grabbed the frog and turned it over. Where a female marsupial
frog would have a pouch, this one had none. Griffith swabbed it,
photographed it, and put it back in the tree.

"You are a beautiful boy," he told the frog.

Amphibians are among the planet's great survivors. The ancestors of today's frogs and toads crawled out of the water some 400
million years ago, and by 250 million years ago the earliest representatives of what became the modern amphibian clades—one includes frogs and toads; a second, newts and salamanders—had
evolved. This means that amphibians have been around not just
longer than mammals, say, or birds; they have been around since
before there were dinosaurs. Most amphibians—the word comes
from the Greek meaning "double life"—are still closely tied to the
aquatic realm from which they emerged. (The ancient Egyptians
thought that frogs were produced by the coupling of land and water during the annual flooding of the Nile.) Their eggs, which have
no shells, must be kept moist in order to develop. There are frogs
that lay their eggs in streams, frogs that lay them in temporary
pools, frogs that lay them underground, and frogs that lay them in
nests that they construct out of foam. In addition to frogs that carry
their eggs on their backs and in pouches, there are frogs that carry
them in their vocal sacs, and, until recently, at least, there were
frogs that carried their eggs in their stomachs and gave birth
through their mouths. Amphibians emerged at a time when all the
land on Earth was part of one large mass; they have since adapted
to conditions on every continent except Antarctica. Worldwide,
more than 6,000 species have been identified, and while the greatest number are found in the tropical rainforests, there are amphibians that, like the sandhill frog of Australia, can live in the desert and also amphibians that, like the wood frog, can live above the
Arctic Circle. Several common North American frogs, including
spring peepers, are able to survive the winter frozen solid.

When, about two decades ago, researchers first noticed that something odd was happening to amphibians, the evidence didn't seem to make sense. David Wake is a biologist at the University of California at Berkeley. In the early 1980s, his students began returning from frog-collecting trips in the Sierra Nevada empty-handed. Wake remembered from his own student days that frogs in the Sierras had been difficult to avoid. "You'd be walking through meadows, and you'd inadvertently step on them," he told me. "They were just everywhere." Wake assumed that his students were going to the wrong spots or that they just didn't know how to look. Then a postdoc with several years of experience collecting amphibians told him that he couldn't find any either. "I said, 'OK, I'll go up with you and we'll go out to some proven places,'" Wake recalled. "And I took him out to this proven place and we found, like, two toads."

Around the same time, other researchers, in other parts of the world, reported similar difficulties. In the late 1980s, a herpetologist named Marty Crump went to Costa Rica to study golden toads; she was forced to change her project because, from one year to the next, the toad essentially vanished. (The golden toad, now regarded as extinct, was actually orange; it is not to be confused with the Panamanian golden frog, which is technically also a toad.) Probably simultaneously, in central Costa Rica the populations of twenty species of frogs and toads suddenly crashed. In Ecuador the jambato toad, a familiar visitor to back-yard gardens, disappeared in a matter of years. And in northeastern Australia, biologists noticed that more than a dozen amphibian species, including the southern day frog, one of the more common in the region, were experiencing drastic declines.

But as the number of examples increased, the evidence only seemed to grow more confounding. Though amphibians in some remote and — relatively speaking — pristine spots seemed to be collapsing, those in other, more obviously disturbed habitats seemed to be doing fine. Meanwhile, in many parts of the world there weren't good data on amphibian populations to begin with, so it was hard to determine what represented terminal descent and what might be just a temporary dip.

"It was very controversial to say that amphibians were disappearing," Andrew Blaustein, a zoology professor at Oregon State University, recalls. Blaustein, who was studying the mating behavior of frogs and toads in the Cascade Mountains, had observed that some long-standing populations simply weren't there anymore. "The debate was whether or not there really was an amphibian population problem, because some people were saying it was just natural variation." At the point that Karen Lips went to look for her first research site, she purposefully tried to steer clear of the controversy.

"I didn't want to work on amphibian decline," she told me. "There were endless debates about whether this was a function of randomness or a true pattern. And the last thing you want to do is get involved when you don't know what's going on."

But the debate was not to be avoided. Even amphibians that had never seen a pond or a forest started dying. Blue poison-dart frogs, which are native to Suriname, had been raised at the National Zoo, in Washington, D.C., for several generations. Then, suddenly, the zoo's tank-bred frogs were nearly wiped out.

It is difficult to say when, exactly, the current extinction event —sometimes called the sixth extinction—began. What might be thought of as its opening phase appears to have started about 50,000 years ago. At that time, Australia was home to a fantastic assortment of enormous animals; these included a wombatlike creature the size of a hippo, a land tortoise nearly as big as a VW Beetle, and the giant short-faced kangaroo, which grew to be ten feet tall. Then all of the continent's largest animals disappeared. Every species of marsupial weighing more than two hundred pounds—there were nineteen of them—vanished, as did three species of giant reptiles and a flightless bird with stumpy legs known as *Genyornis newtoni.*

This die-off roughly coincided with the arrival of the first people on the continent, probably from Southeast Asia. Australia is a big place, and there couldn't have been very many early settlers. For a long time, the coincidence was discounted. Yet thanks to recent work by geologists and paleontologists, a clear global pattern has emerged. About 11,000 years ago, three-quarters of North America's largest animals—among them mastodons, mammoths, giant beavers, short-faced bears, and saber-toothed tigers—began to go extinct. This is right around the time the first humans are believed to have wandered onto the continent across the Bering land bridge. In relatively short order, the first humans settled South America as well. Subsequently, more than thirty species of South American "megamammals," including elephant-size ground sloths and rhinolike creatures known as toxodons, died out.

And what goes for Australia and the Americas also goes for many other parts of the world. Humans settled Madagascar around 2,000 years ago; the island subsequently lost all mammals weighing more than twenty pounds, including pygmy hippos and giant lemurs. "Substantial losses have occurred throughout near time," Ross MacPhee, a curator at the American Museum of Natural History, in New York, and an expert on extinctions of the recent geological past, has written. "In the majority of cases, these losses occurred when, and only when, people began to expand across areas that had never before experienced their presence." The Maori arrived in New Zealand around eight hundred years ago. They encountered eleven species of moas—huge ostrichlike creatures without wings. Within a few centuries—and possibly within a single century—all eleven moa species were gone. While these "first contact" extinctions were most pronounced among large animals, they were not confined to them. Humans discovered the Hawaiian Islands around 1,500 years ago; soon afterward, 90 percent of Hawaii's native bird species disappeared.

"We expect extinction after people arrive on an island," David Steadman, the curator of ornithology at the Florida Museum of Natural History, has written. "Survival is the exception."

Why was first contact with humans so catastrophic? Some of the animals may have been hunted to death; thousands of moa bones have been found at Maori archaeological sites, and man-made artifacts have been uncovered near mammoth and mastodon remains at more than a dozen sites in North America. Hunting, however, seems insufficient to account for so many losses across so many different taxa in so many parts of the globe. A few years ago, researchers analyzed hundreds of bits of emu and *Genyornis newtoni* eggshell, some dating from long before the first people arrived in Australia and some from after. They found that around 45,000 years ago, rather abruptly, emus went from eating all sorts of plants to relying mainly on shrubs. The researchers hypothesized that Australia's early settlers periodically set the countryside on fire— perhaps to flush out prey—a practice that would have reduced the variety of plant life. Those animals which, like emus, could cope with a changed landscape survived, while those which, like *Genyornis,* could not died out.

When Australia was first settled, there were maybe half a million people on Earth. There are now more than 6.5 billion, and it is expected that within the next three years the number will reach 7 billion.

Human impacts on the planet have increased proportionately. Farming, logging, and building have transformed between a third and a half of the world's land surface, and even these figures probably understate the effect, since land not being actively exploited may still be fragmented. Most of the world's major waterways have been diverted or dammed or otherwise manipulated—in the United States, only 2 percent of rivers run unimpeded—and people now use half the world's readily accessible freshwater runoff. Chemical plants fix more atmospheric nitrogen than all natural terrestrial processes combined, and fisheries remove more than a third of the primary production of the temperate coastal waters of the oceans. Through global trade and international travel, humans have transported countless species into ecosystems that are not prepared for them. We have pumped enough carbon dioxide into the air to alter the climate and to change the chemistry of the oceans.

Amphibians are affected by many—perhaps most—of these disruptions. Habitat destruction is a major factor in their decline, and agricultural chemicals seem to be causing a rash of frog deformities. But the main culprit in the wavelike series of crashes, it's now believed, is a fungus. Ironically, this fungus, which belongs to a group known as chytrids (pronounced "kit-rids"), appears to have been spread by doctors.

Chytrid fungi are older even than amphibians—the first species evolved more than 600 million years ago—and even more widespread. In a manner of speaking, they can be found—they are microscopic—just about everywhere, from the tops of trees to deep underground. Generally, chytrid fungi feed off dead plants; there are also species that live on algae, species that live on roots, and species that live in the guts of cows, where they help break down cellulose. Until two pathologists, Don Nichols and Allan Pessier, identified a weird microorganism growing on dead frogs from the National Zoo, chytrids had never been known to attack vertebrates. Indeed, the new chytrid was so unusual that an entire genus had to be created to accommodate it. It was named *Batrachochytrium dendrobatidis*— *batrachos* is Greek for "frog"—or Bd for short.

Nichols and Pessier sent samples from the infected frogs to a mycologist at the University of Maine, Joyce Longcore, who managed to culture the Bd fungus. They then exposed healthy blue poison-dart frogs to it. Within three weeks, the animals sickened and died.

The discovery of Bd explained many of the data that had previously seemed so puzzling. Chytrid fungi generate microscopic spores that disperse in water; these could have been carried along by streams or in the runoff after a rainstorm, producing what in Central America showed up as an eastward-moving scourge. In the case of zoos, the spores could have been brought in on other frogs or on tracked-in soil. Bd seemed to be able to live on just about any frog or toad, but not all amphibians are as susceptible to it, which would account for why some populations succumbed while others appeared to be unaffected.

Rick Speare is an Australian pathologist who identified Bd right around the same time that the National Zoo team did. From the pattern of decline, Speare suspected that Bd had been spread by an amphibian that had been moved around the globe. One of the few species that met this condition was *Xenopus laevis,* commonly known as the African clawed frog. In the early 1930s, a British zoologist named Lancelot Hogben discovered that female *Xenopus laevis,* when injected with certain types of human hormones, laid eggs. His discovery became the basis for a new kind of pregnancy test and, starting in the late 1930s, thousands of African clawed frogs were exported out of Cape Town. In the 1940s and '50s, it was not uncommon for obstetricians to keep tanks full of the frogs in their offices.

To test his hypothesis, Speare began collecting samples from live African clawed frogs and also from specimens preserved in museums. He found that specimens dating back to the 1930s were indeed already carrying the fungus. He also found that live African clawed frogs were widely infected with Bd but seemed to suffer no ill effects from it. In 2004 he coauthored an influential paper that argued that the transmission route for the fungus began in southern Africa and ran through clinics and hospitals around the world.

"Let's say people were raising African clawed frogs in aquariums, and they just popped the water out," Speare told me. "In most cases when they did that, no frogs got infected, but then, on that hundredth time, one local frog might have been infected. Or people might have said, 'I'm sick of this frog, I'm going to let it go.' And certainly there are populations of African clawed frogs established in a number of countries around the world, to illustrate that that actually did occur."

At this point, Bd appears to be, for all intents and purposes, unstoppable. It can be killed by bleach — Clorox is among the donors to EVACC — but it is impossible to disinfect an entire rainforest. Sometime in the last year or so, the fungus jumped the Panama Canal. (When Edgardo Griffith swabbed the frogs on our trip, he was collecting samples that would eventually be analyzed for it.) It also seems to be heading into Panama from the opposite direction, out of Colombia. It has spread through the highlands of South America, down the eastern coast of Australia, and into New Zealand, and has been detected in Italy, Spain, and France. In the United States, it appears to have radiated from several points, not so much in a wavelike pattern as in a series of ripples.

In the fossil record, mass extinctions stand out so sharply that the very language scientists use to describe Earth's history derives from them. In 1840 the British geologist John Phillips divided life into three chapters: the Paleozoic (from the Greek for "ancient life"), the Mesozoic ("middle life"), and the Cenozoic ("new life"). Phillips fixed as the dividing point between the first and second eras what would now be called the end-Permian extinction, and that between the second and third the end-Cretaceous event. The fossils from these eras were so different that Phillips thought they represented three distinct episodes of creation.

Darwin's resistance to catastrophism meant that he couldn't accept what the fossils seemed to be saying. Drawing on the work of the eminent geologist Charles Lyell, a good friend of his, Darwin maintained that the apparent discontinuities in the history of life were really just gaps in the archive. In *On the Origin of Species,* he argued:

> With respect to the apparently sudden extermination of whole families or orders, as of Trilobites at the close of the palaeozoic period and of Ammonites at the close of the secondary period, we must remember what has been already said on the probable wide intervals of time between our consecutive formations; and in these intervals there may have been much slow extermination.

All the way into the 1960s, paleontologists continued to give talks with titles like "The Incompleteness of the Fossil Record." And this view might have persisted even longer had it not been for a remarkable, largely inadvertent discovery made in the following decade.

In the mid-1970s, Walter Alvarez, a geologist at the Lamont Doherty Earth Observatory, in New York, was studying Earth's polarity. It had recently been learned that the orientation of the planet's magnetic field reverses, so that every so often, in effect, south becomes north and vice versa. Alvarez and some colleagues had found that a certain formation of pinkish limestone in Italy, known as the *scaglia rossa,* recorded these occasional reversals. The limestone also contained the fossilized remains of millions of tiny sea creatures called foraminifera. In the course of several trips to Italy, Alvarez became interested in a thin layer of clay in the limestone that seemed to have been laid down around the end of the Cretaceous. Below the layer, certain species of foraminifera—or forams, for short—were preserved. In the clay layer there were no forams. Above the layer, the earlier species disappeared and new forams appeared. Having been taught the uniformitarian view, Alvarez wasn't sure what to make of what he was seeing, because the change, he later recalled, certainly "looked very abrupt."

Alvarez decided to try to find out how long it had taken for the clay layer to be deposited. In 1977 he took a post at the University of California at Berkeley, where his father, the Nobel Prize–winning physicist Luis Alvarez, was also teaching. The older Alvarez suggested using the element iridium to answer the question.

Iridium is extremely rare on Earth's surface but more plentiful in meteorites, which, in the form of microscopic grains of cosmic dust, are constantly raining down on the planet. The Alvarezes reasoned that if the clay layer had taken a significant amount of time to deposit, it would contain detectable levels of iridium, and if it had been deposited in a short time it wouldn't. They enlisted two other scientists, Frank Asaro and Helen Michel, to run the tests, and gave them samples of the clay. Nine months later, they got a phone call. There was something seriously wrong. Much too much iridium was showing up in the samples. Walter Alvarez flew to Denmark to take samples of another layer of exposed clay from the end of the Cretaceous. When they were tested, these samples, too, were way out of line.

The Alvarez hypothesis, as it became known, was that everything—the clay layer from the *scaglia rossa,* the clay from Denmark, the spike in iridium, the shift in the fossils—could be explained by a single event. In 1980 the Alvarezes and their colleagues proposed that a six-mile-wide asteroid had slammed into Earth, killing off not only the forams but the dinosaurs and all the other organisms that went extinct at the end of the Cretaceous. "I can remember working very hard to make that 1980 paper just as solid as it could possibly be," Walter Alvarez recalled recently. Nevertheless, the idea was greeted with incredulity.

"The arrogance of those people is simply unbelievable," one paleontologist told the *Times.*

"Unseen bolides dropping into an unseen sea are not for me," another declared.

Over the next decade, evidence in favor of an enormous impact kept accumulating. Geologists looking at rocks from the end of the Cretaceous in Montana found tiny mineral grains that seemed to have suffered a violent shock. (Such "shocked quartz" is typically found in the immediate vicinity of meteorite craters.) Other geologists, looking in other parts of the world, found small, glasslike spheres of the sort believed to form when molten-rock droplets splash up into the atmosphere. In 1990 a crater large enough to have been formed by the enormous asteroid that the Alvarezes were proposing was found, buried underneath the Yucatán. In 1991 that crater was dated and discovered to have been formed at precisely the time the dinosaurs died off.

"Those eleven years seemed long at the time, but looking back they seem very brief," Walter Alvarez told me. "Just think about it for a moment. Here you have a challenge to a uniformitarian viewpoint that basically every geologist and paleontologist had been trained in, as had their professors and their professors' professors, all the way back to Lyell. And what you saw was people looking at the evidence. And they gradually did come to change their minds."

Today, it's generally accepted that the asteroid that plowed into the Yucatán led, in very short order, to a mass extinction, but scientists are still uncertain exactly how the process unfolded. One theory holds that the impact raised a cloud of dust that blocked the sun, preventing photosynthesis and causing widespread starvation. According to another theory, the impact kicked up a plume of vaporized rock traveling with so much force that it broke through the atmosphere. The particles in the plume then recondensed, generating, as they fell back to Earth, enough thermal energy to, in effect, broil the surface of the planet.

Whatever the mechanism, the Alvarezes' discovery wreaked havoc with the uniformitarian idea of extinction. The fossil record, it turned out, was marked by discontinuities because the history of life was marked by discontinuities.

In the nineteenth century, and then again during World War II, the Adirondacks were a major source of iron ore. As a result, the mountains are now riddled with abandoned mines. On a gray day this winter, I went to visit one of the mines (I was asked not to say which) with a wildlife biologist named Al Hicks. Hicks, who is fifty-four, is tall and outgoing, with a barrel chest and ruddy cheeks. He works at the headquarters of the New York State Department of Environmental Conservation in Albany, and we met in a parking lot not far from his office. From there we drove almost due north.

Along the way, Hicks explained how, in early 2007, he started to get a lot of strange calls about bats. Sometimes the call would be about a dead bat that had been brought inside by somebody's dog. Sometimes it was about a live—or half-alive—bat flapping around on the driveway. This was in the middle of winter, when any bat in the Northeast should have been hanging by its feet in a state of torpor. Hicks found the calls bizarre, but beyond that, he didn't know what to make of them. Then, in March 2007, some colleagues went to do a routine census of hibernating bats in a cave west of Albany. After the survey, they, too, phoned in.

"They said, 'Holy shit, there's dead bats everywhere,'" Hicks recalled. He instructed them to bring some carcasses back to the office, which they did. They also shot photographs of live bats hanging from the cave's ceiling. When Hicks examined the photographs, he saw that the animals looked as if they had been dunked, nose first, in talcum powder. This was something he had never run across before, and he began sending the photographs to all the bat specialists he could think of. None of them could explain it, either.

"We were thinking, Oh, boy, we hope this just goes away," he told me. "It was like the Bush administration. And, like the Bush administration, it just wouldn't go away." In the winter of 2008, bats with the white powdery substance were found in thirty-three hibernating spots. Meanwhile, bats kept dying. In some hibernacula, populations plunged by as much as 97 percent.

That winter, officials at the National Wildlife Health Center, in Madison, Wisconsin, began to look into the situation. They were able to culture the white substance, which was found to be a never before identified fungus that grows only at cold temperatures. The condition became known as white-nose syndrome, or WNS. White nose seemed to be spreading fast; by March 2008, it had been found on bats in three more states—Vermont, Massachusetts, and Connecticut—and the mortality rate was running above 75 percent. This past winter, white nose was found to have spread to bats in five more states: New Jersey, New Hampshire, Virginia, West Virginia, and Pennsylvania.

In a paper published recently in *Science,* Hicks and several co-authors observed that "parallels can be drawn between the threat posed by W.N.S. and that from chytridiomycosis, a lethal fungal skin infection that has recently caused precipitous global amphibian population declines."

When we arrived at the base of a mountain not far from Lake Champlain, more than a dozen people were standing around in the cold, waiting for us. Most, like Hicks, were from the DEC and had come to help conduct a bat census. In addition, there was a pair of biologists from the U.S. Fish and Wildlife Service and a local novelist who was thinking of incorporating a subplot about white nose into his next book. Everyone put on snowshoes, except for the novelist, who hadn't brought any, and began tromping up the slope toward the mine entrance.

The snow was icy and the going slow, so it took almost half an hour to reach an outlook over the Champlain Valley. While we were waiting for the novelist to catch up — apparently, he was having trouble hiking through the three-foot-deep drifts — the conversation turned to the potential dangers of entering an abandoned mine. These, I was told, included getting crushed by falling rocks, being poisoned by a gas leak, and plunging over a sheer drop of a hundred feet or more.

After another fifteen minutes or so, we reached the mine entrance — essentially, a large hole cut into the hillside. The stones in front of the entrance were white with bird droppings, and the snow was covered with paw prints. Evidently, ravens and coyotes had discovered that the spot was an easy place to pick up dinner.

"Well, shit," Hicks said. Bats were fluttering in and out of the mine and in some cases crawling on the ground. Hicks went to catch one; it was so lethargic that he grabbed it on the first try. He held it between his thumb and forefinger, snapped its neck, and placed it in a Ziploc bag.

"Short survey today," he announced.

At this point, it's not known exactly how the syndrome kills bats. What is known is that bats with the syndrome often wake up from their torpor and fly around, which leads them to die either of starvation or of the cold or to get picked off by predators.

We unstrapped our snowshoes and put on helmets. Hicks handed out headlamps — we were supposed to carry at least one extra — and packages of batteries; then we filed into the mine, down a long, sloping tunnel. Shattered beams littered the ground, and bats flew up at us through the gloom. Hicks cautioned everyone to stay alert. "There's places that if you take a step you won't be stepping back," he warned. The tunnel twisted along, sometimes opening up into concert-hall-size chambers with side tunnels leading out of them.

Over the years, the various sections of the mine had acquired names; when we reached something called the Don Thomas section, we split up into groups to start the survey. The process consisted of photographing as many bats as possible. (Later on, back in Albany, someone would have to count all the bats in the pictures.) I went with Hicks, who was carrying an enormous camera, and one of the biologists from the Fish and Wildlife Service, who had a laser pointer. The biologist would aim the pointer at a cluster of bats hanging from the ceiling. Hicks would then snap a photograph. Most of the bats were little brown bats; these are the most common bats in the United States and the ones you are most likely to see flying around on a summer night. There were also Indiana bats, which are on the federal endangered species list, and small-footed bats, which, at the rate things are going, are likely to end up there. As we moved along, we kept disturbing the bats, which squeaked and started to rustle around, like half-asleep children.

Since white nose grows only in the cold, it's odd to find it living on mammals, which, except when they're hibernating (or dead), maintain a high body temperature. It has been hypothesized that the fungus normally subsists by breaking down organic matter in a chilly place and that it was transported into bat hibernacula, where it began to break down bats. When news of white nose began to get around, a spelunker sent Hicks photographs that he had shot in Howe's Cave, in central New York. The photographs, which had been taken in 2006, showed bats with clear signs of white nose and are the earliest known record of the syndrome. Howe's Cave is connected to Howe's Caverns, a popular tourist destination.

"It's kind of interesting that the first record we have of this fungus is photographs from a commercial cave in New York that gets about two hundred thousand visits a year," Hicks told me.

Despite the name, white nose is not confined to bats' noses; as we worked our way along, people kept finding bats with freckles of fungus on their wings and ears. Several of these were dispatched, for study purposes, with a thumb and forefinger. Each dead bat was sexed—males can be identified by their tiny penises—and placed in a Ziploc bag.

At about 7 P.M., we came to a huge, rusty winch, which, when the mine was operational, had been used to haul ore to the surface. By this point, we were almost down at the bottom of the mountain, except that we were on the inside of it. Below, the path disappeared into a pool of water, like the River Styx. It was impossible to go any further, and we began working our way back up.

Bats, like virtually all other creatures alive today, are masters of adaptation descended from lucky survivors. The earliest bat fossil that has been found dates from 53 million years ago, which is to say

12 million years after the impact that ended the Cretaceous. It belongs to an animal that had wings and could fly but had not yet developed the specialized inner ear that, in modern bats, allows for echolocation. Worldwide, there are now more than a thousand bat species, which together make up nearly a fifth of all species of mammals. Most feed on insects; there are also bats that live off fruit, bats that eat fish—they use echolocation to detect minute ripples in the water—and a small but highly celebrated group that consumes blood. Bats are great colonizers—Darwin noted that even New Zealand, which has no other native mammals, has its own bats—and they can be found as far north as Alaska and as far south as Tierra del Fuego.

In the time that bats have evolved and spread, the world has changed a great deal. Fifty-three million years ago, at the start of the Eocene, the planet was very warm, and tropical palms grew at the latitude of London. The climate cooled, the Antarctic ice sheet began to form, and, eventually, about 2 million years ago, a period of recurring glaciations began. As recently as 15,000 years ago, the Adirondacks were buried under ice.

One of the puzzles of mass extinction is why, at certain junctures, the resourcefulness of life seems to falter. As powerful as the Alvarez hypothesis proved to be, it explains only a single mass extinction.

"I think that after the evidence became pretty strong for the impact at the end of the Cretaceous, those of us who were working on this naively expected that we would go out and find evidence of impacts coinciding with the other events," Walter Alvarez told me. "And, of course, it's turned out to be much more complicated. We're seeing right now that a mass extinction can be caused by human beings. So it's clear that we do not have a general theory of mass extinction."

Andrew Knoll, a paleontologist at Harvard, has spent most of his career studying the evolution of early life. (Among the many samples he keeps in his office are fossils of microorganisms that lived 2.8 billion years ago.) He has also written about more recent events, like the end-Permian extinction, which took place 250 million years ago, and the current extinction event.

Knoll noted that the world can change a lot without producing huge losses; ice ages, for instance, come and go. "What the geological record tells us is that it's time to worry when the rate of change is fast," he told me. In the case of the end-Permian extinction, Knoll and many other researchers believe that the trigger was a sudden burst of volcanic activity; a plume of hot mantle rock from deep in the Earth sent nearly a million cubic miles' worth of flood basalts streaming over what is now Siberia. The eruption released enormous quantities of carbon dioxide, which presumably led — then as now — to global warming and to significant changes in ocean chemistry.

"CO_2 is a paleontologist's dream," Knoll told me. "It can kill things directly by physiological effects, of which ocean acidification is the best known, and it can kill things by changing the climate. If it gets warmer faster than you can migrate, you're in trouble."

In the end, the most deadly aspect of human activity may simply be the pace of it. Just in the past century, CO_2 levels in the atmosphere have changed by as much — 100 parts per million — as they normally do in a 100,000-year glacial cycle. Meanwhile, the drop in ocean pH levels that has occurred over the past fifty years may well exceed anything that happened in the seas during the previous 50 million. In a single afternoon, a pathogen like Bd can move, via United or American Airlines, halfway around the world. Before man entered the picture, such a migration would have required hundreds, if not thousands, of years — if, indeed, it could have been completed at all.

Currently, a third of all amphibian species, nearly a third of reef-building corals, a quarter of all mammals, and an eighth of all birds are classified as "threatened with extinction." These estimates do not include the species that humans have already wiped out or the species for which there are insufficient data. Nor do the figures take into account the projected effects of global warming or ocean acidification. Nor, of course, can they anticipate the kinds of sudden, terrible collapses that are becoming almost routine.

I asked Knoll to compare the current situation with past extinction events. He told me that he didn't want to exaggerate recent losses or to suggest that an extinction on the order of the end-Cretaceous or end-Permian was imminent. At the same time, he noted, when the asteroid hit the Yucatán "it was one terrible afternoon." He went on, "But it was a short-term event, and then things started getting better. Today, it's not like you have a stress and the stress is relieved and recovery starts. It gets bad and then it keeps being bad, because the stress doesn't go away. Because the stress is us."

Aeolus Cave, in Dorset, Vermont, is believed to be the largest bat hibernaculum in New England; it is estimated that before white nose hit, more than 200,000 bats—some from as far away as Ontario and Rhode Island—came to spend the winter there.

In late February, I went with Hicks to visit Aeolus. In the parking lot of the local general store, we met up with officials from the Vermont Fish and Wildlife Department, who had organized the trip. The entrance to Aeolus is about a mile and a half from the nearest road, up a steep, wooded hillside. This time, we approached by snowmobile. The temperature outside was about 25 degrees—far too low for bats to be active—but when we got near the entrance we could, once again, see bats fluttering around. The most senior of the Vermont officials, Scott Darling, announced that we'd all have to put on latex gloves and Tyvek suits before proceeding. At first, this seemed to me to be paranoid; soon, however, I came to see the sense of it.

Aeolus is a marble cave that was created by water flow over the course of thousands of years. The entrance is a large, nearly horizontal tunnel at the bottom of a small hollow. To keep people out, the Nature Conservancy, which owns the cave, has blocked off the opening with huge iron slats, so that it looks like the gate of a medieval fortress. With a key, one of the slats can be removed; this creates a narrow gap that can be crawled (or slithered) through. Despite the cold, there was an awful smell emanating from the cave—half game farm, half garbage dump. When it was my turn, I squeezed through the gap and immediately slid on the ice into a pile of dead bats. The scene, in the dimness, was horrific. There were giant icicles hanging from the ceiling, and from the floor large knobs of ice rose up, like polyps. The ground was covered with dead bats; some of the ice knobs, I noticed, had bats frozen into them. There were torpid bats roosting on the ceiling and also wide-awake ones, which would take off and fly by or, sometimes, right into us.

Why bat corpses pile up in some places, while in others they get eaten or in some other way disappear, is unclear. Hicks speculated that the weather conditions at Aeolus were so harsh that the bats didn't even make it out of the cave before dropping dead. He and Darling had planned to do a count of the bats in the first chamber of the cave, known as Guano Hall, but this plan was soon abandoned, and it was decided just to collect specimens. Darling explained that the specimens would be going to the American Museum of Natural History, so that there would at least be a record of the bats that had once lived in Aeolus. "This may be one of the last opportunities," he said. In contrast to a mine, which has been around at most for a few centuries, Aeolus, he pointed out, has existed for millennia. It's likely that bats have been hibernating there, generation after generation, since the end of the last ice age.

"That's what makes this so dramatic—it's breaking the evolutionary chain," Darling said.

He and Hicks began picking dead bats off the ground. Those which were too badly decomposed were tossed back; those which were more or less intact were sexed and placed in two-quart plastic bags. I helped out by holding open the bag for females. Soon it was full and another one was started. It struck me, as I stood there holding a bag filled with several dozen stiff, almost weightless bats, that I was watching mass extinction in action.

Several more bags were collected. When the specimen count hit somewhere around five hundred, Darling decided that it was time to go. Hicks hung back, saying that he wanted to take some pictures. In the hours we had been slipping around the cave, the carnage had grown even more grotesque; many of the dead bats had been crushed and now there was blood oozing out of them. As I made my way up toward the entrance, Hicks called after me: "Don't step on any dead bats." It took me a moment to realize that he was joking.

PART FIVE

The Environment: Small Blessings

ROBERT KUNZIG

Scraping Bottom

FROM *National Geographic*

ONE DAY IN 1963, when Jim Boucher was seven, he was out working the trapline with his grandfather a few miles south of the Fort McKay First Nation reserve on the Athabasca River in northern Alberta. The country there is wet, rolling fen, dotted with lakes, dissected by streams, and draped in a cover of skinny, stunted trees—it's part of the boreal forest that sweeps right across Canada, covering more than a third of the country. In 1963 that forest was still mostly untouched. The government had not yet built a gravel road into Fort McKay; you got there by boat or in the winter by dogsled. The Chipewyan and Cree Indians there—Boucher is a Chipewyan—were largely cut off from the outside world. For food they hunted moose and bison; they fished the Athabasca for walleye and whitefish; they gathered cranberries and blueberries. For income they trapped beaver and mink. Fort McKay was a small fur trading post. It had no gas, electricity, telephone, or running water. Those didn't come until the 1970s and 1980s.

In Boucher's memory, though, the change begins that day in 1963, on the long trail his grandfather used to set his traps, near a place called Mildred Lake. Generations of his ancestors had worked that trapline. "These trails had been here thousands of years," Boucher said one day last summer, sitting in his spacious and tasteful corner office in Fort McKay. His golf putter stood in one corner; Mozart played softly on the stereo. "And that day, all of a sudden, we came upon this clearing. A huge clearing. There had been no notice. In the 1970s they went in and tore down my grandfather's cabin—with no notice or discussion." That was Boucher's first encounter with the oil sands industry. It's an industry that has utterly transformed this part of northeastern Alberta in just the past few years, with astonishing speed. Boucher is surrounded by it now and immersed in it himself.

Where the trapline and the cabin once were, and the forest, there is now a large open-pit mine. Here Syncrude, Canada's largest oil producer, digs bitumen-laced sand from the ground with electric shovels five stories high, then washes the bitumen off the sand with hot water and sometimes caustic soda. Next to the mine, flames flare from the stacks of an "upgrader," which cracks the tarry bitumen and converts it into Syncrude Sweet Blend, a synthetic crude that travels down a pipeline to refineries in Edmonton, Alberta; Ontario; and the United States. Mildred Lake, meanwhile, is now dwarfed by its neighbor, the Mildred Lake Settling Basin, a four-square-mile lake of toxic mine tailings. The sand dike that contains it is by volume one of the largest dams in the world.

Nor is Syncrude alone. Within a twenty-mile radius of Boucher's office are a total of six mines that produce nearly three-quarters of a million barrels of synthetic crude oil a day; and more are in the pipeline. Wherever the bitumen layer lies too deep to be strip-mined, the industry melts it in situ with copious amounts of steam, so that it can be pumped to the surface. The industry has spent more than $50 billion on construction during the past decade, including some $20 billion in 2008 alone. Before the collapse in oil prices in the fall of 2008, it was forecasting another $100 billion over the next few years and a doubling of production by 2015, with most of that oil flowing through new pipelines to the United States. The economic crisis has put many expansion projects on hold, but it has not diminished the long-term prospects for the oil sands. In mid-November, the International Energy Agency released a report forecasting $120-a-barrel oil in 2030—a price that would more than justify the effort it takes to get oil from oil sands.

Nowhere on Earth is more earth being moved these days than in the Athabasca Valley. To extract each barrel of oil from a surface mine, the industry must first cut down the forest, then remove an average of two tons of peat and dirt that lie above the oil sands layer, then two tons of the sand itself. It must heat several barrels of water to strip the bitumen from the sand and upgrade it, and afterward it discharges contaminated water into tailings ponds like the one near Mildred Lake. They now cover around fifty square miles. Last April some five hundred migrating ducks mistook one of those ponds, at a newer Syncrude mine north of Fort McKay, for a hospitable stopover, landed on its oily surface, and died. The incident stirred international attention—Greenpeace broke into the Syncrude facility and hoisted a banner of a skull over the pipe discharging tailings, along with a sign that read WORLD'S DIRTI-EST OIL: STOP THE TAR SANDS.

The United States imports more oil from Canada than from any other nation, about 19 percent of its total foreign supply, and around half of that now comes from the oil sands. Anything that reduces our dependence on Middle Eastern oil, many Americans would say, is a good thing. But clawing and cooking one barrel of crude from the oil sands emits as much as three times more carbon dioxide than letting one gush from the ground in Saudi Arabia. The oil sands are still a tiny part of the world's carbon problem—they account for less than a tenth of one percent of global CO_2 emissions—but to many environmentalists they are the thin end of the wedge, the first step along a path that could lead to other, even dirtier sources of oil: producing it from oil shale or coal. "Oil sands represent a decision point for North America and the world," says Simon Dyer of the Pembina Institute, a moderate and widely respected Canadian environmental group. "Are we going to get serious about alternative energy, or are we going to go down the unconventional-oil track? The fact that we're willing to move four tons of earth for a single barrel really shows that the world is running out of easy oil."

That thirsty world has come crashing in on Fort McKay. Yet Jim Boucher's view of it, from an elegant new building at the entrance to the besieged little village, contains more shades of gray than you might expect. "The choice we make is a difficult one," Boucher said when I visited him last summer. For a long time the First Nation tried to fight the oil sands industry, with little success. Now, Boucher said, "we're trying to develop the community's capacity to take advantage of the opportunity." Boucher presides not only over this First Nation, as chief, but also over the Fort McKay Group of Companies, a community-owned business that provides services to the oil sands industry and brought in $85 million in 2007. Unemployment is under 5 percent in the village, and it has a health clinic, a youth center, and a hundred new three-bedroom houses that the community rents to its members for far less than market rates. The First Nation is even thinking of opening its own mine: it owns 8,200 acres of prime oil sands land across the river, right next to the Syncrude mine where the ducks died.

As Boucher was telling me all this, he was picking bits of meat from a smoked whitefish splayed out on his conference table next to a bank of windows that offered a panoramic view of the river. A staff member had delivered the fish in a plastic bag, but Boucher couldn't say where it had come from. "I can tell you one thing," he said. "It doesn't come from the Athabasca."

Without the river, there would be no oil sands industry. It's the river that over tens of millions of years has eroded away billions of cubic yards of sediment that once covered the bitumen, thereby bringing it within reach of shovels—and in some places all the way to the surface. On a hot summer day along the Athabasca, near Fort McKay for example, bitumen oozes from the riverbank and casts an oily sheen on the water. Early fur traders reported seeing the stuff and watching natives use it to caulk their canoes. At room temperature, bitumen is like molasses, and below 50 degrees F or so it is as hard as a hockey puck, as Canadians invariably put it. Once upon a time, though, it was light crude, the same liquid that oil companies have been pumping from deep traps in southern Alberta for nearly a century. Tens of millions of years ago, geologists think, a large volume of that oil was pushed northeastward, perhaps by the rise of the Rocky Mountains. In the process it also migrated upward, along sloping layers of sediment, until eventually it reached depths shallow and cool enough for bacteria to thrive. Those bacteria degraded the oil to bitumen.

The Alberta government estimates that the province's three main oil sands deposits, of which the Athabasca one is the largest, contain 173 billion barrels of oil that are economically recoverable today. "The size of that, on the world stage—it's massive," says Rick George, CEO of Suncor, which opened the first mine on the Athabasca River in 1967. In 2003, when the *Oil & Gas Journal* added the Alberta oil sands to its list of proven reserves, it immediately propelled Canada to second place, behind Saudi Arabia, among oil-producing nations. The proven reserves in the oil sands are eight times those of the entire United States. "And that number will do nothing but go up," says George. The Alberta Energy Resources and Conservation Board estimates that more than 300 billion bar-

rels may one day be recoverable from the oil sands; it puts th
size of the deposit at 1.7 trillion barrels.

Getting oil from oil sands is simple but not easy. The giant
tric shovels that rule the mines have hardened steel teeth that each
weigh a ton, and as those teeth claw into the abrasive black sand
24/7, 365 days a year, they wear down every day or two; a welder
then plays dentist to the dinosaurs, giving them new crowns. The
dump trucks that rumble around the mine, hauling 400-ton loads
from the shovels to a rock crusher, burn 50 gallons of diesel fuel an
hour; it takes a forklift to change their tires, which wear out in six
months. And every day in the Athabasca Valley, more than a million tons of sand emerge from such crushers and is mixed with
more than 200,000 tons of water that must be heated, typically to
175 degrees F, to wash out the gluey bitumen. At the upgraders,
the bitumen gets heated again, to about 900 degrees F, and compressed to more than 100 atmospheres—that's what it takes to
crack the complex molecules and either subtract carbon or add
back the hydrogen that the bacteria removed ages ago. That's what
it takes to make the light hydrocarbons we need to fill our gas
tanks. It takes a stupendous amount of energy. In situ extraction,
which is the only way to get at around 80 percent of those 173 billion barrels, can use up to twice as much energy as mining, because
it requires so much steam.

Most of the energy to heat the water or make steam comes from
burning natural gas, which also supplies the hydrogen for upgrading. Precisely because it is hydrogen rich and mostly free of impuri-
ties, natural gas is the cleanest-burning fossil fuel, the one that puts
the least amount of carbon and other pollutants into the atmosphere. Critics thus say the oil sands industry is wasting the cleanest fuel to make the dirtiest—that it turns gold into lead. The argument makes environmental but not economic sense, says David
Keith, a physicist and energy expert at the University of Calgary.
Each barrel of synthetic crude contains about five times more energy than the natural gas used to make it, and in much more valuable liquid form. "In economic terms it's a slam dunk," says Keith.
"This whole thing about turning gold into lead—it's the other way
around. The gold in our society is liquid transportation fuels."

Most of the carbon emissions from such fuels come from the tail-
pipes of the cars that burn them; on a "wells-to-wheels" basis, the
oil sands are only 15 to 40 percent dirtier than conventional oil.

But the heavier carbon footprint remains an environmental—and public relations—disadvantage. Last June Alberta's premier, Ed Stelmach, announced a plan to deal with the extra emissions. The province, he said, will spend over $1.5 billion to develop the technology for capturing carbon dioxide and storing it underground —a strategy touted for years as a solution to climate change. By 2015 Alberta is hoping to capture 5 million tons of CO_2 a year from bitumen upgraders as well as from coal-fired power plants, which even in Alberta, to say nothing of the rest of the world, are a far larger source of CO_2 than the oil sands. By 2020, according to the plan, the province's carbon emissions will level off, and by 2050 they will decline to 15 percent below their 2005 levels. That is far less of a cut than scientists say is necessary. But it is more than the U.S. government, say, has committed to in a credible way.

One thing Stelmach has consistently refused to do is "touch the brake" on the oil sands boom. The boom has been gold for the provincial as well as the national economy; the town of Fort Mc-Murray, south of the mines, is awash in Newfoundlanders and Nova Scotians fleeing unemployment in their own provinces. The provincial government has been collecting around a third of its revenue from lease sales and royalties on fossil fuel extraction, including oil sands—it was expecting to get nearly half this year, or $19 billion, but the collapse in oil prices since the summer has dropped that estimate to about $12 billion. Albertans are bitterly familiar with the boom-and-bust cycle; the last time oil prices collapsed, in the 1980s, the provincial economy didn't recover for a decade. The oil sands cover an area the size of North Carolina, and the provincial government has already leased around half of that, including all 1,356 square miles that are minable. It has yet to turn down an application to develop one of those leases, on environmental or any other grounds.

From a helicopter it's easy to see the industry's impact on the Athabasca Valley. Within minutes of lifting off from Fort McMurray, heading north along the east bank of the river, you pass over Suncor's Millennium mine—the company's leases extend practically to the town. On a day with a bit of wind, dust plumes billowing off the wheels and the loads of the dump trucks coalesce into a single enormous cloud that obscures large parts of the mine pit and spills over its lip. To the north, beyond a small expanse of intact forest, a similar cloud rises from the next pit, Suncor's Steepbank mine, and beyond that lie two more, and across the river two more. One evening last July the clouds had merged into a band of dust sweeping west across the devastated landscape. It was being sucked into the updraft of a storm cloud. In the distance steam and smoke and gas flames belched from the stacks of the Syncrude and Suncor upgraders—"dark satanic mills" inevitably come to mind, but they're a riveting sight all the same. From many miles away, you can smell the tarry stench. It stings your lungs when you get close enough.

From the air, however, the mines fall away quickly. Skimming low over the river, startling a young moose that was fording a narrow channel, a government biologist named Preston McEachern and I veered northwest toward the Birch Mountains, over vast expanses of scarcely disturbed forest. The Canadian boreal forest covers 2 million square miles, of which around 75 percent remains undeveloped. The oil sands mines have so far converted over 150 square miles—a hundredth of a percent of the total area—into dust, dirt, and tailings ponds. Expansion of in situ extraction could affect a much larger area. At Suncor's Firebag facility, northeast of the Millennium mine, the forest has not been razed, but it has been dissected by roads and pipelines that service a checkerboard of large clearings, in each of which Suncor extracts deeply buried bitumen through a cluster of wells. Environmentalists and wildlife biologists worry that the widening fragmentation of the forest, by timber as well as mineral companies, endangers the woodland caribou and other animals. "The boreal forest as we know it could be gone in a generation without major policy changes," says Steve Kallick, director of the Pew Boreal Campaign, which aims to protect 50 percent of the forest.

McEachern, who works for Alberta Environment, a provincial agency, says the tailings ponds are his top concern. The mines dump wastewater in the ponds, he explains, because they are not allowed to dump waste into the Athabasca, and because they need to reuse the water. As the thick, brown slurry gushes from the discharge pipes, the sand quickly settles out, building the dike that retains the pond; the residual bitumen floats to the top. The fine clay and silt particles, though, take several years to settle, and when they do, they produce a yogurt-like goop—the technical term is "mature fine tailings"—that is contaminated with toxic chemicals such as naphthenic acid and polycyclic aromatic hydrocarbons (PAH) and would take centuries to dry out on its own. Under the terms of their licenses, the mines are required to reclaim it somehow, but they have been missing their deadlines and still have not fully reclaimed a single pond.

In the oldest and most notorious one, Suncor's Pond 1, the sludge is perched high above the river, held back by a dike of compacted sand that rises more than 300 feet from the valley floor and is studded with pine trees. The dike has leaked in the past, and in 2007 a modeling study done by hydrogeologists at the University of Waterloo estimated that 45,000 gallons a day of contaminated water could be reaching the river. Suncor is now in the process of reclaiming Pond 1, piping some tailings to another pond, and replacing them with gypsum to consolidate the tailings. By 2010, the company says, the surface will be solid enough to plant trees on. Last summer it was still a blot of beige mud streaked with black bitumen and dotted with orange plastic scarecrows that are supposed to dissuade birds from landing and killing themselves.

The Alberta government asserts that the river is not being contaminated—that anything found in the river or in its delta, at Lake Athabasca, comes from natural bitumen seeps. The river cuts right through the oil sands downstream of the mines, and as our chopper zoomed along a few feet above it, McEachern pointed out several places where the riverbank was black and the water oily. "There is an increase in a lot of metals as you move downstream," he said. "That's natural—it's weathering of the geology. There's mercury in the fish up at Lake Athabasca—we've had an advisory there since the 1990s. There are PAHs in the sediments in the delta. They're there because the river has eroded through the oil sands."

Independent scientists, to say nothing of people who live downstream of the mines in the First Nations' community of Fort Chipewyan, on Lake Athabasca, are skeptical. "It's inconceivable that you could move that much tar and have no effect," says Peter Hodson, a fish toxicologist at Queen's University in Ontario. An Environment Canada study did in fact show an effect on fish in the Steepbank River, which flows past a Suncor mine into the

Athabasca. Fish near the mine, Gerald Tetreault and his colleagues found when they caught some in 1999 and 2000, showed five times more activity of a liver enzyme that breaks down toxins—a widely used measure of exposure to pollutants—as did fish near a natural bitumen seep on the Steepbank.

"The thing that angers me," says David Schindler, "is that there's been no concerted effort to find out where the truth lies."

Schindler, an ecologist at the University of Alberta in Edmonton, was talking about whether people in Fort Chipewyan have already been killed by pollution from the oil sands. In 2006 John O'Connor, a family physician who flew in weekly to treat patients at the health clinic in Fort Chip, told a radio interviewer that he had in recent years seen five cases of cholangiocarcinoma—a cancer of the bile duct that normally strikes one in 100,000 people. Fort Chip has a population of around 1,000; statistically it was unlikely to have even one case. O'Connor hadn't managed to interest health authorities in the cancer cluster, but the radio interview drew wide attention to the story. "Suddenly it was everywhere," he says. "It just exploded."

Two of O'Connor's five cases, he says, had been confirmed by tissue biopsy; the other three patients had shown the same symptoms but had died before they could be biopsied. (Cholangiocarcinoma can be confused on CT scans with more common cancers such as liver or pancreatic cancer.) "There is no evidence of elevated cancer rates in the community," Howard May, a spokesperson for Alberta Health, wrote in an e-mail last September. But the agency, he said, was nonetheless conducting a more complete investigation—this time actually examining the medical records from Fort Chip—to try to quiet a controversy that was now two years old.

One winter night when Jim Boucher was a young boy, around the time the oil sands industry came to his forest, he was returning alone by dogsled to his grandparents' cabin from an errand in Fort McKay. It was a journey of twenty miles or so, and the temperature was minus 4 degrees F. In the moonlight Boucher spotted a flock of ptarmigan, white birds in the snow. He killed around fifty, loaded them on the dogsled, and brought them home. Four decades later, sitting in his chief-executive office in white chinos and a white Adidas sport shirt, he remembers the pride on his grandmother's face that night. "That was a different spiritual world," Boucher says. "I saw that world continuing forever." He tells the story now when asked about the future of the oil sands and his people's place in it.

A poll conducted by the Pembina Institute in 2007 found that 71 percent of Albertans favored an idea their government has always rejected out of hand: a moratorium on new oil sands projects until environmental concerns can be resolved. "It's my belief that when government attempts to manipulate the free market, bad things happen," Premier Stelmach told a gathering of oil industry executives that year. "The free-market system will solve this."

But the free market does not consider the effects of the mines on the river or the forest, or on the people who live there, unless it is forced to. Nor, left to itself, will it consider the effects of the oil sands on climate. Jim Boucher has collaborated with the oil sands industry in order to build a new economy for his people, to replace the one they lost, to provide a new future for kids who no longer hunt ptarmigan in the moonlight. But he is aware of the tradeoffs. "It's a struggle to balance the needs of today and tomorrow when you look at the environment we're going to live in," he says. In northern Alberta the question of how to strike that balance has been left to the free market, and its answer has been to forget about tomorrow. Tomorrow is not its job.

MICHAEL SPECTER

A Life of Its Own

FROM *The New Yorker*

THE FIRST TIME Jay Keasling remembers hearing the word "artemisinin," about a decade ago, he had no idea what it meant. "Not a clue," Keasling, a professor of biochemical engineering at the University of California at Berkeley, recalled. Although artemisinin has become the world's most important malaria medicine, Keasling wasn't an expert on infectious diseases. But he happened to be in the process of creating a new discipline, synthetic biology, which — by combining elements of engineering, chemistry, computer science, and molecular biology — seeks to assemble the biological tools necessary to redesign the living world.

Scientists have been manipulating genes for decades; inserting, deleting, and changing them in various microbes has become a routine function in thousands of labs. Keasling and a rapidly growing number of colleagues around the world have something more radical in mind. By using gene-sequence information and synthetic DNA, they are attempting to reconfigure the metabolic pathways of cells to perform entirely new functions, such as manufacturing chemicals and drugs. Eventually, they intend to construct genes — and new forms of life — from scratch. Keasling and others are putting together a kind of foundry of biological components — Bio-Bricks, as Tom Knight, a senior research scientist at Massachusetts Institute of Technology, who helped invent the field, has named them. Each BioBrick part, made of standardized pieces of DNA, can be used interchangeably to create and modify living cells.

"When your hard drive dies, you can go to the nearest computer store, buy a new one, and swap it out," Keasling said. "That's because it's a standard part in a machine. The entire electronics industry is based on a plug-and-play mentality. Get a transistor, plug it in, and off you go. What works in one cell phone or laptop should work in another. That is true for almost everything we build: when you go to Home Depot, you don't think about the thread size on the bolts you buy, because they're all made to the same standard. Why shouldn't we use biological parts in the same way?" Keasling and others in the field, who have formed bicoastal clusters in the Bay Area and in Cambridge, Massachusetts, see cells as hardware, and genetic code as the software required to make them run. Synthetic biologists are convinced that with enough knowledge, they will be able to write programs to control those genetic components, programs that would let them not only alter nature but guide human evolution as well.

No scientific achievement has promised so much, and none has come with greater risks or clearer possibilities for deliberate abuse. The benefits of new technologies—from genetically engineered food to the wonders of pharmaceuticals—often have been oversold. If the tools of synthetic biology succeed, though, they could turn specialized molecules into tiny, self-contained factories, creating cheap drugs, clean fuels, and new organisms to siphon carbon dioxide from the atmosphere.

In 2000 Keasling was looking for a chemical compound that could demonstrate the utility of these biological tools. He settled on a diverse class of organic molecules known as isoprenoids, which are responsible for the scents, flavors, and even colors in many plants: eucalyptus, ginger, and cinnamon, for example, as well as the yellow in sunflowers and the red in tomatoes. "One day a graduate student stopped by and said, 'Look at this paper that just came out on amorphadiene synthase,'" Keasling told me as we sat in his office in Emeryville, across the Bay Bridge from San Francisco. He had recently been named CEO of the Department of Energy's new Joint BioEnergy Institute, a partnership of three national laboratories and three research universities, led by the Lawrence Berkeley National Laboratory. The consortium's principal goal is to design and manufacture artificial fuels that emit little or no greenhouse gases—one of President Obama's most frequently cited priorities.

Keasling wasn't sure what to tell his student. "'Amorphadiene,' I said. 'What's that?' He told me that it was a precursor to artemisinin, an effective antimalarial. I had never worked on malaria. So I got to studying and quickly realized that this precursor was in the general class we were planning to investigate. And I thought, amorphadiene is as good a target as any. Let's work on that."

Malaria infects as many as 500 million of the world's poorest people every year and kills up to 1 million, most of whom are children under the age of five. For centuries, the standard treatment was quinine, and then the chemically related compound chloroquine. At ten cents per treatment, chloroquine was cheap and simple to make, and it saved millions of lives. By the early nineties, however, the most virulent malaria parasite—*Plasmodium falciparum*—had grown largely resistant to the drug. Worse, the second line of treatment, sulfadoxine-pyrimethanine, or SP, also failed widely. Artemisinin, when taken in combination with other drugs, has become the only consistently successful treatment that remains. (Reliance on any single drug increases the chances that the malaria parasite will develop resistance.) Known in the West as *Artemisia annua,* or sweet wormwood, the herb that contains artemisinic acid grows wild in many places, but supplies vary widely and so does the price.

Depending so heavily on artemisinin, while unavoidable, has serious drawbacks: combination therapy costs between ten and twenty times as much as chloroquine, and, despite increasing assistance from international charities, that is too much money for most Africans or their governments. Artemisinin is not easy to cultivate. Once harvested, the leaves and stems have to be processed rapidly or they will be destroyed by exposure to ultraviolet light. Yields are low, and production is expensive.

Although several thousand Asian and African farmers have begun to plant the herb, the World Health Organization expects that for the next several years the annual demand—as many as 500 million courses of treatment per year—will far exceed the supply. Should that supply disappear, the impact would be incalculable. "Losing artemisinin would set us back years, if not decades," Kent Campbell, a former chief of the malaria branch at the Centers for Disease Control and Prevention and director of the Malaria Control Program at the nonprofit health organization PATH, said. "One can envision any number of theoretical public health disasters in the world. But this is not theoretical. This is real. Without artemisinin, millions of people could die."

*

Keasling realized that the tools of synthetic biology, if properly deployed, could dispense with nature entirely, providing an abundant new source of artemisinin. If each cell became its own factory, churning out the chemical required to make the drug, there would be no need for an elaborate and costly manufacturing process, either. Why not try to produce it from genetic parts by constructing a cell to manufacture amorphadiene? Keasling and his team would have to dismantle several different organisms, then use parts from nearly a dozen of their genes to cobble together a custom-built package of DNA. They would then need to construct a new metabolic pathway, the chemical circuitry that a cell needs to do its job—one that did not exist in the natural world. "We have got to the point in human history where we simply do not have to accept what nature has given us," he told me.

By 2003 the team reported its first success, publishing a paper in *Nature Biotechnology* that described how the scientists had created that new pathway, by inserting genes from three organisms into *E. coli,* one of the world's most common bacteria. That research helped Keasling secure a $42.6-million grant from the Bill and Melinda Gates Foundation. Keasling had no interest in simply proving that the science worked; he wanted to do it on a scale that the world could use to fight malaria. "Making a few micrograms of artemisinin would have been a neat scientific trick," he said. "But it doesn't do anybody in Africa any good if all we can do is a cool experiment in a Berkeley lab. We needed to make it on an industrial scale." To translate the science into a product, Keasling helped start a new company, Amyris Biotechnologies, to refine the raw organism, then figure out how to produce it more efficiently. Within a decade, Amyris had increased the amount of artemisinic acid that each cell could produce by a factor of one million, bringing down the cost of the drug from as much as ten dollars for a course of treatment to less than a dollar.

Amyris then joined with the Institute for OneWorld Health, in San Francisco, a nonprofit drug maker, and in 2008 they signed an agreement with the Paris-based pharmaceutical company Sanofi-Aventis to make the drug, which they hope to have on the market by 2012. The scientific response has been reverential—their artemisinin has been seen as the first bona fide product of synthetic biology, proof of a principle that we need not rely on the whims of nature to address the world's most pressing crises. But some people wonder what synthetic artemisinin will mean for the thousands of farmers who have begun to plant the wormwood crop. "What happens to struggling farmers when laboratory vats in California replace farms in Asia and East Africa?" Jim Thomas, a researcher with ETC Group, a technology watchdog based in Canada, asked. Thomas has argued that there has been little discussion of the ethical and cultural implications of altering nature so fundamentally. "Scientists are making strands of DNA that have never existed," Thomas said. "So there is nothing to compare them to. There are no agreed mechanisms for safety, no policies."

Keasling, too, believes that the nation needs to consider the potential impact of this technology, but he is baffled by opposition to what should soon become the world's most reliable source of cheap artemisinin. "Just for a moment, imagine that we replaced artemisinin with a cancer drug," he said. "And let's have the entire Western world rely on some farmers in China and Africa who may or may not plant their crop. And let's have a lot of American children die because of that. Look at the world and tell me we shouldn't be doing this. It's not people in Africa who see malaria who say, whoa, let's put the brakes on."

Artemisinin is the first step in what Keasling hopes will become a much larger program. "We ought to be able to make any compound produced by a plant inside a microbe," he said. "We ought to have all these metabolic pathways. You need this drug: OK, we pull this piece, this part, and this one off the shelf. You put them into a microbe, and two weeks later out comes your product."

That's what Amyris has done in its efforts to develop new fuels. "Artemisinin is a hydrocarbon, and we built a microbial platform to produce it," Keasling said. "We can remove a few of the genes to take out artemisinin and put in a different gene, to make biofuels." Amyris, led by John Melo, who spent years as a senior executive at British Petroleum, has already engineered three microbes that can convert sugar to fuel. "We still have lots to learn and lots of problems to solve," Keasling said. "I am well aware that makes some people anxious, and I understand why. Anything so powerful and new is troubling. But I don't think the answer to the future is to race into the past."

For the first 4 billion years, life on Earth was shaped entirely by nature. Propelled by the forces of selection and chance, the most efficient genes survived, and evolution insured that they would thrive. The long, beautiful Darwinian process of creeping forward by trial and error, struggle and survival, persisted for millennia. Then, about 10,000 years ago, our ancestors began to gather in villages, grow crops, and domesticate animals. That led to stone axes and looms, which in turn led to better crops and a varied food supply that could feed a larger civilization. Breeding of goats and pigs gave way to the fabrication of metal and machines. Throughout it all, new species, built on the power of their collected traits, emerged, while others were cast aside.

By the beginning of the twenty-first century, our ability to modify the smallest components of life through molecular biology had endowed humans with a power that even those who exercise it most proficiently cannot claim to fully comprehend. Human mastery over nature has been predicted for centuries—Francis Bacon insisted on it, William Blake feared it profoundly. Little more than a hundred years have passed, however, since Gregor Mendel demonstrated that the defining characteristics of a pea plant—its shape, its size, and the color of the seeds, for example—are transmitted from one generation to the next in ways that can be predicted, repeated, and codified.

Since then, the central project of biology has been to break that code and learn to read it—to understand how DNA creates and perpetuates life. The physiologist Jacques Loeb considered artificial synthesis of life the goal of biology. In 1912 Loeb, one of the founders of modern biochemistry, wrote that there was no evidence that "the artificial production of living matter is beyond the possibilities of science" and declared, "We must either succeed in producing living matter artificially, or we must find the reasons why this is impossible."

In 1946, the Nobel Prize–winning geneticist Hermann J. Muller attempted to do that. By demonstrating that exposure to X-rays can cause mutations in the genes and chromosomes of living cells, he was the first to prove that heredity could be affected by something other than natural selection. He wasn't entirely sure that people would use that information responsibly, though. "If we did attain to any such knowledge or powers, there is no doubt in my mind that we would eventually use them," Muller said. "Man is a megalomaniac among animals—if he sees mountains he will try to imitate them by pyramids, and if he sees some grand process like evolution, and thinks it would be at all possible for him to be in on that game, he would irreverently have to have his whack at that too."

The theory of evolution explained that every species on Earth is related in some way to every other species; more important, we each carry a record of that history in our body. In 1953 James Watson and Francis Crick began to make it possible to understand why, by explaining how DNA arranges itself. The language of just four chemical letters—adenine, cytosine, guanine, and thymine —comes in the form of enormous chains of nucleotides. When they are joined, the arrangement of their sequences determines how each human differs from all others and from all other living beings.

By the 1970s, recombinant-DNA technology permitted scientists to cut long, unwieldy molecules of nucleotides into digestible sentences of genetic letters and paste them into other cells. Researchers could suddenly combine the genes of two creatures that would never have been able to mate in nature. As promising as these techniques were, they also made it possible for scientists to transfer viruses—and microbes that cause cancer—from one organism to another. That could create diseases anticipated by no one and for which there would be no natural protection, treatment, or cure. In 1975 scientists from around the world gathered at the Asilomar Conference Center, in northern California, to discuss the challenges presented by this new technology. They focused primarily on laboratory and environmental safety and concluded that the field required little regulation. (There was no real discussion of deliberate abuse—at the time, there didn't seem to be any need.)

Looking back nearly thirty years later, one of the conference's organizers, the Nobel laureate Paul Berg, wrote, "This unique conference marked the beginning of an exceptional era for science and for the public discussion of science policy. Its success permitted the then contentious technology of recombinant DNA to emerge and flourish. Now the use of the recombinant DNA technology dominates research in biology. It has altered both the way questions are formulated and the way solutions are sought."

Decoding sequences of DNA was tedious. It could take a scientist a year to complete a stretch that was ten or twelve base pairs long. (Our DNA consists of 3 billion such pairs.) By the late 1980s, automated sequencing had simplified the procedure, and today machines can process that information in seconds. Another new tool

—polymerase chain reaction—completed the merger of the digital and biological worlds. Using PCR, a scientist can take a single DNA molecule and copy it many times, making it easier to read and to manipulate. That permits scientists to treat living cells like complex packages of digital information that happen to be arranged in the most elegant possible way.

Using such techniques, researchers have now resurrected the DNA of the Tasmanian tiger, the world's largest carnivorous marsupial, which has been extinct for more than seventy years. In 2008 scientists from the University of Melbourne and the University of Texas M. D. Anderson Cancer Center, in Houston, extracted DNA from tissue that had been preserved in the Museum Victoria, in Melbourne. They took a fragment of DNA that controlled the production of a collagen gene from the tiger and inserted it into a mouse embryo. The DNA switched on just the right gene, and the embryo began to churn out collagen. That marked the first time that any material from an extinct creature other than a virus has functioned inside a living organism.

It will not be the last. A team from Pennsylvania State University, working with hair samples from two woolly mammoths—one of them 60,000 years old and the other 18,000—has tentatively figured out how to modify that DNA and place it inside an elephant's egg. The mammoth could then be brought to term in an elephant mother. "There is little doubt that it would be fun to see a living, breathing woolly mammoth—a shaggy, elephantine creature with long curved tusks who reminds us more of a very large, cuddly stuffed animal than of a T. Rex.," the *Times* editorialized soon after the discovery was announced. "We're just not sure that it would be all that much fun for the mammoth."

The ultimate goal, however, is to create a synthetic organism made solely from chemical parts and blueprints of DNA. In the mid-1990s, Craig Venter, working at the Institute for Genomic Research, and his colleagues Clyde Hutchison and Hamilton Smith began to wonder whether they could pare life to its most basic components and then use those genes to create such an organism. They began modifying the genome of a tiny bacterium called *Mycoplasma genitalium,* which contained 482 genes (humans have about 23,000) and 580,000 letters of genetic code, arranged on one circular chromosome—the smallest genome of any cell that has been grown in laboratory cultures. Venter and his colleagues then removed genes one by one to find a minimal set that could sustain life.

Venter called the experiment the Minimal Genome Project. By the beginning of 2008, his team had pieced together thousands of chemically synthesized fragments of DNA and assembled a new version of the organism. Then, using nothing but chemicals, they produced from scratch the entire genome of *Mycoplasma genitalium*. "Nothing in our methodology restricts its use to chemically synthesized DNA," Venter noted in the report of his work, which was published in *Science*. "It should be possible to assemble any combination of synthetic and natural DNA segments in any desired order." That may turn out to be one of the most understated asides in the history of science. Next Venter intends to transplant the artificial chromosome into the walls of another cell and then "boot it up," thereby making a new form of life that would then be able to replicate its own DNA — the first truly artificial organism. (Activists have already named the creation Synthia.) Venter hopes that Synthia and similar products will serve essentially as vessels that can be modified to carry different packages of genes. One package might produce a specific drug, for example, and another could have genes programmed to digest carbon in the atmosphere.

In 2007 the theoretical physicist Freeman Dyson, after having visited both the Philadelphia Flower Show and the Reptile Show in San Diego, wrote an essay in the *New York Review of Books* noting that "every orchid or rose or lizard or snake is the work of a dedicated and skilled breeder. There are thousands of people, amateurs and professionals, who devote their lives to this business." This, of course, we have been doing in one way or another for millennia. "Now imagine what will happen when the tools of genetic engineering become accessible to these people."

It is only a matter of time before domesticated biotechnology presents us with what Dyson described as an "explosion of diversity of new living creatures . . . Designing genomes will be a personal thing, a new art form as creative as painting or sculpture. Few of the new creations will be masterpieces, but a great many will bring joy to their creators and variety to our fauna and flora."

Biotech games played by children "down to kindergarten age but played with real eggs and seeds" could produce entirely new species — as a lark. "These games will be messy and possibly dangerous," Dyson wrote. "Rules and regulations will be needed to make sure that our kids do not endanger themselves and others. The dangers of biotechnology are real and serious."

Life on Earth proceeds in an arc—one that began with the big bang and evolved to the point where a smart teenager is capable of inserting a gene from a cold-water fish into a strawberry to help protect it from the frost. You don't have to be a Luddite—or Prince Charles, who, famously, has foreseen a world reduced to gray goo by avaricious and out-of-control technology—to recognize that synthetic biology, if it truly succeeds, will make it possible to supplant the world created by Darwinian evolution with one created by us.

"Many a technology has at some time or another been deemed an affront to God, but perhaps none invites the accusation as directly as synthetic biology," the editors of *Nature*—who nonetheless support the technology—wrote in 2007. "For the first time, God has competition."

"What if we could liberate ourselves from the tyranny of evolution by being able to design our own offspring?" Drew Endy asked the first time we met in his office at MIT, where, until the summer of 2008, he was assistant professor of biological engineering. (That September he moved to Stanford.) Endy is among the most compelling evangelists of synthetic biology. He is also perhaps its most disturbing, because, although he displays a childlike eagerness to start engineering new creatures, he insists on discussing both the prospects and the dangers of his emerging discipline in nearly any forum he can find. "I am talking about building the stuff that runs most of the living world," he said. "If this is not a national strategic priority, what possibly could be?"

Endy, who was trained as a civil engineer, spent his youth fabricating worlds out of Lincoln Logs and Legos. Now he would like to build living organisms. Perhaps it was the three well-worn congas sitting in the corner of Endy's office, or the choppy haircut that looked like something he might have got in a tree house, or the bicycle dangling from his wall, but when he speaks about putting together new forms of life, it's hard not to think of that boy and his Legos.

Endy made his first mark on the world of biology by nearly failing the course in high school. "I got a D," he said. "And I was lucky to get it." While pursuing an engineering degree at Lehigh University, Endy took a course in molecular genetics. He spent his years in graduate school modeling bacterial viruses, but they are complex, and Endy craved simplicity. That's when he began to think about putting cellular components together.

Never forgetting the secret of Legos—they work because you can take any single part and attach it to any other—in 2005 Endy and colleagues on both coasts started the BioBricks Foundation, a nonprofit organization formed to register and develop standard parts for assembling DNA. Endy is not the only scientist, or even the only synthetic biologist, to translate a youth spent with blocks into a useful scientific vocabulary. "The notion of pieces fitting together—whether those pieces are integrated circuits, microfluidic components, or molecules—guides much of what I do in the laboratory," the physicist and synthetic biologist Rob Carlson writes in his new book, *Biology Is Technology: The Promise, Peril, and Business of Engineering Life.* "Some of my best work has come together in my mind's eye accompanied by what I swear was an audible click."

The BioBricks registry is a physical repository, but it is also an online catalogue. If you want to construct an organism or engineer it in new ways, you can go to the site as you would to one that sells lumber or industrial pipes. The constituent parts of DNA—promoters, ribosome-binding sites, plasmid backbones, and thousands of other components—are catalogued, explained, and discussed. It is a kind of theoretical Wikipedia of future life forms, with the added benefit of actually providing the parts necessary to build them.

I asked Endy why he thought so many people seem to be repelled by the idea of constructing new forms of life. "Because it's scary as hell," he said. "It's the coolest platform science has ever produced, but the questions it raises are the hardest to answer." If you can sequence something properly and you possess the information for describing that organism—whether it's a virus, a dinosaur, or a human being—you will eventually be able to construct an artificial version of it. That gives us an alternate path for propagating living organisms.

The natural path is direct descent from a parent—from one generation to the next. But that process is filled with errors. (In Darwin's world, of course, a certain number of those mutations are necessary.) Endy said, "If you could complement evolution with a secondary path, decode a genome, take it offline to the level of information"—in other words, break it down to its specific sequences of DNA the way one would break down the code in a software program—"we can then design whatever we want, and recompile it," which could permit scientists to prevent many genetic diseases. "At that point, you can make disposable biological systems that don't have to produce offspring, and you can make much simpler organisms."

Endy stopped long enough for me to digest the fact that he was talking about building our own children. "If you look at human beings as we are today, one would have to ask how much of our own design is constrained by the fact that we have to be able to reproduce," he said. In fact, those constraints are significant. In theory, at least, designing our own offspring could make those constraints disappear. Before speaking about that, however, it would be necessary to ask two essential questions: What sorts of risk does that bring into play, and what sorts of opportunity?

The deeply unpleasant risks associated with synthetic biology are not hard to imagine: who would control this technology, who would pay for it, and how much would it cost? Would we all have access or, as in the 1997 film *Gattaca*, which envisaged a world where the most successful children were eugenically selected, would there be genetic haves and have-nots and a new type of discrimination—genoism—to accompany it? Moreover, how safe can it be to manipulate and create life? How likely are accidents that would unleash organisms onto a world that is not prepared for them? And will it be an easy technology for people bent on destruction to acquire? "We are talking about things that have never been done before," Endy said. "If the society that powered this technology collapses in some way, we would go extinct pretty quickly. You wouldn't have a chance to revert back to the farm or to the pre-farm. We would just be gone."

Those fears have existed since humans began to transplant genes in crops. They are the central reason that opponents of genetically engineered food invoke the precautionary principle, which argues that potential risks must always be given more weight than possible benefits. That is certainly the approach suggested by people like Jim Thomas, of ETC, who describes Endy as "the alpha Synthusiast." But he also regards Endy as a reflective scientist who doesn't discount the possible risks of his field. "To his credit, I think he's the one who's most engaged with these issues," Thomas said.

The debate over genetically engineered food has often focused on theoretical harm rather than on tangible benefits. "If you build a bridge and it falls down, you are not going to be permitted to design bridges ever again," Endy said. "But that doesn't mean we should never build a new bridge. There we have accepted the fact that risks are inevitable." He believes the same should be true of engineering biology.

We also have to think about our society's basic goals and how this science might help us achieve them. "We have seen an example with artemisinin and malaria," Endy said. "Maybe we could avoid diseases completely. That might require us to go through a transition in medicine akin to what happened in environmental science and engineering after the end of World War II. We had industrial problems, and people said, Hey, the river's on fire—let's put it out. And, after the nth time of doing that, people started to say, Maybe we shouldn't make factories that put shit into the river. So let's collect all the waste. That turns out to be really expensive, because then we have to dispose of it. Finally, people said, Let's redesign the factories so that they don't make that crap."

Endy pointed out that we are spending trillions of dollars on health care and that preventing disease is obviously more desirable than treating it. "My guess is that our ultimate solution to the crisis of health-care costs will be to redesign ourselves so that we don't have so many problems to deal with. But note," he stressed, "you can't possibly begin to do something like this if you don't have a value system in place that allows you to map concepts of ethics, beauty, and aesthetics onto our own existence.

"These are powerful choices. Think about what happens when you really can print the genome of your offspring. You could start with your own sequence, of course, and mash it up with your partner, or as many partners as you like. Because computers won't care. And, if you wanted evolution, you can include random number generators." That would have the effect of introducing the element of chance into synthetic design.

Although Endy speaks with passion about the biological future, he acknowledges how little scientists know. "It is important to unpack some of the hype and expectation around what you can do with biotechnology as a manufacturing platform," he said. "We have not scratched the surface. But how far will we be able to go? That question needs to be discussed openly, because you can't address issues of risk and society unless you have an answer."

Answers, however, are not yet available. The inventor and materials scientist Saul Griffith has estimated that powering our planet requires between fifteen and eighteen terawatts of energy. How much of that could we manufacture with the tools of synthetic biology? Estimates range between five and ninety terawatts. "If it turns out to be the lower figure, we are screwed," Endy said. "Because why would we take this risk if we cannot create much energy? But if it's the top figure, then we are talking about producing five times the energy we need on this planet and doing it in an environmentally benign way. The benefits in relation to the risks of using this new technology would be unquestioned. But I don't know what the number will be, and I don't think anybody can know at this point. At a minimum, then, we ought to acknowledge that we are in the process of figuring that out and the answers won't be easy to provide.

"It's very hard for me to have a conversation about these issues, because people adopt incredibly defensive postures," Endy continued. "The scientists on one side and civil society organizations on the other. And to be fair to those groups, science has often proceeded by skipping the dialogue. But some environmental groups will say, Let's not permit any of this work to get out of a laboratory until we are sure it is all safe. And as a practical matter that is not the way science works. We can't come back decades later with an answer. We need to develop solutions by doing them. The potential is great enough, I believe, to convince people it's worth the risk."

I wondered how much of this was science fiction. Endy stood up. "Can I show you something?" he asked, as he walked over to a bookshelf and grabbed four gray bottles. Each one contained about half a cup of sugar, and each had a letter on it: A, T, C, or G, for the four nucleotides in our DNA. "You can buy jars of these chemicals that are derived from sugarcane," he said. "And they end up being the four bases of DNA in a form that can be readily assembled. You hook the bottles up to a machine, and into the machine comes information from a computer, a sequence of DNA—like T-A-A-T-A-G-C-A-A. You program in whatever you want to build, and that machine will stitch the genetic material together from scratch. This is the recipe: you take information and the raw chemicals and compile genetic material. Just sit down at your laptop and type the letters and out comes your organism."

We don't have machines that can turn those sugars into entire genomes yet. Endy shrugged. "But I don't see any physical reason why we won't," he said. "It's a question of money. If somebody wants to pay for it, then it will get done." He looked at his watch, apologized, and said, "I'm sorry, we will have to continue this discussion another day, because I have an appointment with some people from the Department of Homeland Security."

I was a little surprised. "They are asking the same questions as you," he said. "They want to know how far is this really going to go."

Scientists skipped a step at the birth of biotechnology, thirty-five years ago, moving immediately to products without first focusing on the tools required to make them. Using standard biological parts, a synthetic biologist or biological engineer can already, to some extent, program living organisms in the same way a computer scientist can program a computer. However, genes work together in ways that are staggeringly complex; proteins produced by one will counteract—or enhance—those made by another. We are far from the point where scientists might yank a few genes off the shelf, mix them together, and produce a variety of products. But the registry is growing rapidly—and so is the knowledge needed to drive the field forward.

Research in Endy's Stanford lab has been largely animated by his fascination with switches that turn genes on and off. He and his students are attempting to create genetically encoded memory systems, and his current goal is to construct a cell that can count to 256—a number derived from the mathematics of Basic computer code. Solving the practical challenges will not be easy, since cells that count will need to send reliable signals when they divide and remember that they did.

"If the cells in our bodies had a little memory, think what we could do," Endy said the next time we talked. I wasn't quite sure what he meant. "You have memory in your phone," he explained. "Think of all the information it allows you to store. The phone and the technology on which it is based do not function inside cells. But if we could count to two hundred using a system that was based on proteins and DNA and RNA—well, now, all of a sudden we would have a tool that gives us access to computing and memory that we just don't have.

"Do you know how we study aging?" Endy continued. "The tools we use today are almost akin to cutting a tree in half and counting the rings. But if the cells had a memory, we could count properly. Every time a cell divides, just move the counter by one. Maybe that will let me see them changing with a precision nobody can have today. Then I could give people controllers to start retooling those cells. Or we could say, Wow, this cell has divided two hundred times, it's obviously lost control of itself and become cancer. Kill it. That lets us think about new therapies for all kinds of diseases."

Synthetic biology is changing so rapidly that predictions seem pointless. Even that fact presents people like Endy with a new kind of problem. "Wayne Gretzky once said, 'I skate to where the puck is going to be.' That's what you do to become a great hockey player," Endy told me. "But where do you skate when the puck is accelerating at the speed of a rocket, when the trajectory is impossible to follow? Whom do you hire and what do we ask them to do? Because what preoccupies our finest minds today will be a seventh-grade science project in five years. Or three years.

"We are surfing an exponential now, and, even for people who pay attention, surfing an exponential is a really tricky thing to do. And when the exponential you are surfing has the capacity to impact the world in such a fundamental way, in ways we have never before considered, how do you even talk about that?"

For decades, people have invoked Moore's law: the number of transistors that could fit onto a silicon chip would double every two years, and so would the power of computers. When the IBM 360 computer was released in 1964, the top model came with eight megabytes of main memory, and cost more than $2 million. Today cell phones with a thousand times the memory of that computer can be bought for about a hundred dollars.

In 2001 Rob Carlson, then a research fellow at the Molecular Sciences Institute in Berkeley, decided to examine a similar phenomenon: the speed at which the capacity to synthesize DNA was growing. He produced what has come to be known as the Carlson curve, and it shows a rate that mirrors Moore's law—and has even begun to exceed it. The automated DNA synthesizers used in thousands of labs cost $100,000 a decade ago. Now they cost less than $10,000, and most days at least a dozen used synthesizers are for sale on eBay—for less than a thousand dollars.

Between 1977, when Frederick Sanger published the first paper on automatic DNA sequencing, and 1995, when the Institute for Genomic Research reported the first bacterial-genome sequence, the field moved slowly. It took the next six years to complete the first draft of the immeasurably more complex human genome, and six years after that, in 2007, scientists from around the world began mapping the full genomes of more than a thousand people. The Harvard geneticist George Church's Personal Genome Project now plans to sequence more than a hundred thousand.

In 2003, when Endy was still at MIT, he and his colleagues Tom Knight, Randy Rettberg, and Gerald Sussman founded iGEM— the International Genetically Engineered Machine competition— whose purpose is to promote the building of biological systems from standard parts. In 2006 a team of Endy's undergraduate students used BioBrick parts to genetically reprogram *E. coli* (which normally smells awful) to smell like wintergreen while it grows and like bananas when it has finished growing. They named their project Eau d'E Coli. By 2008, with more than a thousand students from twenty-one countries participating, the winning team—a group from Slovenia—used biological parts that it had designed to create a vaccine for the stomach bug *Helicobacter pylori,* which causes ulcers. There are no such working vaccines for humans. So far the team has tested its creation on mice, with promising results.

This is open-source biology, where intellectual property is shared. What's available to idealistic students, of course, would also be available to terrorists. Any number of blogs offer advice about everything from how to preserve proteins to the best methods for desalting DNA. Openness like that can be frightening, and there have been calls for tighter control of the technology. Carlson, among many others, believes that strict regulations are unlikely to succeed. Several years ago, with very few tools other than a credit card, he opened his own biotechnology company, Biodesic, in the garage of his Seattle home—a biological version of the do-it-your-self movement that gave birth to so many computer companies, including Apple.

The product that he developed enables the identification of proteins using DNA technology. "It's not complex," Carlson told me, "but I wanted to see what I could accomplish using mail order and synthesis." A great deal, it turned out. Carlson designed the molecule on his laptop, then sent the sequence to a company that synthesizes DNA. Most of the instruments could be bought on eBay (or, occasionally, on LabX, a more specialized site for scientific equipment). All you need is an Internet connection.

"Strict regulation doesn't accomplish its goals," Carlson said. "It's not an exact analogy, but look at Prohibition. What happened when government restricted the production and sale of alcohol? Crime rose dramatically. It became organized and powerful. Legitimate manufacturers could not sell alcohol, but it was easy to make in a garage—or a warehouse."

By 2002 the U.S. government had intensified its effort to curtail the sale and production of methamphetamine. Previously, the drug had been manufactured in many mom-and-pop labs throughout the country. Today production has been professionalized and centralized, and the Drug Enforcement Administration says that less is known about methamphetamine production than before. "The black market is getting blacker," Carlson said. "Crystal-meth use is still rising, and all this despite restrictions." Strict control would not necessarily insure the same fate for synthetic biology, but it might.

Bill Joy, a founder of Sun Microsystems, has frequently called for restrictions on the use of technology. "It is even possible that self-replication may be more fundamental than we thought, and hence harder—or even impossible—to control," he wrote in an essay for *Wired* called "Why the Future Doesn't Need Us." "The only realistic alternative I see is relinquishment: to limit development of the technologies that are too dangerous, by limiting our pursuit of certain kinds of knowledge."

Still, censoring the pursuit of knowledge has never really worked, in part because there are no parameters for society to decide who should have information and who should not. The opposite approach might give us better results: accelerate the development of technology and open it to more people and educate them to its purpose. Otherwise, if Carlson's methamphetamine analogy proves accurate, power would flow directly into the hands of the people least likely to use it wisely.

For synthetic biology to accomplish any of its goals, we will also need an education system that encourages skepticism and the study of science. In 2007 students in Singapore, Japan, China, and Hong Kong (which was counted independently) all performed better on an international science exam than American students. The U.S. scores have remained essentially stagnant since 1995, the first year the exam was administered. Adults are even less scientifically literate. Early in 2009, the results of a California Academy of Sciences poll (conducted throughout the nation) revealed that only 53 percent of American adults know how long it takes for Earth to revolve around the sun, and a slightly larger number—59 percent—are aware that dinosaurs and humans never lived at the same time.

Synthetic biologists will have to overcome this ignorance. Optimism prevails only when people are engaged and excited. Why should we bother? Not just to make *E. coli* smell like chewing gum or to make fish glow in vibrant colors. The planet is in danger, and nature needs help.

The hydrocarbons we burn for fuel are believed to be nothing more than concentrated sunlight that has been collected by leaves and trees. Organic matter rots, bacteria break it down, and it moves underground, where, after millions of years of pressure, it turns into oil and coal. At that point, we dig it up—at huge expense and with disastrous environmental consequences. Across the globe, on land and sea, we sink wells and lay pipe to ferry our energy to giant refineries. That has been the industrial model of development, and it worked for nearly two centuries. It won't work any longer.

The industrial age is drawing to a close, eventually to be replaced by an era of biological engineering. That won't happen easily (or quickly), and it will never solve every problem we expect it to solve. But what worked for artemisinin can work for many of the products our species will need to survive. "We are going to start doing the same thing that we do with our pets, with bacteria," the genomic futurist Juan Enriquez has said, describing our transition from a world that relied on machines to one that relies on biology.

"A house pet is a domesticated parasite," he noted. "It is evolved to have an interaction with human beings. Same thing with corn"—a crop that didn't exist until we created it. "Same thing is going to start happening with energy," he went on. "We are going to start domesticating bacteria to process stuff inside enclosed reactors to produce energy in a far more clean and efficient manner. This is just the beginning stage of being able to program life."

BRIAN BOYD

Purpose-Driven Life

FROM *The American Scholar*

> [Darwinism] seems simple, because you do not at first realize all that it
> involves. But when its whole significance dawns on you, your heart sinks into
> a heap of sand within you. There is a hideous fatalism about it, a ghastly and
> damnable reduction of beauty and intelligence, of strength and purpose, of
> honor and aspiration.
> —George Bernard Shaw, *Back to Methuselah* (1912)

EVOLUTIONARY THINKING has lately expanded from the biological to the human world, first into the social sciences and recently into the humanities and the arts. Many people therefore
now understand the human, and even human culture, as inextricably biological. But many others in the humanities—in this, at least,
like religious believers who reject evolution outright—feel that a
Darwinian view of life and a biological view of humanity can only
deny human purpose and meaning.

Does evolution by natural selection rob life of purpose, as so
many have feared? The answer is no. On the contrary, Charles Darwin has made it possible to understand how purpose, like life,
builds from small beginnings, from the ground up. In a very real
sense, evolution creates purpose.

Evolution generates problems and solutions as it generates life.
Rocks may crack and erode, but they do not have problems. Amoe-
bas and apes do. Natural selection creates complex new possibilities, and therefore new problems, as it assembles self-sustaining
organisms piecemeal, cycle after cycle, by generating partial solutions, testing them, and regenerating from the basis of the best solutions available in the current cycle. In time, it can create richer
solutions to richer problems.

*

In *On the Origin of Species* (1859), Darwin showed how new species could evolve through a process of blind variation and selective retention. He transformed at a stroke our understanding of natural design. Living things manifest complex design but can be produced by a mindless process, one that does no more than passively register, in terms of survival and reproduction, the advantages of particular variations. In *The Blind Watchmaker* (1986), Richard Dawkins explains how nature is like a watchmaker who builds intricate mechanisms without forethought, and he thereby overturns the famous argument of the theologian and naturalist William Paley. Paley opens his *Natural Theology: or Evidences of the Existence and Attributes of the Deity, Collected from the Appearances of Nature* (1802), with these words:

> In crossing a heath, suppose I pitched my foot against a *stone,* and were asked how the stone came to be there; I might possibly answer, that, for anything I knew to the contrary, it had lain there forever: nor would it perhaps be very easy to show the absurdity of this answer. But suppose I had found a *watch* upon the ground, and it should be inquired how the watch happened to be in that place; I should hardly think of the answer which I had before given, that, for anything I knew, the watch might have always been there . . . [The precision and intricacy of its mechanism would have forced us to conclude] that the watch must have had a maker; that there must have existed, at some time, and at some place or other, an artificer or artificers, who formed it for the purpose which we find it actually to answer; who comprehended its construction, and designed its use.

Nobody could reasonably disagree, Paley adds, yet this is tantamount to what an atheist does, for "every indication of contrivance, every manifestation of design, which existed in the watch, exists in the works of nature; with the difference, on the side of nature, of being greater and more, and that in a degree which exceeds all computation." As Dawkins notes, we now know that the complexities of natural organisms surpass those of the most intricate watch by far more than science could guess in Paley's time; yet he goes on to show how simple processes of variation and selective retention can, over many cycles, create products with even this degree of design.

Other processes working *within* natural selection have been found to follow the same principle: the human immune system; the synapses in the young human brain (in the neural Darwinism of Gerald Edelman and others); culture (in the work of David Sloan Wilson and others); and invention (in the work of Donald Campbell and David Hull). Such "Darwinian systems," "Darwin machines," or, in Dawkins's term, "universal Darwinism," allow genuine novelty to be achieved without advance knowledge of what will work best in an unpredictable and open world. The common principle of blind variation and selective retention allows for a deeply indeterministic process that explores patches of possibility space in multiple directions and pursues any direction provisionally more promising than others. It tracks through the vastness of the possible in ways that surprisingly often lead to rich solutions by compounding immediate advantages and retaining achieved complexity in the next round of variations. Such Darwinian processes might well occur anywhere we find deeply original novelty.

Darwin's explanation of evolution by natural selection shocked, and still shocks, because it appears to deny purpose. We think of purpose as something prior to decision and action: I want to raise my arm and, unless I am paralyzed or restrained, I do. But in fact purpose emerges slowly, in the species and in the individual. My capacity to move my arm in as many ways as I can depends on things like the evolution of forelegs into arms early in the primate line, the evolution millions of years later of a rotating socket in the shoulder of great apes, to enable swinging in trees, and the further freeing up of arm movements after early hominids became fully bipedal. Babies flail their arms uncontrolledly and purposelessly for months before they can direct them in a particular way for a particular purpose.

Paley's example of the watch assumes a purpose we already understand: the intricate integration of material objects into instruments for telling the time. But humans did not evolve to be capable of constructing multipart mechanisms until within the last 50,000 years or so. Until their manual control reached a high level and their stone toolmaking had had more than 2 million years of refinement, they would have been unable to conceive of such mechanisms or, if confronted with them, unable to recognize anything of their construction or purpose. The idea of telling the time precisely would have been unknown and meaningless to our human ancestors even at the end of the Stone Age.

Purposes can emerge only piecemeal; problems cannot even define themselves until many of the elements are already in place. The position of the sun in its daily sweep can indicate the phases of the day, but nothing more precise. Sundials and the sticks in the ground that preceded them afforded more finely determined divisions of time during daylight and made it possible to imagine coordinating common actions in advance. The first water clocks and sand clocks enabled even tighter coordination. Mechanical clocks and bells to chime the hour or even the quarter-hour took social synchronization still further. Not until European navigation and mapping flourished in the fifteenth century did anyone consider devising a portable clock to ascertain longitude, yet maritime clocks remained highly inaccurate for another three centuries. The first watches, in the early sixteenth century, could register only the hour; and it took more than another century to tell the time down to the minute on a portable mechanism. By Paley's day, the recent invention of the duplex escapement had made it possible to keep time accurately on a pocket watch, but it was only over the course of the nineteenth century that highly accurate timing made possible new degrees of precision measurement, and hence new research options in physics and in psychology. Not until well into the twentieth century did cesium clocks attain the exactitude and reliability of timekeeping needed for quantum physics or space flight.

In the development of both instruments and ever finer standards for measuring time intervals, new purposes have been discovered, each inconceivable a stage or two previously. Purposes arise not in advance, but as possibilities materialize. Of course, when the purpose becomes established, it can *then* be implemented in advance of any particular manifestation: I can choose to move my arm in such a way as to put on a sweater, or to time mental responses down to the millisecond in a psychology laboratory as a measure of the complexity of the neural processing involved. But each of these purposes, although it precedes a definite action, has a long history that precedes it in the species, the culture, and the individual, a history of prior trials and errors, before the purpose could be conceived and fully defined, let alone specified in advance.

Life could become established only when matter organized itself in a way complex enough to sustain and reliably reproduce itself.

Maintaining such a highly improbable and functional arrangement of matter became life's first purpose. As species continued to evolve, so did the purposes of their organs and behaviors. New behaviors, like new organs, begin uncertainly, with slight modifications of existing structure, but become defined over time, and their function or purpose specifiable in advance: a certain kind of spider will spin a certain kind of web to catch a certain range of insects under certain conditions, and so on.

As creatures began to act in more complex and flexible ways, nature evolved emotions to motivate better decisions. Satisfying these emotions—escaping fear, appeasing hunger, fulfilling desire, sustaining love, and so on—became important purposes in themselves for much of the animal kingdom.

As behaviors standardize, as purposes define themselves, social animals can learn to understand not only the actions of other members of their species but even their desires and intentions before they act. Not only do we learn to infer others' intentions but, in social species that benefit from cooperation, we also evolve to empathize with or emotionally react against others' purposes. (Without this, stories would be impossible.)

Yet we should not forget that despite our thinking of purposes as prior to actions, they have emerged over long stretches of evolutionary and individual time. Intentions are efficient routes to objectives clearly defined only after many preliminary stages of variation and selection within animals' evolutionary and individual pasts. Like design, purpose emerges rather than precedes, except in the case of purposes that have developed long enough to become standardized. Only in that sense can purpose be said to precede the particular instance, whether the function of an organ or the intention of an action.

Purposes evolve, and Darwinian processes extend them. Intelligence and creativity are purposes that have emerged over the course of life on Earth. Stephen Jay Gould famously argued that if we could rewind and replay the tape of evolution, humans and human intelligence would not reappear. Quite possibly not; no one disputes that contingency strongly inflects evolution. But as Simon Conway Morris has stressed, certain capacities have evolved again and again, because of the singular advantages they offer: senses, locomotion, minds, emotions, sociality, intelligence, creativity, co-operation, to name those that concern us most. Let's consider two of these, intelligence and creativity.

Intelligence allows us to respond flexibly to circumstances, to solve problems not only according to successful old routines (prior purposes, if you like) but in novel and more or less context-sensitive ways. Because it can sometimes find new solutions, intelligence is highly advantageous—yet not at all easy for evolution to evolve. Although minds have been necessary for all motile creatures, more advanced intelligence has emerged in relatively few lineages, although quite diverse ones: invertebrates like octopi and cuttlefish; vertebrates like crows and parrots among birds and cetaceans and primates among mammals.

Intelligence has large benefits, but it also incurs costs. Out of the pressure to develop social intelligence, humans have grown in self-awareness, so that we can imagine ourselves as others see us in competitive and cooperative scenarios. That ability offers real benefits in anticipating others' actions and reactions, but among its costs is the fact that we can also envisage our own death and absence from the ongoing world. For humans this has raised the question of our purpose in the face of our ultimate lifelessness, one we have answered most frequently by concluding that we continue in some form after death. To judge by grave rituals dating back at least 70,000 years, and the evidence of the fear of death and the hope of immortality in the records of early civilizations, preoccupation with death has loomed large ever since the appearance of a distinctly human culture.

Creativity is the capacity to develop significant and valuable novelty. This seems the most difficult capacity of all for evolution to evolve, and for good reason. Significant and valuable by what criteria?

Human creativity matters for human beings. But creativity hardly matters for evolution. Single-celled organisms reproduce themselves readily, and life can go on—did go on, for billions of years on Earth—with barely more complexity. Life persists through reproduction, through transmitting accumulated complexity to subsequent generations. If inherited design were radically changed each time an organism reproduced, the hard-won gains of natural selection would rapidly be lost. Life can evolve new possibilities only slowly, through variations small enough not to threaten existing evolved functions, accreting functional novelty generation by generation from minor and undirected variation. But although evolution has thereby spawned many new species and even major new forms of life, it does not need or aim for creativity.

Yet organisms vary, even if only through imperfections in reproduction, and conditions change. Over enough time conditions will always alter, including competition with other organisms in the environment. Since any organism can become a source of energy for others, each has to find ways of exploiting others more efficiently and to avoid being exploited by others, including predators, parasites, and pathogens. In species with a wide range of variations, some individuals will be able to exploit changing opportunities or to avoid changing threats more effectively than others. Variation in itself—considerable enough to gain advantage, but not so large as to imperil existing design—therefore offers a measure of security against unpredictable circumstances. For this reason, sex has evolved many times as a way of recombining genes unpredictably but reliably, and hence of generating a range of initially viable variations from which conditions will select. Some species even toggle between reproducing sexually or asexually according to the degree of environmental instability. Sexual recombination therefore ensures a wide and unpredictable range of genetic variation that can cope better with unpredictable circumstances.

Just as natural selection has evolved sex as a means for amplifying genetic variation, it has evolved art in humans as a means for amplifying behavioral variation. Art has been designed by evolution for creativity.

The human immune system and the infant human brain naturally overproduce options to cope with as many unpredictable situations as they can. They then pare back whatever is not activated by experience and regenerate from whatever has been stimulated by experience. These second-order Darwinian processes allow an additional level of flexibility beyond first-order genetic variation, a still more sensitive adjustment to even shorter-term unpredictability.

Art is a subsidiary Darwin machine that generates not natural but "unnatural" variations or options. By "unnatural" I mean only that art involves highly deliberate human choices, both individual and cultural, even if these are themselves ultimately the products of nature.

Creativity as a principle, as a Darwinian process, solves no particular prespecifiable problem; but it offers an additional way of generating new possibilities that *may* prove to solve problems, even significant ones, provided there is a consistent pressure toward a solution—whether over generations, as in natural selection, or over weeks or months or years, as when a storyteller, say, drafts and revises a story, or in only minutes or seconds in the spreading neural activation in a poet or a scientist seeking a new image or idea, in a mind prepared over many years by many trials.

Art did not evolve in order to foster creativity. Evolution has no foresight. It cannot evolve what has only future advantages, but can evolve only what offers benefit *now*. Yet art now does foster creativity. So how did it evolve and why?

New evolutionary solutions themselves often spawn new problems. When our brains allowed us to become superpredators, to dominate our environments and earn the food we needed in much less time than our waking hours, we did not solve the "problem" of spare time, as did other top predators, such as lions, tigers, or bears, by sleeping the extra hours away to conserve energy. Even at rest our large brains consume a high proportion of our energy, and since they offer us most of our advantages against other species and other individuals, we benefit not from resting them as much as possible but from developing them in times of security and leisure. Art as cognitive play, appealing to our appetite for potentially meaningful patterned information, engages our attention in a self-rewarding way and therefore encourages us to strengthen the processing power of our minds in the kinds of information that matter most to us.

Because it appeals to our own cognitive preferences, we have built-in incentives to generate art: its effects should be pleasing in themselves. Since the criteria for success are human preferences, since the testing mechanisms are already in our minds as we sing, or tell a story, or dance, or daub, we can readily adjust our actions to produce more satisfying effects: we can easily select from what we do, as we do it, and try out new variations, or stop when interest flags.

In most societies art has been collective and active, and even in modern cities, dance and song often still remain so. Where art tends to be more individual than communal, those with talent enough to spark the interest of others have a strong extra incentive to develop their skill for the attention, gratitude, and status it can earn them. Although professional artists may not have appeared until agriculture and permanent settlement allowed resources to accumulate and labor to specialize, the quality of some of the earliest art suggests that *some* individuals, long before agriculture, had the luxury of developing singular skills. Creative concentration and feedback during composition could work like a speeded-up version of natural selection, as these artists rapidly generated, discarded, and regenerated new variations.

Even in societies where art has become individualized and professionalized, it remains highly social. Art not only activates our private cognitive preferences but also adjusts and amplifies them through our sociality. From the first, mothers and others engage with infants in multimodal social play involving fine-grained attunement and interaction. We instinctively make learning enjoyable for young children by making it social, by making it play, and by making it art, by appealing to the cognitive preferences that art animates. Throughout life, participation in artistic activities in group settings, whether actively (in performance) or more or less passively (as audiences), continues to amplify the emotional charge of art.

Art's social nature not only motivates our participation, but also provides ready-made models to reduce the costs of invention and increase the benefits of response. Tribal arts like weaving, classic forms like the sonnet, or modern arts like filmmaking all depend on the existence of shared norms to provide prompts and challenges. As the film theorist David Bordwell observes: "Norms help unambitious filmmakers attain competence, but they challenge gifted ones to excel. By understanding these norms we can better appreciate skill, daring, and emotional power on those rare occasions when we meet them."

Art can engender variations through other factors present elsewhere in nature—through randomness, "an intrinsic part of brain function," and "nature's way of exploring unforeseen possibilities" in other domains too, and through copying errors—and through distinctively human factors. Art does not need to start from scratch but can recombine elements already developed in the same or a different art or tradition. Just as sex, by recombining genes, and hybridization, by recombining lineages that have had time to separate, can engender novel forms, art too can readily recombine, from the animal-human blends in cave art to the Minotaurs of the ancient Greeks or the modernist Picasso.

In *any* species, attention diminishes with persistence or reiteration, but humans are especially curious and thus susceptible to boredom. And as the developmental psychologist Michael Tomasello and colleagues have particularly stressed, attention, especially shared or commonly focused attention, has become unprecedentedly important to our species. To attract attention, art explores variation, even in traditional societies, and all the more so in societies where professional art and a highly competitive market for attention act as incentives to discover either new variations within existing forms or entirely novel combinations.

If art is "unnatural" variation, science is "unnatural" selection. Art appeals to our species preferences and our intuitive understandings, often as they have been modified by local culture. Science rejects our species preferences and our intuitive understandings, *even as* modified by local culture. It tests ideas not against human preferences but against a resistant world not designed for humans. Its methods of testing, by logic, observation, and experiment, encourage us to reject ideas, even those that seem self-evident and apparently confirmed repeatedly by tradition.

By exposing itself to falsifying evidence, science makes possible the cumulative retention of only the most rigorously selected ideas. This does not prove them all correct, but it improves on the ratio of tested to untested ideas attainable by any other known procedure. After the winnowing process, although there may still be wrong ideas in what we think of as science, there are far fewer than in any other human domain.

Art could evolve as an adaptation because it appealed to our deep-grained species preferences. Science could not. It appeals to one strong species preference, our curiosity, but it otherwise goes against the grain of our intuitive understanding. Until Galileo, people assumed, with Aristotle, that a heavier object fell more rapidly than a lighter one. Information gathering, invaluable for all kinds of animals and even for plants, has mattered especially for humans, but the knowledge gained has mostly been in the form of heuristics, partly right, but not necessarily so, like our hunches about falling objects or the sun's motion around Earth. And although accurate information is invaluable, indecision is fatal, and no organism can afford the time to search for correct information at a moment when immediate responses are required. It was not possible to devote effort to a time-consuming, difficult-to-imagine, and increasingly resource-expensive process of testing ideas until in Renaissance Italy the right conditions happened to converge: a considerable buffer of security and overproduction; opportunities for intense specialization; and the availability of information and conflicting explanations that the printing press made possible.

Science still calls for qualities that are unnatural. Children are information sponges and soak up what they need to understand, like the basics of their world or their language. They need not be taught how to speak or to play. But they do need slow formal instruction to read, write, or calculate, and they need even more training and the help of externalized information (books, diagrams, models) to master the knowledge on which science builds. If they undergo the intensive training required of scientists, they will still need imagination to find new ways of testing or reexplaining received knowledge. Even for those with training, looking for potential refutations of cherished ideas is both emotionally difficult and imaginatively draining. And whereas art appeals to human preferences, science has to account for a world not built to suit human tastes or talents.

Unnatural selection though it is, science allows for the accumulation of advantageous variations and the rapid "evolution" of complex intellectual and technological design. Art functions very differently, as a form of unnatural variation. Much that it produces is therefore not intrinsically deeply valuable. But some art *is* deeply valuable and speaks profoundly to many people, over long stretches of time or life and across many cultures. Because art is primarily a process of variation — although artists and audiences also select — there is not the same ratcheted accumulation of better design that occurs in science. Hence art of thousands of years ago, like that of Homer or Nok sculptors, can be superior in many ways, as examples of creativity, to most works generated now, simply because Homer or the Nok craftsmen could appeal to preferences that they understood deeply and that have not changed massively since their day.

Religion partakes of elements of both art and science. It could not have begun without our uniquely human understanding of false belief, which develops in individuals during their fourth year—our awareness that another person can have a different understanding of a situation from what we know to be the case, and our concomitant awareness that in other circumstances *we* may not have all we need to understand this or that situation. Our capacity to understand false belief has amplified our curiosity and spurred us to the quest for the deeper knowledge that has led to both religion and science.

Nor could religion have begun without the capacity for storytelling that grew out of our theory of mind and our first inclinations to art, like chant and bodily decoration. Storytelling launched a thousand tales. Those tales most often retold not only involved agents with exceptional powers but also helped to solve problems of cooperation by suggesting that we are continually watched over by spirits who monitor our deeds and punish or reward them.

Religious stories could also allay the unease that arose in us because of our awareness of false belief. The social intelligence out of which our grasp of false belief arose allowed us to imagine being dead and to foresee the world without us. It brought with it a new anxiety about the possible purposelessness of our lives, although this could be allayed to some extent by stories of spirits without bodies as a guarantor of purpose prior to human life or as a promise of continued existence afterward.

Religion and power commandeered art, not entirely but substantially, for millennia. Not that art as play did not persist—between parent and child, or among children, or among adults letting off steam. But where they could, religion and power appropriated toward their own ends art's ability to appeal to human imaginations.

Only when science began to offer alternative, naturalistic explanations of the world did religion and art start to diverge widely again. When science offered a detailed explanation of natural design without the need for a designer—the theory of evolution by natural selection—*that,* more than any other single idea, stripped us of a world made comfortable by a sense of purpose, apparently guaranteed by beings greater than ourselves.

Nevertheless, if we develop Darwin's insight, we can see the emergence of purpose, as of life itself, by small degrees, not from above, but by small increments, from below. The first purpose was the organization of matter in ways complex enough to sustain and replicate itself—the establishment, in other words, of life, or in still other terms, of problems and solutions. With life emerged the first purpose, the first problem, to preserve at least the improbable complexity already reached, and to find new ways of resisting damage and loss.

As life proliferated, variety offered new hedges against loss in the face of unpredictable circumstances, and even new ways of evolving variety, like sex. Still richer purposes emerged with emotions, intelligence, and cooperation, and most recently with creativity itself, pursued naturally, and unnaturally, through human invention, in art, and pursued unnaturally, through challenging what we have inherited, in science.

Art at its best offers us the durability that became life's first purpose, the variety that became its second, the appeal to the intelligence and the cooperative emotions that took so much longer to evolve, and the creativity that keeps adding new possibilities, including religion and science. We do not know a purpose guaranteed from outside life, but we can add as much as we can to the creativity of life. We do not know what other purposes life may eventually generate, but creativity offers us our best chance of reaching them.

PHILIP GOUREVITCH

The Monkey and the Fish

FROM *The New Yorker*

BACK IN THE 1980s and '90s, Greg Carr made a couple of hundred million dollars developing and marketing voice-mail and Internet services. Carr came from Idaho, and he lived in Cambridge, Massachusetts, and in 1998, just before he turned forty, he decided that he would become a full-time philanthropist. He didn't just want to give his money away; he also wanted to give himself to his projects—body and soul. So, for instance, a few years later, when Carr was out walking in Cambridge with a friend, the theater director and critic Robert Brustein, they passed an old building that used to house Grendel's bar, on JFK Street; Brustein said it would be fun to turn that place into a sort of laboratory theater, and Carr fell in love with the idea. He put more than $1 million into converting the place into a proper, ninety-nine-seat theater and began producing plays. "What I would do is spend all summer in Idaho, a lot of it by myself, with stacks of plays, just reading," he said.

Alongside what he called "new, strange" work, Carr read ancient Greek drama, and he became obsessed with Euripides. In contrast to Aeschylus and Sophocles, whom he saw as paragons of the Athenian establishment—"apologists for the current order"—Euripides, he said, "is writing these plays about slaves, and women taken captive in war, and noncitizens, and crazy people, all the outsiders. And he's writing these plays about—well, what if you were an outsider? What would it be like?" The play that really blew Carr away was *The Bacchae,* in which the women of Thebes rebel against the city's Apollonian order (sunshine and rationality) and turn to worshipping Dionysus (night and debauchery). The leader of these women is called Agave, and her son Pentheus is the king of Thebes, and one night, in a Bacchanalian frenzy, the women set upon him, and Agave tears his head off. "And she's holding this bloody head in her hands," Carr told me. "And she kind of looks at him, and she goes, Oh, that's my son. And then she has this moment of recognition, like, Who am I? What have I become? I've been fever-following a god and, um, I don't know who I am anymore. Maybe I've been following the wrong god. What path am I on?"

Carr became obsessed with such moments—the moment when one fever breaks and gives way to a new fever, the moment of self-regard when one calls oneself into question and reverses course. He commissioned a filmmaker, Jessica Yu, to make a documentary about people who experience an "Agave moment"—a terrorist, for instance, and a bank robber, who suddenly saw themselves engaged in action of a kind that they wanted to believe they stood against. In Carr's own life, there was no severed head, no drama worthy of Euripides, but the chapter that was at odds with the way he thinks of himself was, he said, the years he spent as a "crazed businessman"—and after he cashed out he had gone through a long period of not knowing what to do. His theatrical venture, the Market Theatre, belonged to that period. After just two seasons, he shut it down. He had fallen in love with a national park in Africa, which is where we met a few months ago, and he told me this story.

Gorongosa National Park is a wilderness at the southern tip of the Great Rift Valley in central Mozambique, and when Carr showed up there five years ago, it had been all but abandoned to ruin. The park is the size of Rhode Island and was established in 1960 by Portugal, which had dominated Mozambique for nearly five hundred years. For a time, it was one of the top safari parks in Africa: choked with big herds of big game, served by commercial airlines, equipped with a headquarters—Chitengo Camp—that boasted modish accommodations, including a pool, and provided Volkswagen microbuses for exploring the bush. But in 1975 a Marxist liberation movement called FRELIMO drove the Portuguese out of Mozambique, and independence was soon followed by sixteen years of civil war. It was an epoch of appalling national devastation:

1 million Mozambicans killed, 5 million driven from their homes, tens of thousands tortured or maimed, the national infrastructure effectively dismantled, the ground sown with a seeming infinitude of land mines.

Gorongosa District, which includes the park, was the scene of much of the heaviest fighting. Both district and park take their names from Mt. Gorongosa, a 6,000-foot rainforested peak that rises fifteen miles west of the park and that served throughout the civil war as the military and political stronghold of RENAMO, the anti-FRELIMO insurgency, which was sponsored by the white supremacist regimes in South Africa and Rhodesia (now Zimbabwe). Gorongosa was a full-tilt battlefield, a zone of terrifying close-range combat, with tank battles and air raids adding to the maelstrom, as towns and villages were repeatedly overrun by one army, only to be reclaimed by the other, back and forth, year in and year out.

The war was even more unsparing of the wildlife than it was of people, as soldiers from both sides slaughtered game for food and commerce. When the fighting finally ended, the park was left a no man's land. Local trappers desperate for meat, along with organized and heavily armed poaching syndicates, moved into the breach, and the hunt only accelerated, until the fields shimmered with bleaching bones. Between 1972 and 2001, the number of Cape buffalo counted in the park fell from 13,000 to just 15; the wildebeest count fell from 6,400 to 1; hippos went from 3,500 to 44; and instead of 3,300 zebras there were 12. Elephant herds and lion prides, too, were reduced, by 80 to 90 percent. Of hyenas, black and white rhinos, and wild dogs, there were none.

In 2004, Carr said, you could walk or drive all day without seeing any other living thing but some birds. That was when he committed much of his fortune and much of the rest of his working life to resurrecting the park. Now, when we rode out from Chitengo Camp, we routinely saw lions, elephants, any number of species of antelope (oribi, impala, nyala, eland, sable, Lichtenstein's hartebeest, reedbuck, waterbuck, duiker), and sometimes even the rare buffalo. Vervet and green monkeys popped up here and there. Baboons and warthogs were everywhere. In Lake Urema, pods of hippos milled in the shallows, and shoals of giant crocodiles crowded the muddy banks. Once I nearly stepped on a spitting cobra, and later I watched a giant monitor lizard lumbering along the edge of a pond. At night we saw civet cats and water mongoose and listened to the cries of bush babies against the general clamor of bugs and frogs. The bird life was stupendous. Carr, who has never married and has no children, takes a patriarchal pride in every animal he sees in the park. "You give nature half a chance and it's resilient," he said.

Mozambicans generally express surprise on meeting Greg Carr because he likes to dress like a bum and projects none of the grandeur of his wealth. His uniform at Gorongosa is a couple of days' growth of beard, a rumpled short-sleeved shirt or T-shirt, tattered cargo shorts, and Timberland boat shoes with no socks. Carr is happy with the look—it's comfortable and it's disarming. Although he is not in Gorongosa for profit, he considers giving to be a form of entrepreneurship, and he retains a fast belief that private enterprise is the surest instrument for positive change in the world. His idea in Gorongosa is to use his philanthropy to create the conditions and set the example that will attract for-profit ecotourism businesses to the park, insuring its economic self-sufficiency, and benefiting the impoverished rural communities that surround it. He has already spent more than $20 million on revitalizing the park and its environs, and he expects to spend at least as much again before Gorongosa will no longer need him, which is his definition of success.

So far, Carr's team of 120 game scouts has dismantled hundreds of thousands of poachers' snares and gin traps in the park and confiscated nearly a hundred poachers' guns. As a result, more animals have survived and multiplied, and those animals have grown less chary of human presence. But the scouts keep finding more traps and making more arrests, and for every poacher they thwart they have to assume that several more are prospering.

The big animals that the park has lost are not only valuable as tourist attractions but essential to sustaining its ecosystem. If grass is not eaten in sufficient quantities, forest encroaches, and the inevitable dry-season fires that flicker across the savannahs and woodlands rage out of control. There were never fewer than half a dozen—and at times there were more than forty—sizable burns going in the park during the two weeks I spent there in the wicked heat (frequently over 100 degrees at midday) and desiccation of early October. Similarly, in the wetlands, lakes, and rivers that keep the land alive, hippos are needed to churn the muddy bottom and prevent excessive silting; and they are needed, too, to shit in the water to fertilize the rainy-season floods, which begin in December.

So Carr is not just leaving nature to replenish itself. Three years ago, he started bolstering Gorongosa's depleted species with animals donated from parks that can spare them elsewhere in Mozambique and in South Africa. So far he has brought in 180 wildebeest, 139 buffalo, 6 elephants, and 5 hippos, and he figures that he needs at least as many more again of these species before the park has the breeding stock it requires. "We're probably twenty thousand bulk grazers short of what we need to keep the grass down," he said.

Carr is particularly keen to get more zebras, but his chief conservationist says that the subspecies that is endemic to the park can be had only in Zimbabwe, and in the current political crisis it's impossible to get them from there. "We're going to have to get the president of this country to call Mugabe," Carr told me. "That's next year's challenge." And he'd like some rhinos. And he wants predators: hyenas, leopards, maybe some more lions. And he wants Africa's top safari tourism operations to lease concessions—Carr's staff of ecologists, tourism consultants, and engineers have carved Gorongosa into nine huge tracts—and to develop environmentally correct lodges that will generate $30 million a year of business (10 percent of which will go to the park's budget), in addition to park entry fees. And then there is what Carr calls the "greater Gorongosa ecosystem" to attend to—the 10-kilometer-wide buffer zone around the park, where the species in most immediate need of attention is humankind.

In the buffer zone, upward of 30 percent of the population is afflicted with AIDS, and most people subsist on less than a dollar a day, with an average life expectancy of between forty and forty-five years. Here Carr has created hundreds of jobs; he has built two schools and two clinics and a handful of computer centers; he has had wells drilled. He sends a nurse out from his base at Chitengo Camp four days a week to provide basic medical care to nearby villages. He has funded a factory in the regional capital, Vila Gorongosa, where local produce is carefully rendered into fancily packaged dried-fruit snacks. He has sponsored scientific research to develop conservation-minded agricultural practices for the buffer zone, and medical teams to conduct epidemiological studies. He has brought in the U.S. Centers for Disease Control and Prevention to distribute mosquito nets for every one of the 250,000 people in and around the buffer zone—the first large-scale attempt to provide universal malaria protection to a population.

In 2007 Carr signed an agreement with the government of Mozambique to oversee the park and to run community and conservation projects in its buffer zone for the next twenty years. He seeks official advice and consent for each move he makes in Gorongosa, but he recognizes that by giving him charge of the park, the government is basically saying that it recognizes conservation as a necessity that it cannot afford. "I'm not faulting them for this," Carr told me, but he felt the attitude was: "Some dude wants to work on that, let him, because the truth of the matter is, educated Mozambicans, for the most part, want to be in the capital city. The most educated Mozambicans, for the most part, are not clamoring to run a national park."

Carr likes to say that his project is strictly apolitical, but Gorongosa District, where RENAMO sympathies linger, has been largely left behind in Mozambique's postwar recovery, and he understands that his value to the government is not simply as a conservationist. "It wants to show RENAMO it's doing a good thing here," he told me. "Absolutely. And the president's been here three times. He dedicated the school. He dedicated the health clinic. He got his own name on that. I mean, the political aspect of this is key, and the economic aspect of it is key." In fact, Carr told me, "I don't think what I do is that different from what the mayor of any small town faces every single day. You're just juggling. OK, we need better schools, we need a better police department—oh, how's the revenue doing? How are the roads? How's everybody?"

So Carr was excited, a few days after I arrived in early October, that Mozambique's minister of the interior was coming to visit the park with an eye to becoming a minority partner in one of the tourist concessions. "We gotta knock his socks off," he told Rob Janisch, a South African who runs a tented camp, Explore Gorongosa, which is the first private safari operation to come into the park under Carr's scheme. Janisch suggested that the minister be taken to a watering hole called Paradise Pond. Go and watch elephants drinking, he said, "because (a) it cools you off, and (b) it just is cool."

But Carr didn't want to take any chances: the minister didn't have much time, and what if there were no elephants? "Last thing we want is he drives around and says, Oh, I guess they don't have animals yet in Gorongosa." After a minute, he said, "OK, he lands at Chitengo, we do all our greetings, we put him in the heli, we give him the lake tour, where he will absolutely see tons of stuff, we land near the crossing, because there's a hippo pod there, and because that's where—that's a camp he wants, what we call Dingue Dingue. So we need to show him that and say, This is yours."

An hour later, Carr was clean-shaven, in a gingham dress shirt, fresh chinos, and black loafers. At the approach of the minister's plane, a Land Cruiser was dispatched to run up and down the Chitengo airstrip, broadcasting the hunting calls of hyenas to scare off strolling baboons, warthogs, and antelope. The minister, a young-looking man, was dressed in the same fashion as Carr, with a Leatherman bush knife in a holster on his belt. As he and Carr flew over the park in the chopper, he exclaimed over the landscape—"These palms and acacias, so beautiful"—and he told Carr, "You have really saved lives here." Carr said it was his honor, and the work had given his life meaning. The minister said it was a greater honor for Mozambique "that you leave all the comforts and come here." Carr told him, "I am extremely comfortable sleeping in the bush." The chopper banked over Lake Urema, where pelicans, fish eagles, and yellow-billed storks cruised over hippos and crocs, then out over the grasslands, where big herds of antelope galloped below. Elephants appeared as if on cue. "Paradise," the minister said, and Carr, sitting behind him, nodded avidly, cracked a big sideways grin, and gave a double thumbs-up.

Greg Carr had never been involved in conservation work before he came to Gorongosa. He was not an ecologist, or a zoologist, or even much of an outdoorsman. Nor was he an old Africa hand, much less an economist. His only knowledge of the tourism business was as a customer, and he spoke no Portuguese. For that matter, he had never had a conscious interest in making money before he got around to it. "I was taught at an early age to look for deeper meaning in life," he said.

He was the seventh of seven children, the son of a practicing Mormon and a practicing physician. "The Golden Rule and the Beatitudes and Matthew—great stuff," he said. "And then you'd have Dad pitching in with a little rationalism." He never felt compelled to choose between faith and reason, since both had such obvious appeal and such obvious limitations. In politics, too, he considered himself a centrist. The way he put it was: "Conservatives want to make a good person, and liberals want to make a good society. Which of those two do you not want to do?" In college he thought he might major in biological anthropology—Darwinism, paleontology, primatology—even while he saved up, as a freshman at Utah State University in Logan, to spend the next two years in Japan as a Mormon missionary.

Carr allows that he made some converts, but as he tells it, what really excited him in Japan was learning about Zen Buddhism. "I was very much a questioner," he said. And his interests were fluid. When he returned from Japan, he read poetry and majored in history and had no idea what to do with his life. He applied to graduate school, leaving his options open: Asian studies at Stanford, the Kennedy School of Government at Harvard, international relations at Yale, or public policy at Princeton. He got in everywhere. The Kennedy School brochure had some stirring lines from JFK about making a better world, so he went there. In his second year at Harvard, he applied to and was accepted in the Ph.D. program in linguistics—"because I wanted to know, Why do we speak? What is a human?"

That was how his mind worked, with an earnest bent toward the forever debatable. So Carr was as surprised as anyone else during the winter of that year—1985–86—when he developed a sudden and all-consuming conviction that he should be making and selling digital voice-mail systems to telephone companies. "It's a funny thing," he told me. "I just had this idea. I was in my dorm, and I was kind of just looking at my phone, and I was thinking about how little it did, really."

Carr had never studied computer science or telecommunications, let alone business or marketing, but he met an engineer named Scott Jones at MIT, and they started a company called Boston Technology. They maxed out their credit cards buying gadgetry, set up shop in Carr's room, and by summer had a prototype of their product up and running. Carr was twenty-six, Jones twenty-five, and three years later they were tycoons.

In 1992 Carr brought in a CEO and became chairman of the board. He had enough money to support forever the life of boyish wonder that had made him the money to begin with—and he still had to figure out what to do with that life. Only now that meant figuring out what to do with the money. He audited classes at Harvard; for three years in a row, he attended a summer course on Homer, Dante, and Joyce's *Ulysses* taught by a professor called Theoharis Theoharis. He read omnivorously and restlessly; he watched movies; he moved into a grand apartment atop the Charles Hotel in Cambridge—room service suited him—and he suffered from insomnia. He spent most of his time with a friend he met in one of Theoharis's classes, Larry Hardesty, who had moved up to Boston from New Haven as the keyboard player for a rock band that took its name, the Young Man Carbuncular, from a line in *The Waste Land*. Carr and Hardesty started a film production company, Bowerbird Productions, that never produced a film but tried for a time to work with the former navy secretary (now senator) Jim Webb on an adaptation of his Vietnam novel *Fields of Fire*. More successfully, Carr served a stint as publisher (with Hardesty as the managing editor) of a magazine of ideas, the *Boston Book Review*.

It seemed clear to Hardesty that Carr was gratified by his wealth as a measure of achievement, but, he said, "I think he was trying to find something to devote his energy to that would be more satisfying on more levels than success in business had been." Still, in 1996 Carr went back into business as the chairman of the pioneering global Internet service provider Prodigy. Then, after just two years, he sold his share of the company to the Mexican media mogul Carlos Slim and quit all his other for-profit ventures. That was when he decided to commit himself to a life of philanthropy. In 1998 he gave Harvard $18 million to establish the Carr Center for Human Rights Policy at the Kennedy School. In 2000 he cofounded the Museum of Idaho, in Idaho Falls, and created the Market Theatre Company, and the next year he bought the former compound of the Aryan Nations in northern Idaho, which had been confiscated from the Nazi group by court order, and donated the land to a local college as a peace park.

Then a friend introduced him to the Mozambican ambassador to the United Nations, and the ambassador invited Carr to come and help his country. At the same time, Carr began spending a lot of time online, reading compulsively about threats to Earth's biodiversity. He was alarmed by the prospect—widely considered likely among environmental scientists—that as many as a quarter of the species now alive will cease to exist by the end of this century. The problem was generally described as a conflict between humanity and the rest of nature. As usual, Carr didn't see how you could win by taking sides; it made more sense to try to remove the conflict, make people and the ecosystems they lived in serve each other better. Thinking like that, he flew into Mozambique, hired a helicopter, and toured the national parks.

He told me that when he came to Gorongosa he knew he'd found his place. Until then, he said, "I'd cast about a lot—yeah, I think it was a hard time—because I wasn't completely fulfilled by any means, and didn't know what was next. I wanted it to be all-engrossing, challenging, and I didn't want to be the philanthropist that writes a check and comes back next year and says, What did you do with my money?" He said, "I mean, I have to actually be pushing against something." He decided to push against extinction.

In 1968 the Department of Fauna of the Portuguese government hired a young South African ecologist named Ken Tinley to study wildlife conservation in Gorongosa National Park. In establishing the park, the colonial authorities had driven out the African villagers who had lived there for as long as anyone could remember, and they were prepared to displace more people in the name of preserving the wild land. Tinley had been asked by his Portuguese employers to identify the full parameters of the Gorongosa ecosystem with the aim of redrawing the park's boundaries accordingly. He picked up his official Land Rover in the Mozambican capital, Lourenço Marques (now Maputo), and with his wife, Lynne, an artist, and their infant son set out for Chitengo. The drive, which can now be done in about eight hours, took them three days, and once they settled into their cottage, they stayed for six years, until the liberation war drove them out. Tinley then spent several years writing up his research in a Ph.D. thesis, which runs to nearly five hundred pages and remains one of the most comprehensive analyses of an African ecosystem ever produced.

Tinley sought, to the degree that the technology of the time allowed, to record every aspect of the park's natural order: the flora and fauna, of course, but also the geology, the hydrology, the wind patterns, and the intricate complexity that made all the elements of Gorongosa—every living thing and every physical property—inextricably part of the larger system. What emerges from his technical scientific prose and no end of charts, maps, drawings, and diagrams is an image of the park as one big organism, and of Tinley as its anatomist.

Mia Couto, Mozambique's preeminent novelist and also one of its leading ecologists, told me that Tinley was "a genius," and that view is pretty universal among people with sufficient scientific knowledge to plow through his thesis. For the rest of us, Lynne Tinley wrote a memoir of the family's life in Gorongosa and of the life of the park, a book called *Drawn from the Plains,* which distills the essence of her husband's findings with the artful and honest immediacy of the best nature writing. Here, for instance, is how she describes the annual rainy-season scene when Lake Urema overflows and submerges hundreds of square miles of the surrounding grasslands:

> Thousands of water birds arrive on the plains during the flooding to take advantage of the rich shallow waters. Egret, ibis, spoonbills, herons and divers nest in the tall *Acacia albida* (called winterthorns because they bear leaves in winter and shed them in summer) that hang over the water. The ancient, sleazy-looking Marabou stork in his funereal dress waits in the shallows under the trees for fledglings that fall out of the twig platforms above or for fish dropped by the parents during feeding. Crocodile also gather under the nesting trees to pick up the dropped food and to feed on fish attracted to the water enriched by birds' dung.

Lake Urema's yearly flooding was the engine that sustained the teeming animal life of Gorongosa, and even before he arrived in the park Ken Tinley had surmised from maps that the distant, cloud-shrouded massif of Mt. Gorongosa was, in turn, the engine of those floods. His research on the mountain confirmed this hypothesis: the mountain was the main water catchment for the park, the source of roughly half the water that flows into it. The primordial forests on the mountain slopes received two and a half times as much rainfall each year as the valley floor and passed it down through a network of streams and rivers to the lake, providing water essential to human communities on the way. It was obvious to Tinley that the mountain belonged in the park, and as he studied the opposite side of the park, he concluded that the ancient game-migration routes and wetlands that ran east from the park to the mangrove swamps lining the mouth of the Zambezi River delta where it opens into the Indian Ocean were the final piece to complete the Gorongosa system. In his thesis, which he finished in 1977, Tinley laid out a scheme for such an expansion of the park, indicating, as his wife summarizes it, "how the whole area could be planned, in the simplest possible manner, for tourist viewing, for research and education, for wilderness areas, and for the cropping of game as a source of protein for the surrounding tribal peoples."

By then, of course, Gorongosa was a civil-war battleground and the mountain was RENAMO's. But Tinley's dream of a park that ran "from mountain to mangroves" held fast in the lore of the place, and when Greg Carr came along it quickly captured his fancy.

Like Tinley, Carr is what his former Kennedy School professor Herman (Dutch) Leonard calls a "system thinker." Leonard, who also teaches at the Harvard Business School, had worked with Carr on his voice-mail venture, and he told me, "Greg realized that what Boston Technology was trying to do was to become a component of a much larger system"—the national telephone network—"and that if the company didn't think about the rest of the system, if it just designed its piece, it was unlikely to be able to find space in that larger system." In other words, Leonard said, "to him it wasn't just a product, and you go and sell this product. It was an intervention in a pretty big, complicated game. And he was able to simultaneously be focused on the individual tree, if you will, and also keep his eye on the forest and think about the nature of how we operate so that we're going to fit in a natural way in this forest." And speaking of trees and forests, Leonard said, "That's not so different from what he's trying to do in Gorongosa."

Indeed, no sooner had Carr grasped the significance of Mt. Gorongosa than he discovered that the rainforest that covered the mountain—simultaneously capturing the water that fell from the clouds and, through evaporation, replenishing it—was being destroyed, piecemeal but steadily, by local practitioners of slash-and-burn agriculture. People had been living in that forest forever: a few thousand people who speak a distinct language, chi-Gorongosi, and adhere to a distinct spiritual order, which holds the mountain to be sacred and its forests to be inhabited by the spirits of their ancestors, who look out for them. But during the civil war other people began arriving on the mountain, followers of RENAMO and displaced people seeking refuge. After the war, more people arrived from the lowlands at the base of the mountain and began clearing land to grow crops—chopping down old-growth hardwoods, burning them for charcoal, and planting potatoes, or corn, or sometimes marijuana between the stumps. After a few crop cycles, the soil—the accretion of perhaps 10,000 years of compost and sedimentation—was depleted, or after a few rainy seasons it would simply wash away, and these farmers would clear a new patch of forest. The trees they cut did not grow back.

Carr's scientists told him that the mountain's value as a water catchment was in peril: if the killing of trees could not be stopped, the mountain could be lost in just a few years. "To save the park," Carr said, "we have to save the mountain." But he had no authority over the mountain, which stood well outside the buffer zone. It seemed to him that the best solution was Ken Tinley's proposal: to have the mountain, or at least the vital, forested part of it above 700 meters, designated a satellite of the park. After all, the rainforest and the high alpine meadows up there were unique in southern Africa; it was the only place in the region where one could find a certain bird—the green-headed oriole—and no doubt countless species of plants. It was a landscape ripe for ecotourism, dotted with clear pools, riven by deep canyons, flanked by waterfalls. Carr and his team worked their government contacts in Maputo, lobbying hard to have the mountaintop added to the park, and at the same time they sought the permission of the traditional leaders of the mountain communities to start conservation programs and lead tours in their jurisdictions.

Most of Gorongosa Park's mountain water comes from two administrative zones, Canda and Sadjunjira. Canda is the turf of a hereditary chief, known by his Portuguese title of *régulo*, and, as Carr paid court to him, he found he could negotiate with the régulo of Canda: if Carr's people hired Canda people to work in Canda, the régulo didn't mind them running a reforestation program on his side of the mountain, and if they brought tourists to see him, to pay a small visitor's fee, and to conduct a ceremony of respect for the ancestors, they could take the tourists there, too. In Sadjunjira, on the other hand, the traditional keeper of the mountain was a kind of shaman known as the samatenje, and when Carr paid him a visit everything went wrong.

Nobody had told the samatenje of Sadjunjira that Carr would be arriving by helicopter, a vehicle known to the mountain people chiefly as an instrument of war, and the noise of the thing as it descended near his compound outraged his entourage. What's more, the helicopter was cherry red, and in the Gorongosi culture red is the color of violence and conflict, so it is strictly forbidden to appear before the samatenje with any trace of red on one's person. Then the helicopter touched down on the wrong side of a stream that demarcated the sacred ground of the samatenje's domain. And as the helicopter's doors opened, a truly astonishing thing happened: a pale, snakelike lizard, of a sort that nobody had seen before, popped out of a hole in the ground right beside it — a ghastly omen. The only thing that could be worse was for someone to touch the creature, and so that's what happened: a herpetologically inclined member of Carr's party, delighted by what he recognized as his first ever sighting of a blind skink, snatched it up to have a closer look. All the samatenje could say was: Get out of here — go!

"Look," Carr said, when I asked him about that visit. "Everybody is always talking about wanting to save a rainforest. I'm actually trying to do it — trying to see what it takes. And I'm finding out."

Carr spends as much as half of each year in Mozambique these days and the rest of his time in America, where he lives in baronial luxury. He loves his big homes, but in Gorongosa, at Chitengo, he loves an equal and opposite extreme of sparse accommodation. For his first few years there, he slept mostly in a tent and sometimes in the back of a truck. He didn't even bother with mosquito netting, an adventure that resulted in more than one case of malaria. Since Chitengo was rehabilitated two years ago, Carr has taken to staying in one of the tourist bungalows, but when those beds are fully booked — a situation that pleases him — he crashes on a mattress on an office floor. On one such night, as he headed off to sleep, he said, "You've seen my houses. One of the things I've really gotten out of being here is discovering that this is all I need." I pointed out that he had a million acres of game park, a fleet of trucks, a staff of hundreds, and a hotel and restaurant at his command. "Yeah," he said. "But seriously—it's like an eight-year-old boy's dream come true."

The one luxury Carr does allow himself in Gorongosa is to keep a chartered helicopter (a blue one now) on standby—another eight-year-old's dream, but a supremely practical one, as I understood when Carr took me in it to see the mountain. The Chitengo restaurant was full of tourists that morning, and all of them had been happy with the game drives they'd been on. "We've got a tourism product, we're ready for safaris," Carr told me before takeoff. "I can see that happening now. So I can turn my mind to the next big thing—like expanding the park." Two years ago, just before Carr signed his twenty-year agreement with the government, the mountaintop above 700 meters had been added to the park's buffer zone. Carr was still eager to have it brought fully into the park. But RENAMO remained a force on the mountain, and nobody in Maputo wanted to rile up the old guerrillas by appearing to take their turf out from under them.

The helicopter touched down briefly in Vila Gorongosa to pick up the head of Carr's forestry team, Regina Cruz. Our next stop was on the mountain, in a patch of tall grass beside the forested banks of a river, where we continued on foot, climbing through reeds and high grass into the sudden damp, dark cool of forest until we came out on a broad slab of rock at the base of a 300-foot waterfall called Murombodze Falls. It was months since the last heavy rains had replenished the land, and even so, the falls were spectacular—shelf upon shelf of interweaving cascades roaring into a deep, clear pool. "This is a key tourist point," Carr announced. "There are waterfalls all over this mountain. A hiker could spend three or four days hut-hopping here." As with the park, Carr envisioned such tourism paying for the conservation of the mountain. "The thing I never stop marveling over in Africa is the economic scale," he said. "While on Wall Street a couple-billion-dollar merger goes down and no one really bats an eye, you can make a fifty-thousand-dollar business on this mountain and save an entire ecosystem."

In fact, Carr was spending $160,000 a year for Cruz to run sixteen nurseries on the mountain—each of which produces, from locally gathered seed, 1,500 saplings a year, which she and her team plant in denuded areas—and the ecosystem was far from saved. Near the foot of the falls, Cruz showed us one of her nurseries, a simple bamboo shed, where her seedlings were barely six inches tall, and right out in front of it we encountered a lively older woman whom Cruz recognized as an avid illegal tree cutter. "She's been warned, but keeps at it," Cruz said, and Carr said, "So bring the police." Cruz said that the previous week the police had come to arrest two illegal cutters, but she would rather see conservation-education programs on the mountain than cops. "We have to teach the people, because they grow up with nature and they do not understand that one day the trees can disappear," she said. "They say to me, 'The trees come from God.' I say, 'OK, so if God put it and you take it we all have to put God's trees back.' They laugh, but maybe they start to understand."

The question of whether to use persuasion or coercion to save the mountain divides Carr's staff sharply. Franziska Steinbruch, the park's manager of science, told me that she had been studying the deforestation of the Gorongosa water catchment in satellite photographs taken annually since 1972 and that the damage had been increasing exponentially. Her attitude was: the law is the law, no excuses, and the government has to uphold it to defend the rainforest. "If that's not coming as a force from the top, the forest will be gone in two years maximum," she said, and she added, "I'm actually losing hope."

Carr himself takes a hard line when he talks about the mountain, but, as we returned from the falls to the helicopter, we found a young man waiting for us, whom Cruz identified as another illegal tree cutter. He wore a ragged shirt decorated with images of subway stations in Brooklyn, and he bowed his head coyly as Carr lectured him about leaving the trees alone. Then the young man asked for money to buy salt. Carr's anger evaporated. "It's all about poverty," he said, and he told Cruz, "The next time you hire people to work in your nurseries, you should hire this guy—that's a twofer, one less tree cutter and one more tree planter."

From the helicopter, it was easy to see what the worry was. Terrain that appeared from a distance, or even from a slight angle, to be a dense tangle of unbroken rainforest looked from directly overhead like moth-eaten fabric, where small and medium-sized clearings opened up to reveal a field of charred tree stumps interspersed with crops or, at least as often, simply abandoned.

"We need Regina to have a hundred nurseries," Carr had said when we took off. "Because we can do it. We've proved the concept." But as we buzzed over smoldering fires in fresh cuts, and fields of rocks laid bare by the washing away of topsoil, he said, "Another hundred nurseries—that's not enough." To be sure, large swaths of the forest stood solid and inviolate, but these areas only made the despoliation elsewhere more obvious. The damage ran all the way to the cloud-shrouded summit, a wild place of craggy rock pillars and low, tangled vegetation, and as we crossed from Canda to the Sadjunjira side the trashing of the forest only got worse.

On several nights at Chitengo, we had watched Ken Burns's recent documentary miniseries on the history of America's national parks. After the screening of one episode, a visiting safari guide said, "It's interesting seeing those American national parks—how they were set aside in the name of the many for the many. Here it's really the national land being set aside in the name of the people for the very few, to be honest. The masses here aren't going to come see these animals, because really they're afraid of them, and they'd rather have the land to farm." But, the guide said, the land would be useless without the mountain to feed it. So, he said of the park, "why save all this land for a couple of thousand whities if it's just going to disappear? That's why, if you ask Greg, he'll tell you the mountain is the first priority."

Carr rarely expresses discouragement directly; instead, he betrays the feeling by proposing ever more extravagant solutions to the dilemma that's bothering him. By the end of our helicopter survey of the mountain, he had begun talking about planting "enormous avocado orchards" at the foot of the mountain to lure the tree cutters down from their backbreaking assault on the forest. But Cruz didn't see any reason to believe that people wanted to abandon the mountain.

"They say, 'No—it's the white people coming to take our land,'" she said. "'Where will we go? Where will we live? What will we eat?'"

"We'll give them jobs," Carr said. "We can give a lot of them jobs."

Cruz remained unconvinced. She had no difficulty expressing her discouragement. When we landed back in Vila Gorongosa to drop her off, she told Carr, "I'm sorry. I'm very sorry. Really, I never realized that the situation was as bad as that. It's really worse this year."

"We can fight it," Carr told her.

"Yeah," she said. "Maybe."

The mountain was a problem. "It has to be a national park above seven hundred meters," Carr told me. "We have to get to the point where you don't live up there." He said as much on nearly every day of my stay. And nearly every day he flew one or another important visitor over the mountain to make his case. Carr was in the process of trying to sign up the first three safari-lodge operators to run concessions in the park, and he wanted them to understand, as well, the potential value that the mountain could add to Gorongosa tourism. He was encouraged that the minister of the interior had described the harm to the mountain as "an environmental catastrophe." But Carr placed his highest hopes for the mountain in a South African named Dave Law, whose company, Barra Resorts, runs a string of Indian Ocean lodges in Mozambique. Carr wanted Law to build a lodge at Murombodze Falls, and when he took Law to see the spot, he invited a young woman, a Portuguese graduate student who was doing research in the park, to ride along in the helicopter and to visit the swimming hole beneath the spilling water. "A girl in a bikini—who doesn't like that?" he said. Law loved the falls; he loved the whole mountain. Over drinks on the evening after his visit, he spoke of running horseback trips over the summit, of stringing zip lines through the rainforest canopy, of building a spa that looked out from his lodge onto the falls.

It sounded nice. But while Law and Carr were up on the mountain, I had paid a visit to Eugénio Almeida, the régulo of Canda, and he had told me that although he wanted more tourists to visit the waterfall, he didn't want anybody going there, or even taking pictures of it, without first coming to see him at his compound to pay a fee, have a certificate stamped, and conduct a ceremony in honor of a spirit whom he called the Owner. Never mind that the régulo's compound lay nearly two hours' drive away from the falls: the same went for the rest of the mountain, he said. "No one should go on the mountain without my consent—I, the régulo," he said. He was particularly concerned about anyone tampering with the waterfall. "It is a holy place," he said. "Long ago, no one lived very close to that place. In the evening, the Owner would move around the place. Even here we would hear him walking. He made a lot of noise. And he climbed from the waterfall high up in the mountain. Now many people have built their houses around the waterfall. By settling their houses around the waterfall, these people closed the way the Owner would use. He doesn't have any path to move from the waterfall into the mountain."

The régulo, a slender man in his sixties, wore an orange polo shirt and shimmering green trousers, and periodically he tapped a bit of snuff from a plastic vial into his palm, then raised it carefully to his nose and snorted it. "Long ago, things weren't like this chaos we experience these days," he said. The régulo wanted his office to be respected. The problem, he said, was that there were too many people trying to lead the community. "Everyone leads and no one listens to the other," he told me. He blamed the war. If there was lawlessness and destruction on the mountain, it was because people had been addled by what he called "the gun effects." I took that to mean shell shock. "These people inhaled a lot of gunpowder from guns and bombs during the war," he said. "This gunpowder affected people's brains and mixed them up. Good sense was replaced by war-related ideas."

He told me that Greg Carr was his friend—"We are like brothers"—and that the people of Canda were dependent on the Carr Foundation for jobs. In fact, he said, the thirty-six people whom Regina Cruz employed on her forestry team were the only people in all of Canda, an area of some four hundred square miles, who had regular salaried jobs. The rest lived by the hoe; they were subsistence farmers. "Here, in my area, people suffer a lot," he said. So they were also grateful to Carr for building them a school. But none of that made them ready to accept the upper mountain's being turned into a park.

In colonial times, the régulo reminded me, many people had lived in the area that is now the park, and when it became the park the Portuguese burned their homes and scattered them. These were the associations that the word "park" carried, he said. In fact, a number of people had resettled in the park during and after the civil war, and Carr was now working on getting them out. He has budgeted $1 million to resettle some seventy families from the shores of Lake Urema, and he said that there would be no coercion, only enticement: better homes, better social services, better economic opportunities. The régulo saw it differently. He thought it was bad enough that locals could no longer hunt for meat in the buffer zone as everyone used to do in the past. "Now you claim all that area to be yours," he told me. By "you" he meant the *muzungu*—the white man. He explained, "*Muzungu* means tomorrow he cheats us and then he takes our property."

The régulo said that park workers had held a meeting at his house "to persuade people to leave the top of the mountain." When the people up the mountain heard that, he said, they came and protested in his yard. Three times they came—two hundred people, then ninety people, then seventy-five people. "All of them refused to leave those areas on top of the mountain where they live," he said. These "complainers," as he called them, had even accused the régulo personally of selling out to the park, and he didn't like that. He told me that Mozambique's land law was clear, and everyone knew that it said people cannot be removed from where they live—they have to agree to move voluntarily. "This was the only disagreement we have ever had with Mr. Greg Carr," he said, and he added, "We don't want foreigners to control our area. We want to control our land ourselves." He took credit for the recent arrests of illegal tree cutters. "The two men were beaten, then jailed," the régulo told me. After all, he said, "we know that this mountain and its trees attract rain. You see, rain is becoming scarce now, and later no rain will fall at all."

Carr, however, said it was untrue that anybody from his project had told people on the mountain that they would have to move: that was not his policy, and it never would be. In fact, he said, he had insisted, in his agreement with the government, that "traditional people" be allowed to dwell in the park, and he told me that that would apply to the mountain, too, if it was given park status. He placed the blame for the régulo's troubles on agents provocateurs from RENAMO, who had gone around accusing him of being a rich American *muzungu* land grabber in order to make some political hay at a time of municipal elections in the Gorongosa District two years ago. I heard the same thing from people in Maputo who had nothing to do with Carr's project, and Carr provided me with a local newspaper clipping describing the smear campaign.

Carr was much more interested to hear that the régulo had said he wanted to see more tourists on the mountain, and to control deforestation. The logistical details didn't trouble him. "Basically, what you do is, you introduce Dave Law to Eugénio the Chief, and they sort it out," Carr said. "Eugénio wants jobs and he wants tour-

ism money, and Dave Law wants the same thing." Frequently, when
Carr sought to explain his mission, he would recite facts and figures about the global threat to biodiversity and about the glorious
past of the Gorongosa ecosystem. But when he talked about the
how, rather than the why, he could sound more like a ward heeler.
He said, "Economics will solve the problem faster than any public
policy." He said, "Every private-sector entrepreneur has this incredible bias toward letting enlightened self-interest pull things forward, and I fully believe that will happen." And he said, "It's all
about jobs. It's all about jobs. It's all about jobs."

Carr's disarming faith in the power of his own good intentions to
render confrontation insignificant has served him well in Mozambique. Following his initial catastrophic red-helicopter-and-blind-
skink encounter with the samatenje of Sadjunjira, Carr issued the
shaman an invitation to visit Chitengo, gave him a grand welcome
and free use of the bar, and sent him home a buddy. In fact, he
posted a photograph of the two of them together as his head shot
on his Facebook page. The samatenje, however, had never come
around to cooperating with the Gorongosa team, so I went to see
him, too.

Unlike the régulo of Canda, the samatenje of Sadjunjira lived
well up the mountain, far from the last paved road, a four-and-a-
half-hour journey from Chitengo. Along the way I learned from
Inácio Júlio Tomás, a member of the Gorongosa forestry crew, who
had agreed to drive and translate for me, that the samatenje I
thought I was going to meet — the Facebook samatenje — had died
in the past year, and that I was going to meet the new samatenje,
his brother. Tomás wanted me to comprehend the incredible powers that this new samatenje and his other brother, the witch doctor
of Sadjunjira, were believed to possess. There was even a local legend, he said, that the brothers had killed the Facebook samatenje
by working some fratricidal magic on him. It sounded far-fetched
to me, and Tomás allowed that the dead samatenje had spent a
lot of time drinking at the tavern in Vila Gorongosa and fooling
around with the ladies there, which was a more routine cause of
death in the area. "But people here don't believe in AIDS," he told
me, and Tomás himself was not prepared to discount the powers of
the two surviving brothers: their father, he said, had had awesome
powers.

Not long ago, Tomás had had recourse to a witch doctor on account of a lame foot. He still wore the dark green fetish thread the healer had fitted his ankle with as he cast his spells, and Tomás had no complaints about the treatment: he had been up and walking again in no time. It seemed everyone had such stories, even the science-minded. Carr told me that every year at the end of the flood season, in mid-April, Gorongosa Park officially reopens, and all the local shamans convene under a hallowed "miracle tree" at Chitengo to call on the park's lions to show their favor to the park. Lo and behold, after one such ceremony Carr saw a lion stroll by the Chitengo restaurant—an unheard-of sighting in the fenced compound—and he found that five more lions had gathered by the camp gate.

The day before my trip to Sadjunjira, I had made another long drive to visit a network of underground bat caves that were known to be guarded by spirits, one of whom took the form of a leopard. Shortly before we reached our destination, my guide had stopped so that we could stretch our legs and eat a sandwich. It was a stupefyingly hot, still day, but as we stood with our lunch by the tailgate of the car, the trees around us began to rattle in a burst of extreme wind that gathered suddenly into a slender twister, blackened with dust, and ripped between us, snatching a sandwich wrapper, which sailed a good seventy feet into the air and was gone forever. Later, in the caves, I stood on a ledge overlooking a deep black pool of water and a snake rose to the surface—some kind of cobra, at least five feet long—and stared straight at me until I turned away and it sank from sight. My guide had no doubt that both twister and snake were visitations. But as Tomás regaled me with similar stories of the powers of the samatenje and his brother—the brother, he said, was a very good doctor, who could give you medicine that would make you so rich that "you'll go to America and buy a helicopter"—he mentioned in passing that the samatenje had nevertheless lost two children the year before to cholera, by modern medical standards one of the most preventable, and treatable, maladies.

After leaving the last graded road on the way to Sadjunjira, we spent the next hour climbing the mountain in four-wheel drive on a road that was properly a footpath, and a badly washed-out one, too. On either side, large patches of the mountain were stripped bare of trees, and ragged fringes of flame licked up the grassy slopes. Then for long stretches there was no grass to burn: the top-soil was entirely gone, and all that remained was a waste of boulder fields. The last few miles of our approach was made on foot, and for the final hundred yards we had to remove our shoes and socks, as we had entered hallowed ground.

Tomás had been anxious that we were arriving around noon, for fear that we might not find the samatenje at home or sober so late in the day. But he was both—a slight, bearded man, clad in the ragged remains of several denim shirts and matching pants. He squatted at the sight of us, cupped his hands, and clapped them rhythmically—*pock, pock, pock, pock, pock, pock, pock*—finishing with a half-beat flourish. This was a ritual that he repeated frequently throughout our visit, and which we returned in kind each time. He gave us straw mats to sit on in the shade of a vast, fruit-laden mango tree and left us there for well over an hour before he commenced our audience. By then he had assembled an entourage of local elders to join us—because, Tomás explained, such a leader cannot meet alone with a white man lest he later be accused of having betrayed the interests of his community.

Throughout our visit, fires crackled through the surrounding bush, and fine ribbons of sooty ash drifted down around us. A steep flank of the mountain rose behind the samatenje, bare and burning, and I asked him if it had been forest when he was a boy. He said, "Now it's worse, because people are destroying and cutting the trees." The ancestors felt the same way, he said: "They feel that something's going wrong. They're warning the people, but the people don't take care."

So would it be a good idea to plant new trees?

"No," the Samatenje said.

Why?

"Tradition."

What about jobs?

"No."

So no more trees?

"The trees will grow themselves."

What if the police come and arrest tree cutters?

"No problem."

What if tourists come?

"No."

That was what the samatenje had to say. His power, it seemed, lay entirely in refusal. There was more clapping: *pock, pock, pock, pock*. Then we left. As soon as we were outside the shaman's sanctuary and had put our shoes back on, Tomás, who had spoken so admiringly of traditional ways on the journey in, erupted in a tirade against the samatenje's stonewalling of the modern world: his people needed trees, his people needed jobs—to deprive his people of trees and jobs was murder! "They're killing a lot of people," Tomás said.

Back at Chitengo, I found Carr drinking gin and tonics with a new group of safari operators. "I don't think I would ever sort out the micropolitics—or that it really needs it," he told me when I described my excursion, and he said, "I see it more as just a we-give-them-opportunity, we-give-them-jobs, we-may-end-up-saving-Canda-and-losing-Sadjunjira sort of thing." It was the closest I ever heard him come to resignation.

The novelist Mia Couto told me a story when we met in Maputo. During Mozambique's first national elections in 1994, he attended a campaign rally in a small village, where a politician from the city gave a speech. The politician said, "I'm here to save you, and we will bring hospitals, schools"—the usual boilerplate. When he got to the end of it, a villager stood up and said, "We are very happy, very touched, because you came from so far away to save us, and that reminds me of the story of the monkey and the fish." The villager didn't say anything more, and Couto realized that he and the politician were the only people there who didn't know the story. Finally, the politician confessed his ignorance, so the man told him the story, and it went like this: A monkey was walking along a river and saw a fish in it. The monkey said, Look, that animal is under water, he'll drown, I'll save him. He snatched up the fish, and in his hand the fish started to struggle. And the monkey said, Look how happy he is. Of course, the fish died, and the monkey said, Oh, what a pity, if I had only come sooner I would have saved this guy.

"Anyway," Couto said, "this is the traditional point of view. But you cannot just say, This is wonderful and don't touch it. Because it is being touched. You can't avoid bringing modernity. It's happening."

He was right that the story took the fish's point of view. But, I wondered, how could it be told to make the monkey look good? I got my answer the next day, when I met Terezinha da Silva, who runs a women's rights organization in Maputo. Nine years ago, da Silva was a fellow at the Carr Center for Human Rights Policy at Harvard, and later, when Carr told her that he was going to invest in Gorongosa Park, she said, she urged him to please be sensitive to the rights of the peasants. But this past summer she went to Gorongosa District to lead seminars, and she had to call into question all her assumptions about the sanctity of traditional culture. "I was so shocked and depressed with what I heard during the course," she told me. Specifically, she was appalled to hear about a practice that was called "early marriage," but which she properly called "forced unions": impoverished parents in remote rural areas would give their five- or six- or seven-year-old daughters to middle-aged men in exchange for a monthly stipend. She said that she was told of schools that had no girls at all, because they were all married. Mozambique had good laws against such things, but they were often not enforced, out of respect for culture and tradition. And the thing was, da Silva said, she later learned that in the past in these same areas, the cultural tradition was for women not to marry until eighteen.

The traditional position—the fish's position—was: don't save us from ourselves. But what if the tradition itself was corrupt, or if the culture had already been lost? Carlos Pereira, Carr's director of conservation in the park, said of Gorongosa District, "This is a terrible place psychologically. To keep alive, in the morning they had to help RENAMO, in the afternoon they had to help FRELIMO. That was every day. Now imagine this person that had to collaborate with one side and the other side. What does he do in life? That's what he tries to do in life. He says yes on this side, but says yes on that side. He goes in the middle, talks a lot, says nothing. You can never be entirely loyal, you never entirely trust."

"So what are my choices?" Greg Carr said. "Absolutely do nothing, never go anywhere near the mountain in twenty years? OK, fine, what do you have then? A bare, stripped mountain washed away. We lose half of the perennial water system of the park. But I respected them and didn't go there, because there's old RENAMO-FRELIMO wounds and old colonial wounds, so I didn't go there.

OK, so that's Plan A. That's just really bold on my part. What's Plan B? Go there and start talking to them. Gosh, can we put up some nurseries? OK, let's do it. Gosh, can we do tourism? No, we don't want you on the Sadjunjira side doing tourism. OK, we won't. What else am I supposed to do?"

After all, he said, "that Gorongosi culture is gone when the last tree gets cut down up there. And furthermore, you've got people up there—the kids don't have schools, the women are basically slaves, and the régulos are not looking out for their people a lot of the time. We all know that traditional societies look out for themselves. What are my choices? I'm a human-rights guy and a conservation guy trying to do both at the same time. The best idea I've come up with is those nurseries. I like it, I think it was a good thing, I'm proud of myself and my team for doing it."

Carr hasn't given up hope that the mountain will become part of the park, but he told me recently that it might be just as well if the mountain were designated a "forest reserve"—a status that insures higher conservation status but makes greater allowances for human habitation. "That's a change in tactic, not a change in goal," he said, and he added, "But if 'forest reserve' is still so politically sensitive, look, I can keep my goal and continue to change my tactics." At the same time, he has asked the park's conservationists to draw up contingency plans to insure that water keeps coming in even if the mountain is lost. His staff talks about diverting the flow of nearby rivers to feed Lake Urema or drilling wells to supply watering holes for animals. Carr didn't like the prospect, but he liked grappling with the problem. "It makes it interesting for me—a person who could be anywhere in the whole world," he said.

One afternoon at Chitengo, as we sat by the pool, where tourists were splashing, Carr told me that he has made provisions for the park in his will. He had said repeatedly that before he found Gorongosa he had lived in dread of becoming "a dabbler." Now he told me, "The way I see it is this: I may not be a serious person, but Gorongosa is a serious project." When his friends speak of his hunger for purpose, they always tend to mention that he was once a Mormon, although they generally can't say how or why this is significant. But Larry Hardesty took a stab at it. "When you've been that committed to a religious ideal when you're young, I think that

that kind of gives you a taste for what it's like to have some big thing in your life that organizes it and gives it meaning," he said. "I think he was looking for that thing again, trying to replace that monolithic thing at the center of his life."

I told Carr what Hardesty had said, and he said, "I think that's right." Sometimes when he felt that everything was going well in Gorongosa, Carr said to me that he might have to find something else to do with himself in a few years. But then he said, "I have a lot of these weird things in my life where I've gone headlong into something and later I look at it and just go, That was strange. But I get the fever, right? You know, I like getting older and calmer. I don't really want a lot more of those fevers. It's just too much. This is my last fever."

PART SIX

The Environment: Big Blessings

RICHARD MANNING

Graze Anatomy

FROM *OnEarth*

WILL WINTER AND TODD CHURCHILL have a plan. It's simple, it's workable, and if enough people do it, it will shrink our carbon footprint, expand biodiversity and wildlife habitat, promote human health, humanize farming, control rampant flooding, radically decrease the use of pesticides and chemical fertilizers, and —for those of us who still eat the stuff—produce a first-class, guilt-free steak. Their plan: let cows eat grass.

The two men share a background in conventional farming. Winter came of age in the heart of the Midwest, starting his veterinary practice in the 1970s when industrial-strength livestock operations were gaining a full head of steam. "I had a syringe of antibiotics in this hand and a syringe of steroids in the other hand and I thought I was going to go cure the world," he says. "But I left conventional farming practice because it was so crude and so cruel. The cruelty drove me nuts." In the feedlots where cattle are stuffed with corn to produce almost all the beef Americans eat, he explains, "we were weaning [the calves], hot-iron branding them, vaccinating them, castrating them, dehorning them, and shipping them in one day. These are vets doing this, shooting them with ten-way vaccines, giving them ten different diseases in one day."

For Churchill the epiphany was less dramatic but no less wound up in the realities of industrial-scale farming. He grew up on a large, efficient corn and soybean farm near Moline, Illinois. "I spent all my summers as a child ripping out the fences, and we'd bulldoze all the trees and make one big cornfield. And then I thought: where do the birds live? The birds' job is to eat the aphids, but since we don't have trees anymore, we don't have birds, so we have to spray the aphids. Does that really make sense in the long run?"

Winter left his veterinary practice and became a foodie, promoting and distributing raw milk products in Minnesota and working as a consultant for graziers. He is the sort of fellow to have several irons in the fire at all times, and he offered up some free-range pork, his latest venture, when I met him at his Minneapolis home for breakfast.

Churchill became an accountant, also in southeast Minnesota. He had heard the heretical claims of a few contrarian farmers who were finishing beef on grass pastures instead of feedlots. It seemed an anachronism, defying the conventional wisdom that only the feedlot system can yield the economic efficiencies that leave Americans amply supplied with cheap beef and milk. But criticism of that system has escalated exponentially and for a host of different reasons: rapidly rising energy prices, concerns about global warming, and feed costs that leave poor people begging for the grain that Americans use to fatten livestock.

After sampling some grass-fed beef—some of it excellent, some inedible—Churchill decided to go into the business himself. He started the Thousand Hills Cattle Company in Cannon Falls, Minnesota, in 2003. A year later he met Winter, and the boots-and-jeans cattleman wannabe invited the jovial, generation-older vet-foodie to join him. Thousand Hills is now a substantial business, buying, slaughtering, and selling about 1,000 grass-fed cows a year to whole food stores, co-ops, restaurants, and three colleges in the Twin Cities area. Churchill buys his cattle from a small network of regional farmers able to meet his standards for quality.

Thousand Hills is part of an evolving nationwide web. The Denver-based American Grassfed Association was founded by just 8 members in 2003 but now claims 380. Carrie Balkcom, the group's executive director, says companies like Thousand Hills have sprung up throughout the country. At one end of the spectrum are large operations like White Oak Pastures in Georgia, which sells to the Whole Foods Market chain throughout the Southeast and Publix Supermarkets in the Atlanta area. At the other end are hundreds of small farms that sell directly to consumers, says Balkcom. Theo Weening, global meat coordinator for Whole Foods, says grass-fed beef is available in virtually all of his company's stores, and demand is growing.

Grass-fed beef, in other words, is poised to move out of the niche market and into the mainstream—as long as farmers can make it profitable.

Winter points out that for agriculture to be "sustainable," it must have a sustainable business model. "Todd says you have to earn the right to be here next year," he says. This was hard at first for operations like Thousand Springs, but the grass-fed system has matured to the point where significant chunks of the nation's corn-ravaged landscape can be converted into far more sustainable permanent pastures—without a loss in production.

At the heart of this shift lies a humble leap in technology, a fencing material called polywire, which you are sure to notice when you walk into a field with Churchill. A polywire fence is short and flimsy, composed of a single strand that resembles yellow-braided fishing line. This makes cheap, easily movable, and effective electric fences, and it is the key to the whole operation.

Modern grass farmers almost universally rely on something called managed intensive rotational grazing. Polywire fences confine a herd of maybe sixty cows to an area the size of a suburban front lawn, typically for twelve hours. Then the grazier moves the fence, to cycle through a series of such paddocks every month or so. This reflects a basic ecological principle. Left to their own devices in a diverse ecosystem, cows will eat just a few species, grazing again and again on the same plants. As with teenagers at a buffet, cows that eat this way are not acting in their own best nutritional interests. Rotational grazing, which forces them to eat the two-thirds of available forage that they would normally leave untouched, produces much more beef or milk per acre than does laissez-faire grazing.

Quality, of course, is just as important as quantity. I watch Churchill hop a strand of fence to enter a pasture. He whips out a kitchen garlic press to smush a sample of grass, then spreads it on a slide and sticks this into what looks like a miniature telescope. The gadget is a refractometer, and he is testing the grass for sugar content; this varies widely according to a range of conditions, not the least of which is the skill of the grazier. Sugar content is key to quality beef, and it is affected by the mix of grass species, the matching

of species to local climate and soil, the proper selection of comple-
mentary forbs (such as clovers), and proper rotation time.

"When we started with grass-fed, the quality wasn't that great,"
says Theo Weening of Whole Foods. "It has improved a lot in the
last twelve months." And as the quality improves, so does the potential to scale up grass-fed farming, with all its environmental benefits.

Splendor in the Grass

Humans can't eat grass, an assertion that sounds odd, considering
that something like three-quarters of all human nutrition comes
from wheat, rice, and corn, all of which are grasses. But what we
eat is actually their seeds, the dense package of complex carbohydrates that is the specialty of annual grasses. Perennial grasses,
which are more common, devote a larger proportion of their energy to roots, stems, and leaves, and the building block of these is
cellulose.

Humans cannot convert cellulose to protein, but cows can,
thanks to their highly specialized stomachs — rumens, as in *rumi-
nants.* "The rumen is the most magical chemical factory in the
world," Winter says. "It can turn cellulose into meat." It is a highly
specialized biodigester; if it didn't exist, biotechnologists would be
trying to invent it.

The health of the miniature ecosystem inside the rumen parallels the health of the larger ecosystem of the perennial pasture.
The health of that larger system sponsors a rich microbial world
beneath the soil, where lowly creatures like dung beetles and earth-
worms grind away at the task of cycling nutrients. Perennial grasses
build deep roots that can extend more than ten feet below the surface. Shallow-rooted annual crops rapidly deplete trace minerals
such as calcium, magnesium, and iodine, but the roots of peren-
nials act like elevators, lifting these minerals back into the system
and making them available to plants and everything else on up the
food chain.

Winter says almost every grain-fed cow has excessive acid in its
body, largely because of the lack of calcium and magnesium in its
diet. This acidity allows a whole range of parasites and diseases to
gain a foothold. A grass diet neutralizes the problem and passes

the benefits to humans, who, because of our narrow diets, are also short on these same trace minerals.

Grass-fed beef and milk also bring us the benefits of fat. The modern wave of obesity has made us fatophobes, but nutritional research is telling us we are obese and prone to heart disease not because we eat fat but because we eat the wrong kind of fat. Grain-fed beef is especially high in omega-6 fats and cholesterol. Grass-fed beef and dairy products are lower in both and higher in omega-3 fats and conjugated linoleic acid, which reduce the risk of heart disease and are lacking in our diets.

Unlike the industrial feedlot system, which is designed to channel all inputs into a single product—meat protein—grass-fed livestock operations must focus on a series of complex, interlocking ecosystems. By paying attention to these, grass farmers are rewarded with their byproduct: beef or milk. Ideally, health resonates throughout these systems, including the human body.

Economists illuminate this idea with the concept of "externalities"—costs that are invisible to the market. As a former CPA, Churchill knows that being attentive to ecosystem health is a way of bringing externalities into the accounting process. The enterprise has to be profitable, and it is. Churchill and hundreds of farmers like him have found that they can take productive corn and soybean land and convert it to perennial pasture, and in the process make more money than highly subsidized corn and soybean farmers. This flies in the face of the assumptions of agricultural economists worldwide, who have traditionally believed that the highest and best use of the world's most productive lands is row-crop agriculture.

Churchill's balance sheet says otherwise, raising the possibility that entire regions of the globe, including the American heartland, don't need to remain environmental sacrifice zones.

Flat and Flooded

Pouring rain notwithstanding, Francis Thicke wants to show me his herd of eighty Jersey dairy cows grazing a paddock on a summer day in 2008. Iowa farmers like Thicke became accustomed to rain last year. In June floods worked their way down the Mississippi River Basin, beginning in southern Minnesota and eventually inundating vast areas throughout the Corn Belt. The floods killed twenty-four people and caused damages running into the tens of billions of dollars.

It's not considered Minnesota-nice to say so, but this expensive inundation is simply another cost of corn and soybean agriculture. Iowa, with the best and flattest prairie topsoils in the nation, has the most altered landscape of any state; 65 percent of its land is planted to corn and soybeans. The state has less than 1 percent of its native habitat left, almost all of which was tallgrass prairie and oak savannah before European settlement. That earlier system included sinuous streams and riparian areas full of wetlands and flood-catching vegetation, but the thirsty prairie has been flattened and plowed into fields that shed rainwater almost as fast as parking lots do. Many of the fields have been underlain with drainage tiles that speed up the flow of surface water into rivers, exacerbating flooding. A stretch of pure prairie will absorb five to seven inches of rain an hour, meaning that twelve feet of rain in a twenty-four-hour stretch yields no runoff. Normal absorption on corn and soybean land ranges from a half inch to one and a half inches an hour, meaning that comparable rainfall yields catastrophic floods like those of 2008.

Thicke and I shed our drenched raingear and adjourn to his study, where he points to a framed black-and-white photograph. It shows a hill contoured with alternating strips of tilled and untilled land, with a pond at its base. The photo was featured in the 1957 U.S. Department of Agriculture's *Yearbook of Agriculture,* at a time when the department was touting this method of farming as the best way to prevent runoff and erosion. In fact, Thicke tells me, the pond in the photo flooded just about every time it rained hard. He knows this because it was his family's farm.

Thicke earned a Ph.D. in soil science and held an executive job at the Department of Agriculture, but he quit to buy a worn-out, eroded, and marginal farm near Fairfield, Iowa, in 1996. He allowed the tilled fields to revert to grass and started an organic dairy. Erosion stopped almost immediately. In the meantime, Thicke's brother kept the family farm and turned the neat strip plots into permanent pasture for beef and dairy. The pond no longer floods. Converting a significant share of corn and soybean lands to perennial pastures, as the Thickes have done, could go a long way toward eliminating flooding, especially if those pastures are strategically located in areas prone to flooding and erosion.

The traditional argument against farms like the Thickes' is that they cannot match the "efficiency" of industrial-scale grain production. But this argument does not take into account the productivity of managed rotational grazing. Todd Churchill says one reason he can make more money than a subsidized corn farmer is that he can produce about two steers per acre. It takes roughly the same acreage to grow the 3,000 pounds of grain used to finish a single steer in a feedlot.

But can enough land be converted to pasture to make any real difference to the landscape? The swath of destruction that is corn agriculture occupies about 80 million acres, mostly in the Midwest, an area only slightly smaller than California. At least half of that acreage is used to grow corn for livestock. Is it really possible to imagine something so radical as the transformation of 40 million acres of land?

In fact, it's been done before. In the 1970s and 1980s, and at the urging of the federal government, farmers greatly increased the acreage under cultivation, plowing up land that had been idle since the Dust Bowl. This triggered a huge increase in erosion, which in turn triggered the federal Conservation Reserve Program. At present farmers receive about $1.8 billion a year and have converted a total of 34.7 million acres from row crops to grass. But high grain prices have spurred farmers to begin pulling those acres out of the program at an alarming pace, about 2 million acres in the last two years. The grass-fed beef and dairy market offers an opportunity to reverse that flow and at the same time insulate the land from plows driven by high grain prices. Moreover, the $1.8 billion subsidy that farmers receive from taxpayers nets them an average of $51 an acre. Grass farmers can net as much as eight times that amount on converted corn and soybean land.

The Good Earth

All these benefits — more humane livestock farming, healthier humans, fewer floods, a richer and more natural landscape — are powerful arguments for a return to perennial pasture. But then there is the greatest potential benefit of all: a massive reduction of carbon emissions.

Soil is a mix of minerals from Earth's crust, a bit of living matter, and dead and decayed plant mass. Pure prairie builds this matter: the richest of virgin soils in the Midwest once ran to ten feet deep and were about 10 percent organic. What's left of those soils now typically contains less than half that amount of organic matter. But perennial pastures can restore the original richness of the soil in a decade or so.

The heart of all organic matter—about half the total—is carbon, the very stuff of global warming. When we speak of farming's carbon footprint, we generally calculate such things as internal combustion in tractors, transportation, and processing. But this ignores the fact that row-crop farming releases into the atmosphere carbon that has been stored in the soil. Researchers have found that tillage releases not only carbon dioxide but also nitrous oxide and methane (both global warming gases) by triggering the decay and erosion of topsoil. Without exception, all of the tillage systems examined in a recent study published in *Science* contributed to global warming, and the worst offenders were conventionally farmed corn, soybeans, and wheat. Fields of perennial crops in the same study pulled both methane and carbon dioxide from the atmosphere and stashed them in the soil. There is even some evidence that perennial grasslands are often better at sequestering carbon than forests are.

The Rodale Institute has tracked the amount of carbon sequestered by organic farming for nearly thirty years. Working with the Pennsylvania Department of Environmental Protection, it has found that such practices have an effect similar to that of perennial pasture in storing carbon in the soil in the form of organic matter. In 2008 researchers with the project concluded that converting all of the nation's cropland to organic agriculture would suck up enough carbon to offset 25 percent of our total fossil fuel emissions. In other words, we would have a continent-wide carbon sink. If that projection is right, such a shift would yield a net reduction of emissions greater than the Obama administration's target for 2020 and put us well on the way to meeting the target for 2050. True, these numbers are for organic farming, but perennial pastures arguably store even more carbon because of their deeper roots, and they almost eliminate the fossil fuel energy that organic farming uses for tillage.

There are good reasons to approach such landscape-level calculations with caution, however, simply because quantifying carbon sequestration is difficult. There are just too many variables. For instance, Randall Jackson, an ecosystem scientist at the University of Wisconsin at Madison who specializes in the study of pasture systems, says evidence suggests that rising global temperatures are already stimulating microbial activity in soils, which in turn may be increasing decay and adding to the release of global warming gases. But even a skeptic like Jackson is confident that conversion of industrial corn and soybean fields to permanent pastures would head us in the right direction. Simply put, conventional cornfields are a carbon source; pastures are more likely to be a sink. If we can get our beef and milk from a sink instead of a source, we probably ought to do it.

Combining pasture and organic crop farming may further enrich the soil and improve the bottom line. While many of the producers from whom Churchill buys stock are pure grass farmers who have sold their tractors and converted all of their land to managed pastures, many are not. And Francis Thicke's operation is not pure grass but rather an organic dairy. He also grows some grain and sells it.

I talked about this with Fred Kirschenmann, whom I first met in the late 1990s at his 3,500-acre organic farm in North Dakota, an operation he took over in 1976 after getting his Ph.D. in religion. He told me then that try as he might, he could not make organic farming pay without livestock. By bringing cattle into the mix, he gained the manure, controlled weeds through grazing, decreased tillage and energy use, and found a use for low-market-value crops such as grass and alfalfa, which can help build soil and stop erosion. These crops were also once regarded as critical to prudent crop rotation. But on conventional farms, nutrients lost to depletion and erosion are simply replaced by chemical fertilizer.

Kirschenmann still runs his farm, but he also serves as a distinguished fellow at Iowa State University's Leopold Center for Sustainable Agriculture. He told me last year that his earlier assertion about livestock has been strongly borne out. Studies throughout the Midwest have shown sharp increases in profitability when livestock is factored in — not just beef and dairy cattle but free-range chickens, hogs, goats, and sheep. Permanent pasture, in other words, is not an all-or-nothing proposition. The important point is to bring back the grass, and that can be done with full-scale grass-fed dairy and beef operations or by introducing a little bit of pasture here and there on organic or even conventional farms. This flexibility means that the benefits of pasture can evolve incrementally.

The presence and diversity of livestock on grass farms make the whole business a lot more interesting and plain pleasant, says Richard Cates, who teaches the operation of grass systems at the University of Wisconsin at Madison and is himself a grass farmer. He sees interest in his classes rapidly building among farm-raised students who are looking to become free-range people. And that is really the last of the many benefits of grass farming: it may stem the flight of the brightest and best young people from the small farm towns of the Midwest.

True, not every farmer can do it. Grass farming is more intellectually demanding than conventional farming. It deals in complexity and in the end is more an art than a science. But Cates and others believe that's one of its appealing features.

And while it may be complex on the level of an individual farm, the idea seems simple and eminently doable when compared with the vast array of costly technologies on the table for reducing our energy use and combating global warming. To be sure, enlightened federal farm policy could go a long way toward encouraging a shift to grass. Our current system of subsidies encourages industrial agriculture, which is to say that our nation's worst environmental problem is taxpayer funded. What is most impressive, though, is that the solution is a bootstrap operation; it is developing in spite of bad policy. The driver is the market pull, and it is gaining strength. Everyone I spoke to for this story agreed that demand for grass-fed products is well ahead of supply. Todd Churchill says the market could easily support ten times the number of grass farmers now in business. We tend to associate exponential growth with environmental harm, but in this case it would be good news, an economic solution designed by nature's economy and scaled up by market demand.

BURKHARD BILGER

Hearth Surgery

FROM *The New Yorker*

TWO MEN WALKED INTO A BAR called the Axe and Fiddle. It was a Thursday night in early August, in the town of Cottage Grove, Oregon, and the house was full. The men ordered drinks and a vegetarian Reuben and made their way to the only seats left, near a small stage at the back. The taller of the two, Dale Andreatta, had clear blue eyes and a long, columnar head crowned with gray hair. He was wearing a pleated kilt festooned with pockets and loops for power tools, and he spoke in a loud, unmodulated voice, like a clever robot. His friend, Peter Scott, was thinner and more disheveled, with a vaguely biblical look. He had long brown hair and sandaled feet, sun-baked skin and piercing eyes.

None of the locals paid them any mind. Cottage Grove, like much of Oregon, is home to hippies and hillbillies in equal measure. At the Axe and Fiddle, lumbermen from the local Weyerhaeuser and Starfire mills sat side by side with former Hoedads —free-living tree planters who'd reforested large tracts of the Bitterroot and Cascade mountains. The bar was flanked by a bookstore and, a few doors down Main Street, a store that specializes in machine guns. "I can't imagine that his market's that big," the bookstore's owner told me. "I mean, how many machine guns does a guy need?"

The featured act at the bar that night was a burlesque troupe from New York called Nice Jewish Girls Gone Bad. Just how they'd landed in the Oregon woods wasn't clear, but they stuck stubbornly to their set list. They sang a song about gefilte fish ("Fear Factor for Jews") and danced suggestively to Yiddish hip-hop. They promised to put the whore back in hora, and when that met with only polite applause—"Look it up on Wikipedia"—they asked for a show of hands from local Jews. There were five. Finally, near the end of the show, one of the performers—a spindly comedian with thick black glasses and a T-shirt that said FREAK—peered out from under the spotlight and fixed her eyes, a little desperately, on Peter Scott. "Do you have a job?" she said, almost to herself.

Scott said no, then yes.

"That sounds fishy. What is it you do?"

Scott fidgeted for a second, then mumbled, "I make stoves for Africa."

"You what?"

"I make stoves for Africa."

Scott was being modest. In the small but fanatical world of stove-makers, he is something of a celebrity. ("Peter is our rock star," another stovemaker told me.) For the past seven years, under the auspices of the German aid agency GTZ, Scott has designed or built some 400,000 stoves in thirteen African countries. He has made them out of mud, brick, sheet metal, clay, ceramic, and discarded oil drums. He has made them in villages without electricity or liquid fuel, where meals are still cooked over open fires, where burns are among the most common injuries and smoke is the sixth leading cause of death. In the places where Scott works, a good stove can save your life.

He and Andreatta were in Cottage Grove for Stove Camp. A mile or two from the Axe and Fiddle, a few dozen engineers, anthropologists, inventors, foreign aid workers, and rogue academics had set up tents in a meadow along a willowy bend in a fork of the Willamette River. They spent their days designing and testing wood-burning stoves, their nights cooking under the stars and debating thermodynamics. Stove Camp was a weeklong event hosted by the Aprovecho Research Center—the engineering offshoot of a local institute, education center, and environmental collective. Now in its tenth year, the camp had become a kind of hippie Manhattan Project. It brought together the best minds in the field to solve a single, intractable problem: How do you build cheap, durable, clean-burning stoves for 3 billion people?

A map of the world's poor is easy to make, Jacob Moss, a Stove Camper who works for the Environmental Protection Agency and started its Partnership for Clean Indoor Air, told me. Just follow the smoke. About half the world's population cooks with gas, kerosene, or electricity, while the other half burns wood, coal, dung, or other solid fuels. To the first group, a roaring hearth has become a luxury—a thing for camping trips and Christmas parties. To the second group, it's a necessity. To the first group, a kitchen is an arsenal of specialized appliances. To the second, it's just a place to build a fire.

Clean air, according to the EPA, contains less than 15 micrograms of fine particles per cubic meter. Five times that amount will set off a smoke alarm. Three hundred times as much—roughly what an open fire produces—will slowly kill you. Wood smoke, as sweet as it smells, is a caustic swirl of chemical agents, including benzene, butadiene, styrene, formaldehyde, dioxin, and methylene chloride. Every leaf or husk adds its own compounds to the fire, producing a fume so corrosive that it can consume a piece of untreated steel in less than a year. The effect on the body is similar. Indoor smoke kills a million and a half people annually, according to the World Health Organization. It causes or compounds a long list of debilities—pneumonia, bronchitis, emphysema, cataracts, cancers, heart disease, high blood pressure, and low birth weight—and has been implicated in a number of others, including tuberculosis, low IQ, and cleft palate, among other deformities.

A well-made stove can easily clear the air by piping the smoke out through a chimney or burning the fuel more efficiently. Yet most appliance manufacturers see no profit in making products for people who can't pay for them. And most aid agencies have found easier ways to help the poor—by administering vaccines, for instance. Stovemakers are a chronically underfunded bunch, used to toiling in the dusty margins of international development. Aside from a few national programs in Asia and the Americas, their projects have tended to be small and scattershot, funded, a few thousand stoves at a time, by volunteers and NGOs. "We've been watering this rock for a long time," Dean Still, the head of Aprovecho, told me.

Lately, though, the rules have changed. As global temperatures have risen, the smoke from Third World kitchens has been upgraded from a local to a universal threat. The average cooking fire produces about as much carbon dioxide as a car, and a great deal more soot, or black carbon—a substance seven hundred times as warming. Black carbon absorbs sunlight. A single gram warms the atmosphere as much as a 1,500-watt space heater running for a week. Given that cooking fires each release 1,000 or 2,000 grams of soot in a year, and that 3 billion people rely on them, cleaning up those emissions may be the fastest, cheapest way to cool the planet.

In June 2009, the sweeping Waxman-Markey climate bill was passed by the U.S. House of Representatives. Hidden among its 1,400 pages was a short section calling on the EPA to identify ways to provide stoves to 20 million households in five years. The bill made no mention of how or where the stoves might be built or who might pay for them. But there was talk of carbon-credit subsidies, international cofinancing, and major appliance manufacturers entering the fray.

The engineers of Stove Camp, in other words, found themselves suddenly blinking in the spotlight—like a band of raccoons caught digging through a scrap heap. "Kill a million and a half people and nobody gives a damn," one government official told me. "But become part of this big climate thing and everyone comes knocking at your door."

The entrance to Stove Camp was marked by a piece of weathered plywood hung on a rusty railroad trestle, with the words FRED'S ISLAND spray-painted on it. The place wasn't technically an island—it was bordered by the river on two sides and the railroad on the third—but it did belong to a retired carpenter named Fred Colgan. When I arrived on a Sunday evening, he and Aprovecho's Dean Still showed me an old trailer where I could sleep, a few yards from the tracks. "Wait till that timber train comes through at four in the morning," Colgan said. He gripped an imaginary bedstead and rattled his head up and down. "If you see giant rats in the middle of the night, you haven't had too much to drink. We're infested with nutria."

Still laughed. "A nutria is a rodent," he said. "Entirely harmless."

"It's a rat the size of a cocker spaniel."

Before Colgan and his wife, Lise, bought the island four years ago, it had belonged to a slaughterhouse and meatpacking operation, which left its buildings scattered across the grounds. Colgan offered the use of them to Still in 2006, after the stove program outgrew its original facilities a few miles up the road. The research center now has seven employees and a rotating cast of volunteers, who spend their time testing and developing stoves for projects worldwide. Their offices occupy a ramshackle complex along the river, with a wooden corral to one side and a labyrinth of labs, workshops, and storage rooms in back. Still holds his stove meetings in the meat locker, where the carcasses used to hang.

"Here's the deal," he told us one morning. "The world is absolutely littered with failed stoves. At the UN they laugh at us when we say that we have another project. So if we keep on blowing it, we're in trouble." He peered at the bleary-eyed campers, about thirty strong, gathered around mismatched Formica tables. Jacob Moss, the EPA official, sat next to a pulmonologist from the National Institutes of Health; Peter Scott had recently returned from Uganda, and others had worked on projects in Haiti, Honduras, Mexico, Malawi, Peru, India, and China. "It ain't easy," Still told them. "But it ain't impossible. We're going to be offered opportunities. But if there's going to be money for twenty million stoves, we have to be ready. And we have to not screw this up."

Still, who is fifty-seven, is one of the presiding spirits of the stove community. He has a large, ruddy face and a mop of white hair, a wide walrus mustache, and dark eyebrows that curve high above his eyes, giving him a look of perpetual, delighted surprise. Decades of living and working in hardscrabble villages have instilled an improbable ebullience in him, and a correlative roundness of form. I once compared him to Buddha when I was talking to Scott, who quickly corrected me. "Dean's a mystic Episcopalian," he said. "The only thing Buddhist about him is his girth." It's true that Still keeps plastic statues of St. Francis, the Virgin Mary, and the Archangel Michael glued to the dashboard of his truck. (Michael's flaming sword, he says, reminds him that "sometimes to make something good happen you have to kick people in the ass.") But when I was with him, they were joined by well-foxed copies of William James and a book of Mad Libs. When it comes to stoves, he said, any spiritual guidance will do.

Earlier that summer, Still had flown to London to accept an Ashden Award for Sustainable Energy, presented by Prince Charles. To prepare for the ceremony and press interviews, he'd had to buy his first suit since his wedding twelve years ago, and the Ashden foundation had given him a week's worth of elocution lessons. ("The English, geez, they're so unconsciously imperialist.") Still has a clear but indecorous way of talking, with an old hippie's loitering rhythms and self-questioning asides. Although he has trained a generation of stove designers and built one of the world's premier stove-testing labs, his science is mostly self-taught and he's uncomfortable playing the expert. "I'm just the mouth," he told us, waving his hands at the engineers in the room. "These guys are the brains." They just needed a little prodding now and then.

He turned to the whiteboard behind him and scrawled out some bullet points with a pink marker. "This is now the definition of a good stove, according to Waxman-Markey," he said. "1. Reduces fuel use by more than fifty percent. 2. Reduces black carbon by more than sixty percent. 3. Reduces childhood pneumonia by more than thirty percent. 4. Affordable ($10 retail or less). 5. Cooks love it. 6. Gets funded." The last three weren't in the bill, Still admitted, but no stove could succeed without them. And none had ever met all six criteria at once.

"So this is what we have to do this week, my dears," he said. "Save the damn world." He grinned. "I mean, you didn't want an easy problem, did you?"

Building a stove is simple. Building a good stove is hard. Building a good, cheap stove can drive an engineer crazy. The devices at Aprovecho looked straightforward enough. Most were about the size and shape of a stockpot, with a cylindrical combustion chamber and a cooking grate on top. You stuck some twigs in the chamber, set them on fire, and put your pot on the grate—nothing to it. Yet one stove used a pound of wood to boil a gallon of water, and another used two pounds. Fire is a fickle, nonlinear thing, and seems to be affected by every millimeter of a stove's design—the size of the opening, the shape and material of the chamber, the thickness of the grate—each variable amplifying the next and being amplified in turn, in a complex series of feedback loops. "You've heard of the butterfly effect?" one engineer told me. "Well, these stoves are full of butterflies."

Like science and religion, stove design is riven into sects and disciplines. Some engineers use only low-cost materials like mud or

brick; others dabble in thermoelectric generators and built-in fans
—cleaner and more efficient, but also more expensive. Most stoves
are built for combustion: they consume the wood and reduce it to
ash. But a few are designed for gasification instead. These stoves
heat the wood until it releases its volatile compounds, which are
ignited in the air. (All that's left of the wood afterward is its carbon
skeleton, which can be burned separately as charcoal or used as a
fertilizer.) Gasifiers can be remarkably clean-burning, but they're
also finicky. Because the fire burns at the top of the stove, rather
than rising up from a bed of coals at the bottom, its flames are easily stifled when new fuel is added, turning the stove into a smoke
bomb.

In the vestibule of the Aprovecho building, Still had set up a
small "Museum of Stoves" on facing wall racks. Its contents came
from more than a dozen countries, in an odd menagerie of shapes
and sizes: an elegant clay chulha from India, a squat steel Jiko from
Kenya, a painted coal-burner from China, like an Easy-Bake oven.
Most were better than an open fire, yet all had failed the test in
some way—too flimsy or inefficient or expensive or unstable or
unclean or hard to use. "We still haven't cracked the nut," Peter
Scott said.

Scott had come to Stove Camp to build a better injera stove. Injera is the spongy pancake that Ethiopians eat with almost every
meal. The batter is usually made of an ancient grain called teff and
fermented until it's bubbly and tart. It's poured onto a ceramic
griddle, or *mitad,* then set over an open fire or a concrete hearth.
In Ethiopia, injera is often cooked by women's cooperatives in
kitchens that may have forty or fifty smoky, inefficient stoves running simultaneously—one reason that the country has lost more
than 90 percent of its forests since the early 1960s. "In the north,
people will travel hundreds of kilometers to get wood, then double
back to bring it to market," Scott told me. A good stove, he figured,
could cut that fuel use in half.

For the past several months, Scott and his kilt-wearing friend
Dale Andreatta—a mechanical engineer from Columbus, Ohio,
who often did stove projects pro bono—had been collaborating
on a prototype. It had an efficient ceramic combustion chamber
shaped like a miniature fireplace, with a round griddle perched
above it like a tabletop. Scott had tried using a traditional mitad,
since local cooks would much prefer it, but the ceramic wouldn't

heat evenly, so he'd switched to steel instead. Steel conducts heat much more efficiently than ceramic, and it's often used for the plancha griddles in tortilla stoves. Injera, though, is an unforgiving dish. Its batter is thin and watery, so it can't be moved around like a tortilla, and any hot spots in the griddle will burn it. "The Ethiopians are unbelievably particular," Scott said. "If the injera doesn't have the exact size of bubble in the batter, they'll say it's garbage."

Luckily, Scott was used to improvising under much rougher conditions. His years in the African bush had left him, at forty, as sober and sinewy as Still was gregarious and stout. Scott had lived in mud huts in Swaziland, battled intestinal infections in Zambia, and been robbed by bandits in Uganda. When he first went to Africa for a stove project in 2002, he was taken hostage on his third day, in an Internet café in Pretoria. "They tied us up, laid us on the ground, a gun at the back of the head," he told me. "I had a strong premonition that I was going to die. But I didn't die. So after that I didn't worry too much about my own safety." He went on to build stoves for refugees in the Congo, tobacco-curing barns in Malawi and Tanzania, and institutional stoves throughout eastern and southern Africa. In 2006 he became the first Aprovecho member to receive an Ashden Award. (As a Canadian, he told me, he was excused from the elocution lessons.)

Over the next few days, I'd periodically find Scott and Andreatta skulking around the Aprovecho workshops and laboratories, looking for tools or discussing metallurgy. Their preliminary tests had not been encouraging: the griddle was 200 degrees hotter at the center than at the edge. When I asked Andreatta how it was going, he lifted an eyebrow. "The optimist thinks the glass is half full," he said. "The pessimist thinks the glass is half empty. The engineer knows the real truth: that the glass is twice as large as it should be for optimum utilization of resources."

When Aprovecho was founded in the late 1970s, building stoves was a good deal less complicated. "Appropriate technology" was the byword then. Grounded in the teachings of Gandhi and the economist E. F. Schumacher, the philosophy held that poor countries are best served by low-cost, low-tech, local development. Better to teach villagers to make a stove than to give them stoves that they can't afford to repair or replace.

Aprovecho took the idea a step further. "We wanted to work as an inverse Peace Corps," Ianto Evans, one of the founding members, told me. Evans was an architect and ecologist who'd done research and volunteer work in Guatemala and was then teaching at Oregon State. Instead of exporting American know-how to the Third World, he and a small group of artists and academics decided, they would try to teach Americans to live more sustainably. "We would bring in villagers from Kenya or Lesotho, have them stay with us, and teach us what they knew—everything from cooking to growing things to assessing how much is too much." They would build a model Third World village in the Oregon woods.

In 1981, with the help of a Canadian foundation, the group bought forty acres of second-growth timber five miles west of Cottage Grove. The land lay on a south-facing slope at the end of a logging road. It was rough, marginally fertile ground, wet year-round and often freezing in the winter. But the new owners spared no effort in improving it. They deep-tilled the soil and enriched it with compost. They planted pear, apple, and quince trees, a grape arbor, and a bamboo grove. They built a library, a workshop, an adobe hut, and passive solar cabins, and, to top it all, a giant tree house thirty feet above the ground. To neighbors or passersby they might have seemed like squatters, yet they were ambitious, industrious, self-serious folk. Aprovecho, in Spanish, means "I make good use of."

Deforestation was the issue of the moment, and Evans believed that stoves were an ideal solution. A few years earlier, at a research center in Quetzaltenango, Guatemala, he and a team of local craftsmen had tested a variety of designs and materials and brought in cooks to try them out. "Any fool can do technical things," he told me. "But if people don't want it, don't bother." The team eventually hit upon a mixture of sand, clay, and pumice that was stable and freely available. They cast it into a massive hearth, about waist-high, carved out a firebox, burners, and interior channels to direct the heat, then added a chimney for the smoke. They dubbed it Lorena, after the Spanish words for mud and sand, *lodo* and *arena*.

The Lorena never made many inroads to American kitchens, but it was an immediate hit internationally. While Evans was still testing it, a United Nations representative saw the stove and persuaded him to publish the design. "The facts are stunning," Evans wrote in a 1979 book on the Lorena. "Data from several sources indicate that improved stoves—and of these the Lorena stove appears to have advantages over the others—can save one-half to three-quarters or more of the wood normally used in cooking." Projects for the Peace Corps, World Bank, USAID, and the governments of Senegal and Lesotho followed, often inspiring others in turn. In some areas, the Lorena was so popular that its name became a generic term: it simply meant "improved stove."

In one sense, though, it was no improvement at all. The Lorena was good at removing smoke and preventing burns (no small things). It was handsome, easy to use, and helped warm the house. What it didn't do was save fuel—at least compared with a well-tended open fire. Its thick walls, rather than concentrating the heat, absorbed it: the stove warmed the room because it wasn't warming the food. Studies later found that the Lorena used up to twice as much wood as an open fire and took up to three times as long to boil a pot of water. "It sounds funny, but there are still people making Lorenas today," Dean Still says. "They don't understand the difference between insulation and a heat sink."

By the time Still arrived at Aprovecho in the summer of 1989, funding for stoves had dried up. The Lorena, as it turned out, was only one of hundreds of well-meaning but misconceived projects worldwide. There were mud stoves that dissolved in the rain, designer stoves that worked only with a certain pot, portable stoves that fell over when you stirred cornmeal mush on them. In 1983 the Indian government launched a national program that distributed some 35 million stoves across the subcontinent. The units came in various designs from local manufacturers; most were neither sturdy nor especially efficient. Several years later, when a doctoral student from Berkeley surveyed the results in Andhra Pradesh, she found a single stove still in use—as a bin for grain.

"They were good-hearted people," Still says of his predecessors at Aprovecho. "But they were idealistic artists. They were farmers and architects and artisans more than they were engineers." Still didn't seem, on the face of it, much better qualified. Before coming to Aprovecho, he'd worked in a trauma ward in Illinois, lived in a trapper's cabin in Colorado, and served as a security guard on an ocean freighter. He had owned a gas station, worked as a janitor in a synagogue, earned a master's degree in clinical psychology but never used it professionally—"Not one day," he says. Instead, he built a seagoing catamaran with two friends and crisscrossed the Pacific in it. Then he sold the boat, moved to Baja, built a thatched hut by the Sea of Cortez, and stayed there for nine years. "My idea was this," he says. "Can Dean learn to sit under a tree and be contented?"

The answer was no. But his wanderings left him oddly suited to building stoves. He was a skilled carpenter and designer, used to improvising with cheap materials. He was intimately familiar with the needs and hazards of life in developing countries. And he was a born community organizer. His parents, Douglas and Hanna Still, were political activists in the heroic sixties mold. They'd worked with César Chávez in California, the Black Panthers in Chicago, and Martin Luther King, Jr., in the South. (King and Still's father, who was a Presbyterian minister, spent a night in jail together in Albany, Georgia, after a protest.) By the age of thirteen, Dean was tagging along to a civil rights rally in Milwaukee wearing a BLACK POWER T-shirt among crowds of bellowing racists. At sixteen he was among the rioters at the 1968 Democratic National Convention in Chicago, narrowly escaping arrest. His parents always encouraged him to be a freethinker, he says. "So when I was in seventh or eighth grade I told them, 'You're right. School is just a training ground for cogs. I'm going to quit and have adventures.'"

Still was first drawn to Aprovecho by its work in sustainable agriculture and forestry. But it was the stoves that kept him there. Not long after he arrived in 1989, he met a local inventor named Larry Winiarski—a mild, bespectacled, dumpling-shaped man in his forties, perennially clad in overalls. Winiarski had a doctorate in engineering from Oregon State and had worked for the EPA for thirteen years, analyzing the heat discharge from power plants. It didn't take him long to spot the Lorena's inadequacies.

Working as a volunteer, Winiarski sketched out ten principles of stove design and began to build prototypes with Evans and other Aprovecho members. The new devices, which they called rocket stoves, for the powerful roar of their draft, were the physical opposite of the Lorena. They were small and lightweight, so that little heat was wasted on warming the stove itself. They had vertical combustion chambers that acted as chimneys, mixing the wood's volatile gases with air so that the rockets burned more efficiently. And they had well-insulated walls that forced the hot gases through narrow gaps around the pot, heating it as quickly as possible.

"Larry is one of those rare people in my life, when you ask him a question about stoves he's almost always right," Still says. "He just really, really understands fire." Aprovecho went on to build a number of rocket stoves and to publicize them in books and newsletters, but the group's loyalty still lay with the Lorena. "People were basically ignoring Larry when I showed up," Still told me. "Hippies love earthen structures." The community's open-air kitchen, for instance, was dominated by a clay bread oven that took hours to heat up and consumed great quantities of firewood. With Winiarski's guidance, Still conducted an experiment. Next to the bread oven, he built a simple rocket stove. It was made of a 55-gallon drum, laid horizontally, with a 33-gallon drum inside it and a rocket combustion chamber below. The new stove looked nothing like a traditional bread oven, yet it was hot within fourteen minutes on the strength of a few twigs. An hour later, when the bread was done, the clay oven was still warming up. "That's what won people over," Still says.

Over the next few years, Still and Winiarski built ever more elaborate devices for the community: room heaters, water heaters, jet-pulse engines, wood-fired refrigerators. They were just tinkering, mostly, in the absence of funding for more ambitious work. Aprovecho, by then, was in turmoil. Evans was evicted from the property in the early 1990s after a dispute over the community's finances. Then county inspectors declared the tree house and other structures illegal, and everything had to be torn down, rebuilt, and reorganized. "It was a hippie nightmare," Still says.

To Peter Scott, who came to Aprovecho in 1997, the situation seems not uncommon. "People in environmental communities tend to be escaping from normal society," he told me. "If things were great where they came from, they wouldn't have left. And that sort of opens us up to the pain of the world and what's happening to it. We're all a little crazy, maybe." Scott was twenty-eight when he showed up in Oregon and already a veteran activist. In British Columbia, where he was born, he had stood in front of bulldozers on logging roads, climbed old-growth trees to spare them the axe, and acted in an environmental theater troupe. An article on solar cookers in *Mother Jones* first led him to Aprovecho, he says — that and memories of a trip to the Congo and the denuded landscapes there. "I'm here to save the forests of Africa by building stoves!" he remembers declaring on his first day. Dean Still just laughed and told him to go pick some vegetables.

Even more than new designs, Still began to realize, stovemakers needed data—to win back their credibility with reliable laboratory and field research. In 2000, when the stove lab was just a toolshed in the woods, Aprovecho built its first emissions detector and began testing Winiarski's designs. By 2004, Still had grants from the EPA and the Shell Foundation to test stoves from other programs. By 2006, when the lab moved to Fred's Island, it had half a million dollars in funding and a staff of scrappy young engineers. (Nordica MacCarty, the lab manager, runs her jury-rigged Datsun on French-fry oil from a local diner. Karl Walter, the electronics designer, once built an airplane by hand and flew it to New York.) The research center now supports itself in good part with sales of microprocessor-controlled portable emissions detectors, designed and built inhouse. The hippie commune has become a quality-control center.

Early in October, Still and I flew to Guatemala to visit the world's longest-running stove study. The village of San Lorenzo, where it's based, is in the remote western highlands, close to 9,000 feet above sea level. It feels like one of the world's forgotten places—its houses, made of mud and straw, cling to terraces that look out over plunging valleys and volcanic peaks—yet its cooks are among the most closely observed in the world. Walk into many local kitchens and you'll find, attached to the walls or in the children's clothes, an array of electronic sensors and transmitters. Some measure particle emissions; others are motion detectors or carbon monoxide monitors. Next to the chimney, on top of the stove, is a piece of black duct tape with a small silver disk beneath it. Plug the disk into a Palm Pilot, and it will tell you exactly when and for how long that stove was used in the previous month.

In seventeenth-century England, when a stovemaker wanted to test a new design, he'd soak a piece of coal in cat's urine and throw it into the fire. If the stench went up the chimney with the smoke, the design was deemed a success. Stove-testing is more of a numbers game now: minutes to boil, grams of fuel, milligrams of black carbon. Yet the practical effects of those numbers aren't always clear—especially on the emissions side. "We have no idea how low you have to go before you get the majority of the health benefits," Jacob Moss told me. "Is it peak exposures you want to get rid of, or is pollution a steady-state thing? Rocket stoves still have a whole slew of emissions that are an order of magnitude higher than EPA standards." Cutting them in half, or even by two-thirds, may not be enough, he said.

The study that Still and I observed was aimed squarely at such uncertainties. Its detectors were the work of Kirk Smith, a professor of global environmental health at Berkeley and one of the world's leading authorities on indoor air pollution. Seven years ago, Smith and a team of students, researchers, and Guatemalan collaborators began tracking more than five hundred local families, all with pregnant mothers or infants less than four months old. The families were divided at random into two groups. Half were given plancha stoves with chimneys; the other half continued to cook over open fires. (After two years, when the first phase of the study was over, the second group got stoves as well.) Every week Smith's team would give the families a medical checkup and download the data from its sensors. In this way, they could track their pollution exposure and its effects in real time. "My wife likes to say that most men spend their lives watching women cook," Smith says. "Her husband has managed to make a career of it."

Smith is a rumpled sixty-two-year-old with tousled gray hair and eyelids as heavy as a basset hound's—he seems both tireless and perpetually short of sleep. When Still and I drove up to his site with him from Guatemala City, he spoke absorbingly, and almost continuously, for six hours about public health. (Last June, for a vacation, he took his wife and daughter to Chernobyl.) San Lorenzo is a six-hour flight plus layover on the red-eye from San Francisco, followed by a vertiginous trek, by truck or multicolored bus, up whipsawing mountain roads. For three years, Smith made the trip every month. His funders left him little choice, he told us.

"I'd go to an air pollution conference and show them my measurements, and they'd say, 'Good Lord, these are orders of magnitude higher than in our cities! And these are the most vulnerable populations in the world. Just go out and fix it!'" Instead of funding stove projects, though, they'd pass him along to the next agency. "So I'd go across the street to the international health meeting," Smith went on, "and they'd say, 'Well, Mr. Smith, you have a pretty convincing problem, but we have seven dollars a year per capita. Do you really expect us to take a dollar out of our budget for vaccines? We need to be damn certain that we can make a difference.'" The pharmaceutical companies had dozens of randomized trials to back up their claims. What did Smith have?

San Lorenzo is his answer. The study, which was funded by the NIH in 2001, now generates so much information that Smith needs two full-time workers to enter it into computers. On the morning after we arrived, Still and I joined the team on their rounds through the village. While Still scrutinized the stoves and suggested ways to improve them (he and Smith were hatching plans for a more efficient "hyper plancha"), I sat and watched the women cook. Diminutive and shy, in their bright embroidered blouses and tapestry skirts, they quietly answered questions as their children clutched their legs or peeked out from behind door frames. The houses were low-ceilinged and bare, with earthen floors, corrugated roofs, and a tree stump or two for furniture. Some had sheaves of Indian corn drying from the rafters or raised eaves that allowed a little light to leak in. A field hand in San Lorenzo makes about twenty dollars a week, Smith said— "Truth be told, they haven't recovered since Cortez." But in most of the houses with stoves, at least the air was clear. In those with open fires it hung so thick and noxious that the walls were blackened, the joists and beams shaggy with creosote. It was like sitting inside a smoker's lung.

Near the end of our rounds, we paid a visit to Angela Jiménez, a small, sharp-featured woman who was part of Smith's original control group. Jiménez is thirty-five and has five children, including four-month-old twins. When we walked in, she was simmering a pot of corn for tortillas and sautéing a *recado de pescado*—a thin brown sauce made with dried fish and cornmeal, ground together on a slab of volcanic rock. Smith's team had given her a stove six years earlier, but she hadn't bothered to maintain it. The clay tiles and steel griddle were pocked with holes, and smoke was billowing into the room. On the wall behind the stove, the team had hung a poster explaining the dangers of carbon monoxide, but the words were too covered in soot to be legible.

We were getting ready to leave when Jiménez's nine-year-old son, Wilder, lurched in with his baby sister, Milvia, in his arms. She was tightly bundled in blankets, with a blue-and-white knit cap on. Her face was covered in dried phlegm and she was crying hard, with a steady, wheezing cough. Jiménez lifted her up and laid her against her shoulder. Her daughter had been sick for eight days, she told us, and was running a fever. "You should take her to the clinic," Smith said. "Eight days is a long time at that age." Jiménez looked at him with hooded eyes and turned back to the stove. If she went to the clinic, they'd just send her to the hospital, she said. "And that's where people go to die."

Smith later prevailed upon Jiménez to let his team drive her to his clinic, where a physician gave both infants a diagnosis of severe pneumonia. Milvia was hypoxic: her lungs were so full of fluid that they couldn't get enough oxygen into her blood. Her twin brother, Selby, was even sicker: his blood was only 82 percent oxygenated, and his lungs made crackling noises under a stethoscope. "He could pass away tonight," Smith said. Pneumonia is the leading killer of children worldwide, and San Lorenzans are especially susceptible to it. They're so malnourished that their height at eighteen months is already two standard deviations below the norm. And their immune systems are further weakened by the toxins in wood smoke. On average, Smith has found, the children in the village get pneumonia every other year.

"So this is the bottom line," he told me that night, bringing up a graph on his laptop. "This is seventeen years of applying for grants, seven years of research, three and a half million dollars, and me coming down here for a week of every month." Thanks to his electronic sensors, Smith knew his subjects' cooking habits in microscopic detail. He knew when they lit the stove but left the room while it was burning. He knew how much smoke was in the air when they were cooking and how much carbon monoxide was in their breath. And by combining such data with their weekly medical records he could show, for the first time, how the risk of disease increased with exposure—what epidemiologists call a dose-response curve.

"For groups like the Gates Foundation and USAID, the metric is cost-effectiveness," Moss had told me. "How many people are you going to save with a hundred million dollars? That's what they want from this field, and they don't have it yet." Until now. Smith had data on half a dozen diseases that a decent stove could help prevent (it could lower blood pressure about as much as a low-salt diet, for instance). But the most dramatic numbers were for pneumonia. The graph on his laptop had an *x* axis for exposure and a *y* axis for disease. In between, the data followed a steeply rising curve. The children who inhaled the least smoke were between 65 and 85 percent less likely to contract severe pneumonia than those who inhaled the most.

"Those numbers are as good as for any vaccine," Smith said. The plancha stoves cost about a hundred dollars each, yet they were a bargain in public health terms. "In our country, we pay forty thousand dollars per year of life saved," Smith said. "Even if you take the lower end of the benefit, this would cost at most a few hundred dollars per life-year. It's a no-brainer." In a country like India, he and a team of coauthors later estimated in an article in *The Lancet*, stoves could save more than 2 million lives in ten years.

Smith's data may be good enough for the Gates Foundation, but the harder part will be convincing local villagers. Most of the San Lorenzans liked their stoves and maintained them well enough. But they considered the smoke from cooking more of an annoyance than a threat. (In Africa, some even welcome it as a defense against flies and mosquitoes.) "These kinds of correlations just aren't that easy to make," Smith said. "Think of cigarettes. They kill one out of two smokers prematurely—no war has ever had that effect. Yet famous scientists have died saying there is no connection." To imagine cooking as harmful is an even greater leap. "It's not cyanide," Still said. "They can always think of an eighty-nine-year-old who's been cooking over an open fire all her life. And Grandma's doing just fine."

The best examples of this insouciance in San Lorenzo were the wood-fired saunas that most of the villagers used. The tradition dated back to the ancient Mayans, who would heat rocks over an outdoor fire and carry them into a stone bathhouse. The modern version, known as a *chuj*, was just a mud-caked hut about the size of a large doghouse. It had an open fire inside, a pallet to lie on, and a blanket to seal the door. A *chuj* was essentially a human smokehouse, yet the same villagers who swore by their plancha stoves—including Vincente Tema, one of Smith's Guatemalan staff—took sauna baths once or twice a week for half an hour. (The baths were especially good for pregnant women, they said.)

When I asked Tema if I could try his *chuj*, Smith shrugged. I might want to take a carbon monoxide monitor with me, he said.

The experience wasn't altogether unpleasant—there are worse things, apparently, than becoming a giant slab of bacon. But by the time I stumbled out, sixteen minutes later, my head was swimming. When Smith later downloaded the monitor's data at his office, it showed the carbon monoxide in the *chuj* spiking to 500 parts per million, then abruptly leveling off. The program wasn't designed to show levels any higher than that, he explained. "Oh, buddy," Still said, staring at the screen. "If you'd gone to a thousand for ten minutes, you'd be in a coma now."

Stories like these were a source of endless frustration to stovemakers. The trouble with tradition, they'd found, is that it can be remarkably thickheaded. Ignore it, and your shiny new stove may get turned into a flowerpot. Cater to it, and you may end up with a new version of the same old problem. The campers in Cottage Grove spent half their time agonizing over cultural sensitivity ("We're highly dominated by elderly white engineering types," a stovemaker who'd worked in Uganda told me. "So you get a lot of preposterous ideas that'll never fly in the kitchen") and the other half grousing about "design drift." Too many stoves start out as marvels of efficiency, they said, and are gradually modified into obsolescence. Once the engineer is gone, the local builder may widen the stove's mouth so it can burn larger sticks, only to draw in too much cold air. Or he'll make the stove out of denser bricks, not realizing that the air pockets in the clay are its best insulation. The better the stove and the tighter its tolerances, the easier it is to ruin.

"When we first got into this, we had this utopian vision of working with local communities to build locally grown stoves," the EPA's Jacob Moss told me. "We've moved away from that—I won't say a hundred and eighty degrees, but maybe a hundred and sixty. I don't really listen to small stove projects anymore. When I hear Dean say that one millimeter can make a nontrivial difference, it's inconceivable to me that all these local stovemakers can make all these stoves efficiently. You have to work in a different way."

Three years ago, on a taxi ride in southern China, Still had a glimpse of the future. He was working as a consultant for the EPA at the time, passing through the city of Kunming, when he spotted some odd little stoves for sale on a street corner. He shouted for the driver to stop and stepped outside to examine one. "It was like Shangri-La," he told me. The stove was meant for burning coal, so its design was all wrong for wood, but it was sturdy, compact, and cleanly manufactured. More important, its combustion chamber was made of a hard yet miraculously light and porous clay—a combination that stovemakers had been scouring the Earth to find. "There, in this two-dollar coal burner, was everything needed to make the world's perfect rocket stove," Still says.

The stove had a telephone number printed on it, so Still called it on his cell phone. Two months later he was visiting the factory where the stove was built, in eastern China. Within two years, the factory was producing a stove to Aprovecho's specifications. Sold under the name StoveTec, it isn't much to look at: a hollow clay tube clad in green sheet metal, with an opening in front and a pot support on top. But it incorporates all ten rocket-design principles with a consistency that only mass production can offer. The StoveTec uses about half as much wood as an open fire, produces less than half as much smoke, and sells for eight dollars wholesale. In the United States, where it retails for five times as much, it has been especially popular among Mormons and survivalists.

Still's stove is a kind of proof of principle. It shows that an efficient, user-friendly stove can be mass-produced at a cost that even the very poor can afford. But it also shows what's missing. The StoveTec isn't suited to some dishes—tortillas, chapatis, heavy porridges—and its life expectancy is less than two years. While it's much less smoky than an open fire, it can't quite meet the Waxman-Markey standards.

The search for the perfect stove continues, in other words. Not long before Stove Camp, I visited a company called Envirofit, in Fort Collins, Colorado. Envirofit's laboratories are housed at Colorado State University in a converted power plant from the 1930s. On the morning of my tour, half a dozen experiments were going on simultaneously. One glass case held nine stoves, all furiously burning pellets fed to them by an automatic hopper. Across the room, the smoke was being parsed into its chemical components by a rack of blinking machinery. (Wood smoke may not be cyanide, as Still put it, but hydrogen cyanide turns out to be one of its trace elements.) On a catwalk upstairs, a programmer was modeling green and yellow flames on his computer, while a biologist down the hall was subjecting live human lung cells to wood smoke. "We grow them in the basement, but they're fully functional," I was told. "They even produce phlegm."

Envirofit's CEO, Ron Bills, is a former executive of Segway, Yamaha, and Bombardier. His new company is technically a nonprofit, yet Bills believes that stovemakers, for too long, have treated the developing world as a charity ward instead of a business opportunity. "A lot of the poor—call them emerging consumers—get inundated with crummy stuff," he told me. "So we're going back to Henry Ford." Envirofit's first new product was essentially a rebranded version of Aprovecho's stove, made by the same Chinese factory with a few improvements in durability and design. In July, however, the company unveiled a new model. It was shaped like an ordinary rocket stove, though much more stylish, and had a major innovation at its core: a durable metal combustion chamber. Made of an alloy developed together with Oak Ridge National Laboratory in Tennessee, it could withstand the caustic fumes of a wood fire for more than five years, yet cost only three dollars a unit to produce. The Envirofit combustion chamber could be shipped for a fraction of the cost of a fully built stove and adapted to local designs and cooking traditions. It was mass production and appropriate technology rolled into one.

"That's the goose that laid the golden egg right there," Bills told me. "That's the Intel inside." He had nothing against groups like Aprovecho, he said. They could continue to hold their Stove Camps and sell their stoves made out of clay. "But Henry Ford didn't stop with the Model T. If we are going to make an impact in my lifetime, it has to be done at scale. And when you have a three-billion-product opportunity, what is enough scale? One million, two million, five million? I like to dream big." Thanks, hippies, he seemed to be saying. Now, please step aside.

On the last day of Stove Camp, I stumbled out of bed late, in search of coffee—the timber train having catapulted me awake, as usual, four hours earlier. Aprovecho was as busy as a science fair. The pulmonologist from NIH was putting the finishing touches on a rocket stove made from an oil drum. A Norwegian designer was running emissions tests on a little tin gasifier. And another camper was watching emission measurements unspool across a laptop. "Look at that!" he shouted. "It's flat-lining! There's almost no particulate matter!" On the whiteboard next door, the words SAVE THE WORLD had long since been erased and replaced with mathematical equations.

Scott and Andreatta were in the far corner of the workshop, probing their injera stove with an infrared thermometer. Their week had been a succession of setbacks and breakthroughs. When their first prototype, with its steel griddle, had too many hot spots, Scott had suggested that they try aluminum. It conducted heat even better than steel and was considerably cheaper. A few e-mails to Ethiopia had confirmed that the metal could be locally cast from recycled engine blocks. By the next morning, Andreatta had roughed out a plywood mold for the griddle and they'd taken it to a foundry in Eugene. But the design proved too complicated to cast—it had radiating fins along the bottom to distribute the heat. So they'd settled on something simpler.

The new griddle was a third of an inch thick and flat on both sides. Andreatta had put a ceramic baffle beneath it to temper and diffuse the flames, but he still had his doubts. The melting point of aluminum is 1,220 degrees Fahrenheit—about half as high as the peak temperature inside a rocket stove. If they weren't careful, the griddle would dissolve before their eyes. Andreatta switched on his LED headlamp and peered at the infrared thermometer. For now, the griddle was holding steady at 433 degrees—just 5 degrees short of the target temperature. Better yet, the center was less than 25 degrees hotter than the outer edge. "Even Ethiopian women don't get it in that range," Scott said.

Still strolled by, wearing a T-shirt with a giant longhorn beetle on it. He had a groggy grin on his face, as if he'd just woken up to a redeemed and revitalized world. Sometimes he saw the stove community more as Ron Bills seemed to see it—as a gathering of undisciplined hobbyists, engaged in the equivalent of building iPods out of toothpicks and aluminum foil. But this wasn't one of those days. Earlier that summer, a research group under Vijay Modi, a professor of mechanical engineering at Columbia, had surveyed cooks in Uganda and Tanzania who had tested a variety of improved stoves. In both studies, the StoveTec/Envirofit design had won the highest rating, beating out the most recent Envirofit stove in the Tanzanian study. "My people, they aren't always very smart," Still had told me. But they were inventive, resourceful, and doggedly resilient. And after thirty years of trial and error and endless field research, they understood fire very, very well.

The injera stove was the kind of project that might always fall to them. "What is the market for an improved cookstove, really?" Still said. "People hope that it's big, but we have an eight-dollar stove and it's not easy to sell. Everyone forgets that poor people are really poor." In Africa, where less than a quarter of the population has electricity and the most efficient technologies are beyond reach, an open fire can still seem hard to beat, if only because it's free. "But you know what? We're going to do it," Still said. "A lot of people think that if you don't make a whole lot of money at something it can't be good. I think those people are wrong. If you want to do what poor people need, and you really don't stop, you're not going to be rich. Not unless you're a lot smarter than I am."

Just before we broke camp the next morning, Scott came to find me in the meat locker: the prototype was ready for its first pancake. He and Andreatta had hoped to cook true injera bread for the occasion, but they couldn't find the time—or the teff—to make a proper sourdough. So they'd settled for Aunt Jemima. "This is our first test," Scott said, holding up a pitcher of pancake batter. "People of the world, cut us some slack." Then he poured it onto the hot griddle.

Over the next three months, the stove would go through more rounds of fiddling and redesign. The aluminum would prove too conductive for real injera and get swapped out for a traditional mitad. To get the ceramic to heat evenly, the baffles beneath it would have to be removed. At one point, in Addis Ababa, Scott would nearly abandon the project, only to have an Ethiopian cook make some key suggestions. Yet the result would be even better than it seemed on this sunny August morning: the world's first successful rocket injera stove—twice as efficient and many times more durable than those it was meant to replace.

As the batter hit the griddle, it spread into a circle that nearly reached the edge. Within a minute, it was bubbling up evenly across its surface. "Yeah, baby!" Scott said. "If we'd tried that last

Friday, it would be blackened char in the middle." He slid a spatula under the batter and tried to flip it, leaving half on the griddle but the rest well browned. He stared at the pancake. "We can't really fucking believe it," he said. "I mean, these designs usually take months and you're still scratching your head." The stove was almost ready, he thought. Now they just had to convince a few million Ethiopians.

EVAN OSNOS

Green Giant

FROM *The New Yorker*

ON MARCH 3, 1986, four of China's top weapons scientists—each a veteran of the missile and space programs—sent a private letter to Deng Xiaoping, the leader of the country. Their letter was a warning: decades of relentless focus on militarization had crippled the country's civilian scientific establishment; China must join the world's *xin jishu geming*, the "new technological revolution," they said, or it would be left behind. They called for an élite project devoted to technology ranging from biotech to space research. Deng agreed and scribbled on the letter, "Action must be taken on this now." This was China's "Sputnik moment," and the project was code-named the 863 Program, for the year and month of its birth.

In the years that followed, the government pumped billions of dollars into labs and universities and enterprises on projects ranging from cloning to underwater robots. Then, in 2001, Chinese officials abruptly expanded one program in particular: energy technology. The reasons were clear. Once the largest oil exporter in East Asia, China was now adding more than two thousand cars a day and importing millions of barrels; its energy security hinged on a flotilla of tankers stretched across distant seas. Meanwhile, China was getting nearly 80 percent of its electricity from coal, which was rendering the air in much of the country unbreathable and hastening climate changes that could undermine China's future stability. Rising sea levels were on pace to create more refugees in China than in any other country, even Bangladesh.

In 2006 Chinese leaders redoubled their commitment to new energy technology; they boosted funding for research and set targets for installing wind turbines, solar panels, hydroelectric dams, and other renewable sources of energy that were higher than goals in the United States. China doubled its wind power capacity that year, then doubled it again the next year, and the year after. The country had virtually no solar industry in 2003; five years later, it was manufacturing more solar cells than any other country, winning customers from foreign companies that had invented the technology in the first place. As President Hu Jintao, a political heir of Deng Xiaoping, put it in October 2009, China must "seize preemptive opportunities in the new round of the global energy revolution."

A China born again green can be hard to imagine, especially for people who live here. After four years in Beijing, I've learned how to gauge the pollution before I open the curtains; by dawn on the smoggiest days, the lungs ache. The city government does not dwell on the details; its daily air-quality measurement does not even tally the tiniest particles of pollution, which are the most damaging to the respiratory system. Last year the U.S. Embassy installed an air monitor on the roof of one of its buildings, and every hour it posts the results to a Twitter feed, with a score ranging from 1, which is the cleanest air, to 500, the dirtiest. American cities consider anything above 100 to be unhealthy. The rare times in which an American city has scored above 300 have been in the midst of forest fires. In these cases, the government puts out public health notices warning that the air is "hazardous" and that "everyone should avoid all physical activity outdoors." As I type this in Beijing, the embassy's air monitor says that today's score is 500.

China is so big—and is growing so fast—that in 2006 it passed the United States to become the world's largest producer of greenhouse gases. If China's emissions keep climbing as they have for the past thirty years, the country will emit more of those gases in the next thirty years than the United States has in its entire history. So the question is no longer whether China is equipped to play a role in combating climate change but how that role will affect other countries. David Sandalow, the U.S. assistant secretary of energy for policy and international affairs, has been to China five times in five months. He told me, "China's investment in clean energy is extraordinary." For America, he added, the implication is clear: "Unless the U.S. makes investments, we are not competitive in the clean-tech sector in the years and decades to come."

One of the firms that are part of the 863 Program is Goldwind Science and Technology Company. It operates a plant and a laboratory in a cluster of high-tech companies in an outlying district of Beijing called Yizhuang, which has been trying to rebrand itself with the name E-Town. (China has been establishing high-tech clusters since the late 1980s, after scientists returned from abroad with news of Silicon Valley and Route 128.) Yizhuang was a royal hunting ground under the last emperor, but as E-Town it has the sweeping asphalt vistas of a suburban office park, around blocks of reflective-glass buildings occupied by Nokia, Bosch, and other corporations. Local planning officials have embraced the vocabulary of a new era; E-Town, they say, will be a model not only of e-business but also of e-government, e-community, e-knowledge, and e-parks.

When I reached Goldwind, the first thing I saw was a spirited soccer game underway on a field in the center of the campus. An artificial rock-climbing wall covered one side of the glass-and-steel research center. I met the chairman, Wu Gang, in his office on the third floor, and I asked about the sports. "We employ several coaches and music teachers," he said. "They do training for our staff." A pair of pushup bars rested on the carpet beside his desk. At fifty-one years old, Wu is tall, with wire-rim glasses, rumpled black hair, and the broad shoulders of a swimmer. ("I can do the butterfly," he said.) For fun he sings Peking opera. Wu said that he had not been a robust child: "My education was very serious. Just learning, learning, learning. I wanted to jump out of that!"

Wu integrates his hobbies into his work life in the manner of a California entrepreneur. He once led seventeen people, including seven Goldwind employees, on a mountaineering expedition across Mt. Bogda, in the Tian Shan range, in western China. "We Chinese are very weak in this field—teamwork," Wu said. He recently put his workers on a five-year self-improvement regimen; among the corporate announcements on Goldwind's website, the company now posts its inhouse sports reports. ("All the vigorous and valiant players shot and dunked frequently," according to a recent basketball report on a game between factory workers.)

Wu was born and raised in the far western region of Xinjiang, home to vast plains and peaks that create natural wind tunnels, with gusts so ferocious that they can sweep trains from their tracks. In the 1980s, engineers from Europe began arriving in Xinjiang in order to test their wind turbines, and in 1987 Wu, then a young engineer in charge of an early Chinese wind farm, worked alongside engineers from Denmark, a center of wind power research. He immersed himself in the mechanics of turbines — "Where are their stomachs, and where are their hearts?" he said. In 1997 state science officials offered him the project of building a 600-kilowatt turbine, small by international standards but still unknown territory in China. Many recipients of government research funding simply used the money to conduct their experiments and move on, but some, like Wu, saw the cash as the kernel of a business. He figured that every dollar from the government could attract more than ten dollars in bank loans: "We can show them, 'This is money we got from the science ministry.'"

Wu saw little reason to start from zero: Goldwind licensed a design from Jacobs Energie, a German company. Manufacturing was not as simple. Early attempts were a "terrible failure," Wu said. "Whole blades dropped off." He shook his head. "The main shafts broke. It was really very dangerous."

Goldwind shut down for three months. The company eventually solved the problems, and, with the help of 863 and other government funding, it expanded into a full range of large and sophisticated turbines. Many of them were licensed from abroad, but as they were built in China, they sold for a third less than European and American rivals. Goldwind's sales doubled every year from 2000 to 2008. In 2007 Wu took the company public, and garnered nearly $200 million.

China has made up so much ground on clean tech in part through protectionism — until recently, wind farms were required to use turbines with locally manufactured parts. The requirement went into effect in 2003; by the time it was lifted, six years later, Chinese turbines dominated the local market. In fact, the policy worked too well: China's wind farms have grown so fast that according to estimates, between 20 and 30 percent aren't actually generating electricity. A surplus of factories was only part of the problem: local bureaucrats, it turned out, were being rewarded not for how much electricity they generated but for how much equipment they installed—a blunder that is often cited by skeptics of China's efforts.

They have a point; many factories are churning out cheap, unreliable turbines because the government lacks sufficient technical standards. But the grid problem is probably temporary. China is already buying and installing the world's most efficient transmission lines—"an area where China has actually moved ahead of the U.S.," according to Deborah Seligsohn, a senior fellow at the World Resources Institute. In the next decade, China plans to install wind power equipment capable of generating nearly five times the power of the Three Gorges Dam, the world's largest producer.

After I met with Wu Gang, the company's director of strategy and global development, Zhou Tong, an elegant woman in her thirties, handed me a hard hat and walked me next door to the turbine-assembly plant, an immaculate four-story hangar filled with workers in orange jumpsuits piecing together turbine parts that were as big and spacy-looking as Airstream trailers. The turbines were astonishing pieces of equipment— so large that some manufacturers put helicopter pads on top—and the technical complexity dispelled any lingering image I had of Chinese factories as rows of unskilled workers stooped over cheap electronics. Wandering among the turbines, we passed some Ping-Pong tables, where a competition was underway, and stopped in front of a shiny white dome that looked like the nose of a passenger jet. It was a rotor hub—the point where blades intersect—and it was part of Goldwind's newest treasure, a turbine large enough to generate 2.5 megawatts of electricity, its largest yet. "Wow, this is a 2.5!" Zhou exclaimed. "I saw the first one in Germany. This is the first one I've seen here." Wu was set to unveil the new turbine at a press conference the next day. A flatbed truck, loaded with turbine parts and idling in the doorway, was bound for wind farms throughout Manchuria.

The prospect of a future powered by the sun and the wind is so appealing that it obscures a less charming fact: coal is going nowhere soon. Even the most optimistic forecasts agree that China and the United States, for the foreseeable future, will remain ravenous consumers. (China burns more coal than America, Europe, and Japan combined.) As Julio Friedmann, an energy expert at the Lawrence

Livermore National Laboratory, near San Francisco, told me, "The decisions that China and the U.S. make in the next five years in the coal sector will determine the future of this century."

In 2001 the 863 Program launched a "clean coal" project, and Yao Qiang, a professor of thermal engineering at Beijing's Tsinghua University, was appointed to the committee in charge. He said that its purpose is simple: to spur innovation of ideas so risky and expensive that no private company will attempt them alone. The government is not trying to ordain which technologies will prevail; the notion of attempting to pick "winners and losers" is as unpopular among Chinese technologists as it is in Silicon Valley. Rather, Yao sees his role as trying to insure that promising ideas have a chance to contend at all. "If the government does nothing, the technology is doomed to fail," he said.

Grants from the 863 Program flowed to places like the Thermal Power Research Institute, based in the ancient city of Xi'an in the center of China's coal country. "The impact was huge," Xu Shisen, the chief engineer at the institute, told me over lunch recently. "Take our project, for example," he said, referring to an experimental power plant that, if successful, will produce very low emissions. "Without 863, the technology would have been delayed for years."

After lunch a pair of engineers took me to see their laboratory: a drab eight-story concrete building crammed with so many pipes and ducts that it felt like the engine room of a ship. We climbed the stairs to the fourth floor and stepped into a room with sacks of coal samples lining the walls like sandbags. In the center of the room was a device that looked like a household boiler, although it was three times the usual size, and pipes and wires bristled from the top and the sides. It was an experimental coal gasifier, which uses intense pressure and heat to turn coal dust into a gas that can be burned with less waste, rather than burning coal the old-fashioned way. With a coal gasifier, it is far easier to extract greenhouse emissions so that they can be stored or reused instead of floating into the atmosphere. Gasifiers have been around for decades, but they are expensive — from $500 million to more than $2 billion for the power-plant size — so hardly any American utilities use them. The researchers in Xi'an, however, set out to make one better and cheaper.

One of the engineers, Xu Yue, joined the gasifier project in 1997. A team of ten worked in twelve-hour shifts, conducting their experiments around the clock. "There was a bed there," he said, pointing to the corner of a soot-stained control room. (The image of China as a nation of engineers toiling for pennies is overstated; Xu Yue works hard, but he earns around $100,000 a year.) Beyond salaries, everything about the lab was cheaper than it would have been in the United States, from the land on which it was built to the cost of heating the building, and when the gasifier was finished it had a price tag one-third to one-half that of the equivalent in the West.

When Albert Lin, an American energy entrepreneur on the board of Future Fuels, a Texas-based power-plant developer, set out to find a gasifier for a pioneering new plant that is designed to spew less greenhouse gas, he figured that he would buy one from GE or Shell. Then his engineers tested the Xi'an version. It was "the absolute best we've seen," Lin told me. (Lin said that the "secret sauce" in the Chinese design is a clever bit of engineering that recycles the heat created by the gasifier to convert yet more coal into gas.) His company licensed the Chinese design, marking one of the first instances of Chinese coal technology's coming to America. "Fifteen or twenty years ago, anyone you asked would have said that Western technologies in coal gasification were superior to anything in China," Lin said. "Now, I think, that claim is not true."

The 863 Program took much of its shape from the American research system used by the National Institutes of Health and the Department of Defense: the government appointed panels of experts, who drew up research priorities, called for bids, and awarded contracts. In 1987 the government gave it an initial budget of around $200 million a year. That figure was small by Western standards, but the sum went far in China, according to Evan Feigenbaum, an Asia specialist at the Council on Foreign Relations, who studied the program. When I mentioned to Xu Shisen, the coal engineer in Xi'an, that American scientists are dubious of top-down efforts to drive innovation, he suggested that the system is more competitive than outsiders imagine. "It is very intense — like a presidential election," he joked, and he sketched out the system: "Normally, each project will have five to eight contenders — some less, some more—but there is a broad field of innovators. A lot of companies are doing the same thing, so everyone wants to have a breakthrough." He went on, "It's not possible to have a flawless system, but it makes relatively few mistakes. It combines the will of the state with mass innovation."

R & D expenditures have grown faster in China than in any other big country—climbing about 20 percent each year for two decades, to $70 billion last year. Investment in energy research under the 863 Program has grown far faster: between 1991 and 2005, the most recent year on record, the amount increased nearly fifty-fold.

In America things have gone differently. In April of 1977, President Jimmy Carter warned that the hunt for new energy sources, triggered by the second Arab oil embargo, would be the "moral equivalent of war." He nearly quadrupled public investment in energy research, and by the mid-1980s the United States was the unchallenged leader in clean technology, manufacturing more than 50 percent of the world's solar cells and installing 90 percent of the wind power.

Ronald Reagan, however, campaigned on a pledge to abolish the Department of Energy, and once in office, he reduced investment in research, beginning a slide that would continue for a quarter-century. "We were working on a whole slate of very innovative and interesting technologies," Friedmann, of the Lawrence Livermore lab, said. "And, basically, when the price of oil dropped in 1986, we rolled up the carpet and said, 'This isn't interesting anymore.'" By 2006, according to the American Association for the Advancement of Science, the U.S. government was investing $1.4 billion a year—less than one-sixth the level at its peak, in 1979, with adjustments for inflation. (Federal spending on medical research, by contrast, nearly quadrupled during that time, to more than $29 billion.)

Scientists were alarmed. The starkest warning came in 2005, from the National Academies, the country's top science advisory body, which released *Rising Above the Gathering Storm,* a landmark report on U.S. competitiveness. It urged the government to boost investment in research, especially in energy. The authors—among them Steven Chu, then the director of the Lawrence Berkeley National Laboratory and now the secretary of energy, and Robert Gates, the former CIA director and now the secretary of defense

—wrote, "We fear the abruptness with which a lead in science and technology can be lost—and the difficulty of recovering a lead once lost, if indeed it can be regained at all."

They called for a new energy agency that could spur the hunt for "transformative" technologies. It would inject money into universities and companies and would be called the Advanced Research Projects Agency-Energy, or ARPA-E, modeled on DARPA, the Defense Department unit that President Eisenhower founded in response to Sputnik. (DARPA went on to play a significant role in the invention of the Internet, stealth technology, and the computer mouse, among other things.) ARPA-E, they hoped, would shepherd new energy inventions from the lab to the market, bridging the funding gap that is referred to in engineering circles as the "valley of death." Congress approved the idea in 2007, but President George W. Bush criticized it as an "expansion of government" into a role that is "more appropriately left to the private sector." He never requested funding, and the idea fizzled.

Other plans withered as well. In January 2008, the Bush administration withdrew support for FutureGen, a proposed project in Illinois that would have been the world's first coal-fired, near-zero-emissions power plant. The administration cited cost overruns, saying the price had climbed to $1.8 billion, but an audit by the Government Accountability Office later discovered that Bush appointees had overstated the costs by $500 million. House Democrats launched an investigation, which concluded, "FutureGen appears to have been nothing more than a public-relations ploy for Bush Administration officials to make it appear to the public and the world that the United States was doing something to address global warming." An internal Energy Department report had warned that canceling the project would set back the advance of carbon-storage technology by "at least 10 years." An e-mail between officials emphasized that Bush's secretary of energy, Samuel Bodman, "wants to kill" FutureGen "with or without a Plan B." (Bodman denies that costs were overstated.)

After FutureGen foundered, China broke ground on its own version: GreenGen. If it opens as planned in 2011, China will have the most high-tech low-emissions coal-fired plant in the world.

Two summers ago, a truckload of Beijing municipal workers turned up in my neighborhood and began unspooling heavy-duty black power lines, which they attached to our houses in preparation for a campaign to replace coal-burning furnaces with electric radiators. Soon the Coal-to-Electricity Project, as it was called, opened a small radiator showroom in a storefront around the corner, on a block shared by a sex shop and a vendor of funeral shrouds. My neighbors and I wandered over to choose from among the radiator options.

Two-thirds of the price was subsidized by the city, which estimates that it has replaced almost 100,000 coal stoves since the project began five years ago, cutting down on sulfur and dust emissions. I settled on a Marley CNLS340, a heater about the size of a large suitcase, manufactured in Shanghai. It had a built-in thermostat preprogrammed to use less electricity during peak day hours and then store it up at night, when demand was lower—a principle similar to the "smart meters" that American utilities plan to install in the next decade.

Neighbors began cutting their electricity bills by climbing up to their rooftops and installing solar water heaters—simple pieces of equipment with a water tank and a stretch of glass tubing to be heated by the sun. (China, which produces 50 percent of the world's solar heaters, now uses more of them than any other country.) And in the hardware stalls of the raucous covered market nearby, where the inventory ranges from live eels to doorbells, coiled high-efficiency light bulbs began crowding out traditional bulbs for sale. The government, it turned out, had instituted a 30-percent wholesale subsidy on efficient bulbs. Without anybody really noticing, China sold 62 million subsidized bulbs in ten months.

When Hu Jintao called on China to adopt a "scientific concept of development" in 2003, he was making a point: China's history of development at all costs had run its course. And in ways that were easy to overlook, China had embarked on deep changes.

In the summer of 2005, Edward Cunningham, a Ph.D. student researching energy policy at MIT, was traveling in the Chinese countryside when he noticed something peculiar: the government was allowing the price of coal to rise sharply after decades of controls. "I said, 'How the hell?'" he recalled. "'That can't be right. Maybe this is just some freak anecdotal evidence.'" It was in fact a pivotal change: manipulating the price of coal had always insured that Chinese utilities would produce ever more electricity, but the unhappy side effect was that utilities needed to build nothing more efficient than the cheapest, dirtiest plants. Coal prices had begun to rise, however, and that would leave power plants no choice but to install cleaner, more efficient equipment. Cunningham, now a postdoctoral fellow at Harvard, said that the effect had broad consequences. "We are going to see a huge amount of learning that we have not seen in the U.S."

Learning, in technology terms, is another way of saying "reducing cost." The more a technology is produced, the cheaper it becomes, and that can lead to change as revolutionary as dreaming up an invention in the first place: Henry Ford invented neither the automobile nor the assembly line. He simply perfected their combination to yield the world's first affordable cars.

In the same way, technology that is too expensive to be profitable in the West can become economical once China is involved; DVD players and flat-screen televisions were luxury goods until Chinese low-cost production made them ubiquitous. So far, many of the most promising energy technologies—from thin-film solar cells to complex systems that store carbon in depleted oil wells—are luxury goods, but the combination of Chinese manufacturing and American innovation is powerful; Kevin Czinger, a former Goldman Sachs executive, called it "the Apple model." "Own the brand, the design, and the intellectual property," he said, and then go to whoever can manufacture the technology reliably and cheaply. A few years ago, Czinger began looking at the business of electric cars. Detroit was going to move slowly, he figured, to avoid undermining its main business, and U.S. startups, including Tesla and Fisker, were planning to sell luxury electric cars for more than $80,000 each. Czinger had something else in mind.

"These cars should be far simpler and far cheaper than anything that's manufactured today," he told me when we met last spring in Beijing. At fifty, Czinger has brown hair swept back, sharp cheekbones, and an intensity that borders on the unnerving. ("Kevin Czinger was the toughest kid to play football at Yale in my thirty-two years as head coach," Carm Cozza, the former Yale coach, wrote in a memoir. "He was also the most unusual personality, probably the outstanding overachiever, maybe the brightest student, and definitely the scariest individual.")

In the spring of 2008, Czinger signed on as the CEO of Miles Electric Vehicles, a small electric-car company in Santa Monica that was looking to expand, and he went searching for a Chinese partner. He ended up at Tianjin Lishen Battery Joint-Stock Company. A decade ago, Japan dominated the world of lithium-ion batteries — the powerful lightweight cells that hold promise for an electric-car future — but in 1998 the Chinese government launched a push to catch up. Lishen received millions in subsidies and hundreds of acres of low-cost land to build a factory. The company grew to $250 million in annual sales, with customers including Apple, Samsung, and Motorola. Last year the 863 Program gave Lishen a $2.6-million grant to get into the electric-car business. That is when Czinger showed up. "We hit it off immediately," Qin Xingcai, the general manager of Lishen, said.

Czinger, who by now was heading up a sister company called Coda Automotive, added components from America and Germany and a chassis licensed from Japan. If all goes as planned, the Coda will become the first mass-produced all-electric sedan for sale in the United States next fall, with a price tag, after government rebates, of about $32,000. The Coda looks normal to the point of banal, a Toyota-ish family car indistinguishable from anything you would find in a suburban cul-de-sac. And that's the point; its tagline, "A model for the mainstream," is a jab at more eccentric and expensive alternatives.

The race to make the first successful electric car may hinge on what engineers call "the pack" — the intricate bundle of batteries that is the most temperamental equipment on board. If the pack is too big, the car will be too pricey; if the pack is too small or of poor design, it will drive like a golf cart. "Batteries are a lot like people," Phil Gow, Coda's chief battery engineer, told me when I visited the Tianjin factory, a ninety-minute drive from Beijing. "They want to have a certain temperature range. They're finicky." To explain, Gow, a Canadian, who is bald and has a goatee, led me to one of Lishen's production lines, similar to the car-battery line that will be fully operational next year. Workers in blue uniforms and blue hairnets were moving in swift precision around long temperature-controlled assembly lines, sealed off from dust and contamination by glass walls.

The workers were making laptop batteries — pinkie-size cylinders, to be lined up and encased in the familiar plastic brick. The system is similar for batteries tiny enough for an iPod or big enough for a car. Conveyor belts carried long, wafer-thin strips of metal into printing press–like rollers, which coated them with electrode-active material. Another machine sandwiched the strips between razor-thin layers of plastic and wound the whole stack together into a tight "jelly roll," a cylinder that looked, for the first time, like a battery. (Square cell-phone batteries are just jelly rolls squashed.)

A slogan on the wall declared VARIATION IS THE BIGGEST ENEMY OF QUALITY. Gow nodded at it gravely. A bundle of batteries is only as good as its weakest cell; if a coating is five-millionths of a meter too thin or too thick, a car could be a lemon. The new plant will have up to 3,000 workers on ten-hour shifts, twenty hours a day. "When you get down to it, you can have ten people working in China for the cost of one person in the U.S.," Mark Atkeson, the head of Coda's China operations, said.

It was easy to see China's edge in the operation. Upstairs, Gow and Atkeson showed me America's edge: their prototype of the pack. For two years, Coda's engineers in California and their collaborators around the world have worked on making it as light and powerful as possible—a life of "optimizing millimeters," as Gow put it. The result was a long, shallow aluminum case, measured to fit between the axles and jam-packed with 728 rectangular cells, topped with a fiberglass case. It carried its own air conditioning system to prevent batteries from getting too cold or too hot. Rattling off arcane points, Gow caught himself. "There's hundreds of things that go into it, so there's hundreds of details," he said. "It's really a great field for people with OCD."

Czinger, in that sense, had found his niche. By November he was crisscrossing the Pacific, leading design teams on both sides; in the months since we first met, he had grown only more evangelical in his belief that if Americans would stop feeling threatened by China's progress on clean technology, they might glimpse their own strengths. Only his American engineers, he said, had the garage-innovation culture to spend "eighteen hours a day for two years to develop a new technology." But only in China had he discovered "the will to spend on infrastructure, and to do it at high speed." The result, he said, was a "state-of-the-art battery facility that was, two years ago, an empty field!"

America has a tradition of overestimating its rivals, and China is a convenient choice these days. But, as with Japan's a generation ago, China's rapid advances in science and technology obscure some deeper limitations. In 2004 a group of U.S.-based Chinese scientists accused the 863 Program of cronyism, of funneling money into pet projects and unworthy labs. (A proverb popular among scientists goes, "Pavilions near the water receive the most moonlight.") When critics published their complaints in a Chinese-language supplement to the journal *Nature,* the government banned it. Less than two years later, Chen Jin, a star researcher at Shanghai Jiaotong University, who had received more than $10 million in grants to produce a Chinese microchip to rival Intel's, was discovered to have faked his results. It confirmed what many Chinese scientists said among themselves: the Chinese science system was riddled with plagiarism, falsified data, and conflicts of interest.

After the Chen Jin scandal, the 863 Program made changes. It began publishing tenders on the web to invite broader participation, and to cut down on conflicts of interest, it started assigning evaluators randomly. But those measures couldn't solve a larger problem: the system that allowed China to master the production of wind turbines and batteries does not necessarily equip China to invent the energy technology that nobody has yet imagined. "Add as many mail coaches as you please, you will never get a railroad," the economist Joseph Schumpeter once wrote. Scale is not a substitute for radical invention, and the Chinese bureaucracy chronically discourages risk. In 1999 the government launched a small-business innovation fund, for instance, but its bureaucratic DNA tells it to place only safe bets. "They are concerned that, given that it's a public fund, if their failure rate is very high the review will not be very good and the public will say, 'Hey, you're wasting money,'" Xue Lan, the dean of the school of public policy at Tsinghua University, told me. "But a venture capitalist would say, 'It is natural that you'll have a lot of failures.'" Financing is not the only barrier to innovation. As an editorial last year in *Nature* put it, "An even deeper question is whether a truly vibrant scientific culture is possible without a more widespread societal commitment to free expression."

The Obama administration is busy repairing the energy legacy of its predecessor. The stimulus package passed in February put more than $38 billion into the Department of Energy for renewable-energy projects—including $400 million for ARPA-E, the agency that Bush opposed. (It also allocated $1 billion toward reviving FutureGen, though a final decision is pending.)

In announcing the opening of ARPA-E in April 2009, Obama vowed to return America's investment in research and development to a level not seen since the space race. "The nation that leads the world in twenty-first-century clean energy will be the nation that leads in the twenty-first-century global economy," he said. "I believe America can and must be that nation."

An uninspiring version of that message is gaining currency in Congress; it frames American leadership as manifesting not so much the courage to seize the initiative as the determination to prevent others from doing so. Senator Charles Schumer, one of several lawmakers who have begun to cast China's role in environmental technology as a threat to American jobs, has warned the Obama administration not to provide stimulus funds to a wind farm in Texas because many of the turbines would be made in China. ("We should not be giving China a head start in this race at our own country's expense," Schumer said in a statement.) Senators John Kerry and Lindsey Graham, in an op-ed in the *Times*, vowed not to "surrender our marketplace to countries that do not accept environmental standards" and suggested a "border tax" on clean-energy technology.

The larger fact, however, is that no single nation is likely to dominate the clean-energy economy. Goldwind, Coda, and the Thermal Power Research Institute are hybrids of Western design and Chinese production, and no nation has yet mastered both the invention and the low-cost manufacturing of clean technology. It appears increasingly clear that winners in the new-energy economy will exploit the strengths of each side. President Obama seems inclined toward this view. When he visited Beijing in November, he and Hu Jintao cut several deals to share energy technology and know-how, which will accelerate progress in both countries. This was hardly a matter of handing technology to China; under one of the deals, for instance, the Missouri-based company Peabody Energy purchased a stake in GreenGen so that it can obtain data from, and lend expertise to, a cutting-edge Chinese power plant.

More important, the two presidents reignited hopes that climate negotiations in Copenhagen, which had been heading for failure, might reach a meaningful compromise. Days after the Beijing summit, China and the United States provided specific targets for controlling emissions. Their pledges were far from bold and left both sides open to criticism: China's emissions, after all, will continue to grow over time, and cuts pledged by the United States still fall far short of what scientists say is required to avert the worst effects of warming. Yet the commitments, for all their weakness, serve a crucial function: they prevent each side from using the other as a foil to justify inaction.

For the United States and China, the climate talks boil down to how much money the rich world will give poorer nations to help them acquire the technology to limit emissions and cope with the droughts, rising sea levels, and other effects caused by those who enjoyed two hundred years of burning cheap fossil fuels. Without sharing costs and technology, it is not at all clear, for instance, that China will invest in the holy grail of climate science: funneling greenhouse gases underground. The process, known as carbon capture and storage, or CCS, is so difficult and expensive that nobody has yet succeeded in using it on a large scale. Like electric cars and coal gasification, CCS would be cheaper to develop in China than in the United States, but China is not interested in paying for it alone. As long as a Chinese citizen earns less than one-seventh what his counterpart in America earns, China is unlikely to back down on the demand that it should be paid to slow down its economy and invest even more in energy technology. And on that point the sides remain far apart.

In November I was spending much of my time at Tsinghua University, a center of clean-tech research, seeing a string of new energy projects that might or might not succeed someday. (My favorite, science aside, is a biofuel based on the process of producing Chinese moonshine.) In a giant, bustling convention hall across town, models in slinky evening gowns and white fur stoles arrayed themselves around mockups of wind turbines as if they were hot rods. Beijing was so overrun with visiting MacArthur geniuses and Nobel laureates and Silicon Valley eminences, all angling to influence China's climate-change policy, that I had to triage conferences.

Traffic alone made it hard to get around. This year China overtook the United States as the world's largest car market, and much of Beijing is gridlocked every day. (Impossibly, the number of cars in the city is expected to double in seven years.) In desperation, I decided to buy an electric bicycle. China has put 100 million of them on the road in barely ten years, an unplanned phenomenon that, energy experts point out, happens to be a milestone: the world's first electric vehicle to go mass market. The 863 Program noticed, and last year it added a program to build a micro-electric car, inspired by bicycle components, for commuters. Researchers at Tsinghua did just that, by attaching four electric-bike motors to a chassis. "We call it the Hali," Ouyang Minggao, the Tsinghua professor in charge of it, told me. They took the name from the Chinese translation of "Harry Potter." The car is tiny and bulbous and is being road-tested near Shanghai.

Hunting for an e-bike, I ended up at a string of shops near the Tsinghua campus, where each storefront offered a competing range of prices and styles to a clientele dominated by students and young families. I settled on a model called the Turtle King—a simple contraption, black and styled like a Vespa, with a 500-watt brushless motor and disc brakes. Built of plastic to save weight, it was more akin to a scooter than to a bicycle, and it ran on a pair of lead-acid batteries, similar to those under the hood of a car. The salesman said that the bike would run for twenty to thirty miles, depending on how fast I went, before I would need to plug its cord into the wall for eight hours or lug the batteries inside to charge. With a top speed of around twenty-five miles per hour, it would do little for the ego, but, at just over $500, it was worth a try.

The manager rang up the sale, and I chatted with two buyers who were students at the Beijing University of Aeronautics and Astronautics. "You must have tons of these in the U.S., because you're always talking about environmental consciousness," one of them, an industrial-design major wearing a Che Guevara T-shirt, said. Not really, I told him; American drivers generally use bikes for exercise, not transportation. He looked baffled. Around his campus and others in Beijing, electric bikes are as routine as motorcycles are in the hill towns of Italy.

I eased the Turtle King down over the curb and accelerated to full speed, such as it was. I threaded through an intersection clotted with honking traffic, and the feeling, I discovered, was sublime. The Turtle King was addictive. I began riding it everywhere, showing up early for appointments, flush with efficiency and a soupçon of moral superiority. For years people had abandoned Beijing's bicycle lanes in favor of cars, but now the bike lanes were alive again, in an unruly showcase of innovation. Young riders souped up their bikes into status symbols, pulsing with flashing lights and subwoofers; construction workers drove them like mules, laden down to the axles with sledgehammers and drills and propane tanks; parents with kids' seats on the back drifted through rush-hour traffic and reached school on time. Before long I was coveting an upgrade to a lithium-ion battery, which is lighter and runs longer. (Lithium-ion batteries have sparked interest in electric bikes in the West. They are a high-minded new accessory in Paris, and more than a few have even turned up in America.)

As a machine, the Turtle King was in desperate need of improvement. The chintzy horn broke the first day. The battery never went as far as advertised, and it was so heavy that I narrowly missed breaking some toes as it crashed to the ground on the way into the living room. Soon the sharp winter wind in Beijing was testing my commitment to transportation al fresco. And yet, for all its imperfections, the Turtle King was so much more practical than sitting in a stopped taxi or crowding onto a Beijing bus that it had become what all new-energy technology is somehow supposed to be: cheap, simple, and unobtrusive enough that using it is no longer a matter of sacrifice but one of self-interest.

GEORGE BLACK

India, Enlightened

FROM *OnEarth*

AT SEVEN O'CLOCK on a late February morning the scene is much as you'd imagine it, much as you've seen it, perhaps, in a score of earnest documentaries or in the highlight reels at this year's Oscars. In the slums of Delhi, a man pedals a bicycle laden with milk cans along a narrow, dusty lane swarming with people. A family of five squeezes into the back of a green-and-yellow autorickshaw meant for three. Somnolent cows lie in a field of garbage beside the railroad tracks, where an endless line of rusted coal cars rolls past, off to fuel some factory or power plant south of the city. Half-naked children squat among the discarded plastic bags and food wrappers to do their morning business. No one really knows how many people live here in the hutments of Okhla; 80,000 perhaps. And not a slumdog millionaire in sight.

Up on the footbridge that crosses the tracks, an old man in a filthy dhoti and turban shuffles past a torn notice advertising jobs for "marketing and tele-calling executives." The ad is a dispatch from another India, the new India, which lies, both literally and figuratively, on the other side of the tracks. You can't see it now, but it's out there somewhere in the smog, which is backlit to an opaque yellow-brown by the rising sun. As the day progresses and some of the haze burns off, shapes will start to emerge: the forest of cranes, the towers of blue steel and reflective glass, the billboards with words like Vodafone and Airtel and Intelenet, the Center Stage Mall and the Spice World Mall, and the call centers of the New Okhla Industrial Development Authority (NOIDA). It's Chinese workers who fill the shelves at Walmart with Christmas tree ornaments and socket wrench sets, but it's Indian workers in places like this who answer those phone queries about your credit card bill or your airline reservation, each call helping to build India's $11-billion-a-year outsourcing industry.

In this new India, faucets run all day. Lights burn all night. Shiny new cars zip into the center of Delhi on the DND Flyway. Water, energy, mobility: three defining elements of the escape from poverty. Such modest goals, but for most Indians still so painfully hard to achieve.

Looking out at NOIDA, at the towers and the smog, I wondered whether India could have it both ways. Could more than a billion people have the prosperity without the environmental havoc, in a country that is already struggling with the impact of a changing climate? Prime Minister Manmohan Singh had seemed to suggest as much in a speech launching India's National Action Plan on Climate Change in June 2008. Rapid economic growth was nonnegotiable, Singh said, if people were "to discard the ignominy of widespread poverty." At the same time, he promised that India would follow "a path of ecologically sustainable development." In seeking to reconcile these two goals, he pointed to the country's "civilizational legacy, which treats Nature as a source of nurture and not as a dark force to be conquered and harnessed to human endeavour."

What did that have to do with NOIDA? Perhaps one hint lay in a passage from Aravind Adiga's best-selling novel, *The White Tiger,* which won the Man Booker Prize four months after Singh's speech. The sardonic antihero writes to the Chinese premier, Wen Jiabao: "Apparently, sir, you Chinese are far ahead of us in every respect, except that you don't have entrepreneurs. And our nation, though it has no drinking water, electricity, sewage system, public transportation, sense of hygiene, discipline, courtesy, or punctuality, does have entrepreneurs. Thousands and thousands of them."

Could this new entrepreneurial spirit be harnessed to provide India's poor with the three essentials that NOIDA takes for granted—water, energy, and mobility? There seemed only one way to test the proposition: to embark on a journey that would give me a sampling of this astoundingly diverse and complicated country, from its mountains to its deserts and back again to the city. And the prime minister's speech seemed to suggest where I should start looking for answers—by going to the river that is the cradle of India's civilizational legacy.

The Water Tower of Asia

The Hindu pilgrimage town of Rishikesh in the Himalayan foothills, where John, Paul, George, and Ringo spent the early months of 1968 in thrall to the Maharishi Mahesh Yogi, sits on the banks of the Ganges, which Indians call the Ganga.

On the outskirts of town, an imposing line of high-tension electricity pylons marched southward from the mountains, carrying power to Delhi and the cities of the plain. Nearby was a poster advertising Ambuja Cement. A heroically muscled man, chin raised, gaze fixed on the future, clutched a gigantic dam under his arm. Ambuja is a private corporation, but the artwork suggested Soviet-era socialist realism.

Coal still accounts for 55 percent of India's energy mix, but hydro supplies 26 percent. That's an unusually high proportion—China, for all the publicity about Three Gorges, generates only about 7 percent of its power from dams—and India's climate plan assumes that hydro will continue to expand steadily. The plan also speaks at some length about the potential for large-scale solar power, since most of the country has clear, sunny skies for 250 to 300 days a year. But solar is expensive, and hydro, despite the huge economic and environmental cost of dams, remains the cheapest of all conventional energy sources.

On the other side of Rishikesh, a few miles upstream, an intense young woman named Priya Patel sat cross-legged in the garden of an ashram and showed me a map of the headwaters of the Ganga. Small rectangular symbols marked the site of proposed hydroelectric projects. Patel is the unofficial leader of the Ganga Ahvaan, a campaign to stop them.

There is already one colossal dam on the upper river at Tehri, which came into operation in 2006 and produces about 2,400 megawatts. (By way of comparison, the Hoover Dam generates about 2,000 megawatts, and Tehri is about 100 feet higher.) The new dams, impoundments, and diversion tunnels on Patel's map would add another 5,000 megawatts to the mix. I counted about two dozen new sites, more or less equally divided between the Bhagirathi and the Alaknanda, the two rivers that come together to form the main stem of the Ganga at the small town of Devaprayag. Patel said that the first of these structures, the 380-megawatt Bhai-ronghati I, would be built just eight or nine miles below the Gangotri glacier, where the Bhagirathi originates in an ice cave. The diversion tunnels and proposed minimum flows would dry up miles of riverbed, she said, and to make matters worse, all these massive engineering works were being planned in one of the highest-risk earthquake zones in the world. When a 6.6-magnitude quake hit the Bhagirathi valley in 1991, the greatest number of casualties occurred in a village that sits on top of one of the new tunnels.

"But surely they must have done an environmental impact assessment?" I asked.

She smiled without humor and enumerated some of the assurances that had been given by the National Thermal Power Corporation, including one that promised that "no historical, religious, or cultural monuments" would be affected by the dams. Of course, the Ganga itself is the sacred core of India's national identity, but the irony of this seemed to have escaped the government.

Later I made the bumpy three-hour drive upriver along a tortuous corniche hundreds of feet above the Ganga until I reached the confluence at Devaprayag. The town is built on a narrow, triangular point of rocks that ends in a *ghat*—the ubiquitous riverside steps where Hindus gather to wash, bathe, worship, and burn their dead. The Bhagirathi, a foaming torrent colored turquoise by silt from the Gangotri glacier, rushed in from the west. From the east, the Alaknanda was an unbroken slick of emerald between sheer cliffs. But the waters were much lower than usual, people said. It had been a strange winter, unusually warm and raining only once, a brief downpour a few days before I arrived. Peaches that normally fruited in April were ripe in February.

It was the second day of the festival of Mahashivaratri, a celebration of Lord Shiva, the Destroyer, which is one of the most important events in the Hindu calendar. Pilgrims and priests had gathered on the lower steps of the ghat, knee-deep in the water, one foot in turquoise, the other in green. The wall behind them was scrawled with Hindi graffiti. Translated, it said, "Dam Is Murderer of Ganga."

The flow of India's sacred river is of much more than local concern. Fully one-fifth of all humanity depends for its survival on the great rivers that are born among the glaciers of the Himalayas, which some people call the water tower of Asia. But even as the downstream demand for water increases, the upstream supply is contracting, because the glaciers are melting, and rapidly.

Before leaving Delhi for the mountains I'd talked to Syed Iqbal Hasnain, India's best-known glaciologist. A jovial, white-haired, grandfatherly man, he punctuated his gloomy observations with improbable bursts of laughter.

"The Ganga system is about 60 to 70 percent snow and ice," he told me. "There are more than eight hundred glaciers in the Ganga basin. The Gangotri is the big one. It used to cover more than 250 square kilometers [about 100 square miles], but now it's breaking up in many places. You will see blocks of dead ice that are no longer connected to the main ice body. I'm afraid that if the current trends continue, within thirty or forty years most of the glaciers will melt out." He chuckled.

No one could fail to notice the changes in the Himalayan weather, Hasnain said: "The monsoons are being affected by climate change. We are not getting the westerlies, which bring snow in the wintertime. Crops like potatoes, peas, and apples are growing at higher altitudes now. At lower elevations the temperatures are no longer suitable.

"There's also the atmospheric 'brown cloud,' a layer of dust particles three kilometers thick, which is warming the glaciers and creating all these anomalies," he went on. "And black soot is being deposited on the white ice of the Tibetan plateau." Together the soot and dust reduce the albedo (from the Latin *albus*, or white) — the amount of solar radiation reflected back into the atmosphere. Instead it is absorbed by the darkened ice. The dust is mainly from fossil fuel emissions, with China the principal culprit. Most of the soot comes from cooking fires on the Indian side, a seemingly trivial source that in fact generates huge amounts of highly polluting "black carbon." I was surprised when Hasnain told me that even the firewood and kerosene burned by the growing numbers of pilgrims to the Gangotri temple and nearby ashrams have a significant impact on the glacier.

The government misreads, or perhaps chooses to misread, these symptoms, Hasnain complained. "Because the glaciers are melting, a lot of water is flowing downstream," he said. "They think, the water is coming, people are happy, so why rake up all these issues of climate change?"

The melting also poses a direct threat to the new hydropower projects, he said. More glacial melt means more silt, and more silt means clogged turbines and incapacitated dams. No one was thinking about that either. "There's a total disconnect," Hasnain said, "between those who are designing these power projects and what is happening on the headwaters." He laughed again.

He said that measuring the precise extent of glacier loss was not easy, and the government's climate action plan had used this shortage of hard data to justify a disturbingly agnostic view of the problem. All the plan says is that "it is too early to establish long-term trends" and that there are "several hypotheses" about the reasons for the great melting. Part of the difficulty is that outside monitors are not welcome in areas that border on China and Pakistan: a matter of national security. You can figure out a certain amount by satellite imagery—even by looking at Google Earth—and it's not hard to measure the distance by which a particular glacier has advanced or receded. But the critical issue is what glaciologists call mass balance, the most sensitive indicator of the impact of climate change, and measuring this requires getting up into the high peaks and taking ice-core samples. Hasnain said he had begun to work with the celebrated glaciologist Lonnie Thompson of Ohio State University. "He's the leader in the ice-core business," Hasnain said. "So in four or five years we may have a credible database." He was no longer laughing now.

Under the Desert Sun

The Ganga Ahvaan campaign was launched in an unlikely place: the former palace of Maharaja Gaj Singh Ji of Marwar-Jodhpur, on the edge of the Thar Desert in western Rajasthan, India's most drought-stricken state. It is also the largest, with 56 million people in an area slightly larger than New Mexico. If India ever realizes its ambition of building affordable, large-scale solar power installations, this will be one of the prime locations.

The palace, a sprawling sandstone complex a few miles outside the ancient city of Jodhpur, is home to the Jal Bhagirathi Foundation. An odd name, I remarked to its director, Kanupriya Harish, considering that we were out in the desert, that *jal* means water, and that the Bhagirathi River is hundreds of miles away in the Himalayas.

Not really, she said with a smile, because the word *Bhagirathi* also has another significance. In Hindu legend, a king named Bhagirath had to do penance for several centuries so that the goddess Ganga would forgive the sins of his ancestors. Finally she granted his wish and decided to come down to Earth, taking the form of a great river.

"Anything that is very hard to do is called *Bhagirath prayah,*" Harish explained. "A very difficult task. And since there is nothing harder than to find water in the desert, that is how we got our name."

Finding enough water is a problem in most of India, and it's getting harder all the time. In theory there should be enough for everyone, since the overall precipitation levels are tremendous. But most of the rain falls in the three-month monsoon season, and in recent years it has been more and more concentrated into a small number of intense downpours. In a single twenty-four-hour period in 2005, for instance, Mumbai got 39 inches of rain. But if the water isn't captured in a timely fashion, it's lost, and India contrives to lose it in myriad ways, including profligate irrigation, degraded infrastructure, and a failure to treat and reuse wastewater. Rajasthan isn't Mumbai, of course; much of the Thar Desert gets only about four inches of rain a year. But the same weather patterns are apparent, Harish said. If the meager rainfall is spread out over several weeks, people can get by, more or less, scratching out a bare subsistence by cultivating lentils, millet, and a poor variety of sesame seeds. However, such rain as there is in the Thar comes more often now in a single, violent burst. "Most of the work we're doing now is actually an adaptive strategy to climate change," Harish said.

In Delhi, Ramaswamy Iyer, a former government official who drafted India's first national water plan in 1987, had told me he saw three basic options for dealing with the water crisis. You can increase the supply by massive engineering projects—dams, canals, the interlinking of major river systems. You can leave it to the market to supply and price water as it would any other commodity. Or you can treat water as a community resource, making decisions at the local level and educating people about conservation. The Indian government has relied heavily on the first two strategies; Kanupriya Harish favors the third.

In earlier times, she said, people in the Thar developed all sorts of creative techniques for harvesting and conserving water. But the collective memory of these skills began to dissipate after independence, when the expectation grew (though it was often ill-founded) that the government would come in, lay a pipe, and solve the problem. The critique of government was a thread that ran through almost every conversation I had in India: no matter how grand various schemes might look on paper, most were beset by bureaucratic inertia, crippling inefficiencies, and a culture of corruption that allowed budgeted funds to drain away into private pockets like water into the desert sands.

Local people could act with much greater agility, Harish said, and communities could develop entrepreneurial skills along the way. The trick was to revive the old, forgotten techniques and combine them with the smartest of the new technologies. She snapped open her laptop and launched into a brisk PowerPoint presentation to show me the spectrum of possibilities. At one end, a woman dug a hole in the sand to collect seepage. At the other, an improbable high-tech structure, shaped like a pyramid, offered a way to harvest Rajasthan's scarcest resource—water—by using its most abundant—the desert sun.

That sun beat down mercilessly all the next day. It was still winter, and by the standards of the Thar it was not especially hot—95 degrees or so. We drove west through a landscape of rolling dunes and spiny scrub, innumerable camel carts, wild peacocks scurrying across the baking sand. Along the way we saw many of the water-harvesting structures Harish had described in her PowerPoint. There were *beris* and *tankas* and *talabs*—shallow and deep wells, ground-level and rooftop storage tanks, ponds large and small. Women walked away from a water hole with heavy pitchers on their heads, climbing uphill through a grove of thorny khejri trees. Harish told me that these are still zealously protected by members of the local Bishnoi caste—the original tree-huggers, who sacrificed their lives in a massacre in 1730 rather than allow the khejri to be cut down by the local maharaja.

As we drew closer to the border with Pakistan, the land was dotted with white salt flats, left behind by the evaporation of last year's monsoons. At one point a man in camouflage fatigues waved us off the one-lane road and ordered us to loop around across the salt flats. For hundreds of yards ahead the roadway was occupied by a long column of battle tanks.

I asked my taciturn driver if he thought India and Pakistan would go to war.

"Inevitable," he grunted, seeming almost to relish the idea.

In the village of Trisingadi Sodha, members of the local water users' association, the jal sabha—village elders with gold earrings and brilliant turbans of red, white, yellow, and purple—garlanded us with oleanders and daubed our foreheads with vermilion. We walked with them to a large pond, perhaps a hundred yards across, which collected rain that was channeled from a jagged line of sandstone hills five miles away. The pond held enough water year-round for 10,000 people. One of the men pointed out a flock of migratory Siberian cranes poking around in the muddy shallows on the far side. Later, as we sipped sweet *masala chai,* the elders brought out their records for inspection—dog-eared notebooks with minutes and decisions from their monthly meetings, signed in neat Hindi script or with thumbprints, careful entries of money spent and received. Some of the jal sabhas charged monthly fees to water users, Harish said. Others sold it by the tankerful. This income financed the necessary maintenance, with each village devising its own system—posting a guard by the pond, for example, to keep away would-be defecators, or perhaps training a young man to keep the pumps and filters in good working order.

"It's a challenge to get the women involved," Harish said. "This is still a very feudal area."

"But what about you?" I asked.

She twinkled. "For most of these communities I've ceased to be a woman. They think that I'm a man."

Something significant was happening here, it seemed to me. The jal sabha was blending traditional principles of community organization with a newer entrepreneurial spirit. In the process, India might ease some of the historic tension between village and city. Gandhi believed that the village was India's beating heart; Nehru, the first prime minister after independence, thought its future lay in the cities. Here was a way to maintain the integrity of the village while building the modest, incremental prosperity that might make it unnecessary for people to migrate to places like the slums of Okhla.

This was not the only way in which Gandhi's vision was being updated. Forty miles to the south, in the straggling village of Roopji Raja Beri, a surreal sight confronted us. I recognized the flattened foil dome, which somewhat resembled a silvery mushroom cap, from Harish's PowerPoint slide. It looked as if an alien spacecraft had set down among the sand dunes, but it was a "water pyramid," only the second of its kind in India. The technology was Dutch; a team of engineers had come here for six weeks to install it, and the inauguration ceremony had taken place just five days earlier.

The entrepreneur in charge was an imposing, barrel-chested villager named Prem Ram, a twenty-year veteran of the Indian army. He said that the water shortage in Roopji Raja Beri had grown so severe that people had come to blows. The groundwater was so salty that you could literally burn your tongue, and there had not been a decent rain since 2003. "The natural order is breaking," he said.

Prem Ram seemed as proud of his pyramid as he was of his military service. He opened a vent for me to look inside, but a fierce surge of heat and humidity drove me back before I could catch more than a glimpse of the glittering pool of fresh water. The strange structure used the power of the sun to function as a combined distillation and desalination plant, he explained. And since it ran entirely on solar energy, the operating costs were close to zero. The brackish groundwater was pumped in from a nearby holding tank; once inside the pyramid, it was distilled through evaporation. He thumped a meaty fist on one of the sloping sides, and shimmering streaks of fresh water ran down the interior walls. The salt that was left behind provided an added source of income for the villagers, who subsisted otherwise by selling the milk from their scrawny herds of cattle, goats, and water buffalo.

One day, if the costs of the technology come down, if the government bureaucracy becomes more efficient, if the high-tech entrepreneurs from Hyderabad and Bangalore put up the start-up capital, there may be other strange sights here in the Thar Desert: gigantic solar farms, perhaps, each one capable of feeding as much power into the national grid as all the new dams on the Ganga. But it will be equally important for India to think small and local and to focus on the entrepreneurial village culture that is emerging in obscure places like Trisingadi Sodha and Roopji Raja Beri. To

Americans, living off the grid may imply a hair-shirt lifestyle choice, a yurt among the Oregon redwoods. To Indians, paradoxically, it may be a pathway to the national mainstream.

Let There Be Light

"Turn left at the monkeys," Sumant Dubey said to my taciturn driver. We were 200 miles north of the water pyramid now. We drove another five minutes or so along the highway, dodging homicidal Tata trucks, until a narrower road turned off into the scrubby Aravalli Hills. At the intersection, a large, dusty lot served as an informal truck stop, and hundreds of monkeys were rooting around for scraps and handouts.

I had been introduced to Dubey, a cheerful, round-faced young man, a few days earlier in Delhi by Leena Srivastava, executive director of The Energy and Resources Institute (TERI). Headed by Rajendra Pachauri, the Nobel Prize–winning chairman of the Intergovernmental Panel on Climate Change, TERI is an unusual hybrid of scholarship, science, policy analysis, and grassroots activism. It has its own university, and its labs specialize in off-the-grid renewable energy technologies.

There are still 400 million people in rural India without access to electricity, Srivastava said, and progress was slow because the government insisted on trying to meet their needs by the conventional means of extending the national grid. "We're still throwing good money after bad," she said. "It's easy enough to set up the infrastructure, even with our monetary constraints, but it's much harder to actually get the electricity flowing through the wires. So distributed generation is something that we need to pursue very aggressively."

We'd been joined now by Dubey's boss, Akanksha Chaurey, an expert on solar photovoltaics. "People are beginning to recognize that a better way to go is the smart mini-grid and micro-grid," she said.

I asked her what she meant by this, and she said, "Multiple small-scale, interconnected power plants in rural areas, serving isolated communities. They may generate power by solar, or small hydro, or biomass. Solar is the easiest, although it's also the most expensive."

What Srivastava said next echoed Kanupriya Harish's point about creating sustainable connections between rural India and the economic mainstream. "In the last two or three years we've proven that if you provide people with the energy they need to run their businesses, you can create new linkages, things like agricultural retail networks where rural people have direct access to urban markets," she said. Take refrigeration: 60 percent of fresh produce is lost before it gets to market; provide affordable electricity, and the local economy can be transformed. The key, she explained, was to identify entrepreneurs, or "franchisees"—individuals who are known and trusted in their communities, who can make sure that the business model is sound, that the bills are paid, that those miniature power plants remain in good working order. It was much the same vision as that expressed by the jal sabhas in the Thar Desert, only this time to provide energy rather than water.

Chaurey told me about a new TERI program called Lighting a Billion Lives. The name seemed stunningly ambitious—a *billion* lives?—but that didn't seem to faze her. It did, however, raise the question that bedevils any local initiative in a country as vast and complex as India. Can it be replicated? In the jargon of development, is it scalable? The water-harvesting structures in the desert were designed to be scalable in horizontal fashion, so to speak. A solar pyramid creates fresh water; villagers from miles around come to see how it works (these days they may even hear about it by cell phone), and they want one too. The model that TERI was promoting worked vertically as well as horizontally: not only did you show the villages what worked, but you showed the government too, and Srivastava said that on a good day it might even sit up and take notice.

Lighting a Billion Lives was launched last year at a ceremony in which Rajendra Pachauri presented Prime Minister Singh with a hand-held solar lantern. The gift was rich in symbolism: it took the power of the sun and the large vision of international climate science and linked them in one direction to the national government and in the other to India's 638,365 villages, all through a simple device that would illuminate the humblest hut in the Aravalli Hills of Rajasthan. The only question was, how would people afford it? Even a solar lantern costs about $80, more than most Indians earn in a month.

Dubey, who was responsible for the implementation of the program in Rajasthan, told the driver to follow the narrow blacktop that branched off the Jaipur highway. We passed through a small town, kept going for a few more miles, then turned off onto a one-lane dirt road, and finally bumped along a narrow, rutted track of dried mud until we reached a two-story concrete house at the edge of a field of wheat and gram, the following season's worth of *chapatis, parathas,* and *nans.* The house was modest enough, but by the standards of the hamlet—one of several that make up the village of Badgujran, which has 5,000 people—it was a mansion. The paunchy, middle-aged man who lived there was the most prosperous person in Badgujran, as well as its solar entrepreneur. He introduced himself as Mahavir Singh.

I asked him why his fields were so green—a rarity in Rajasthan. He removed the cover of his deep tube well, where a rope descended into unfathomable darkness. Five years ago it was 250 feet deep, he said; now he had to go down 800 feet to reach water. It hadn't rained in eight months; the monsoons had ended several weeks earlier than usual. In the old days, wells like these were excavated by gangs of lower-caste workers who charged 150 rupees—three dollars—for every foot they dug. Now a truck came with heavy equipment and did the job in a day. But it was expensive. A well like this costs two lakh rupees (Indians count in lakhs, multiples of 100,000, not in millions). Four thousand dollars, in other words, making the tube well something that only the wealthiest farmers could afford. Poorer ones had to sell their land or leave it uncultivated and work for men like Singh as field laborers.

I was surprised to see that the village had some half-hearted electricity poles, although many of the wires were trailing on the ground. The power reached only four or five houses, Singh said, though others had strung up illegal wiring to feed off the current, as poor people have done since electricity was invented. But the supply was dependent on the whims of load-shedding, when a utility shuts down secondary lines like these during hours of peak demand. In Badgujran, that meant that the juice might start to flow at useless times—at ten at night or four in the morning, while the village slept. And most people couldn't afford it anyway. As Leena Srivastava had said, putting up the infrastructure was the easy part.

Until this year, that left no option but kerosene lamps, still the basic source of light in 68 million Indian homes. Village huts have uneven floors, no windows, walls full of holes and cracks. Kerosene lamps—usually no more than a bottle with a crude wick—burn black and smoky. Children knock them over, the wind blows them down. Smoke inhalation and kerosene fires are among the leading causes of child mortality. Furthermore, Chaurey had told me, in terms of the intensity of carbon emissions, kerosene is the dirtiest of all fossil fuels. Lighting a Billion Lives is TERI's alternative; the program is now up and running in thirty-three villages in Rajasthan and is expanding rapidly nationwide.

A young woman named Sunita took me up to the roof of Singh's house to show me how it worked. Like many Rajasthani women, she was a walking rainbow: orange sari, lime green scarf spangled with silver, bright bangles and earrings, fingers and toes painted with intricate patterns of henna, a cherry red cell phone. She showed me the small array of solar panels, which stood next to a stack of dung cakes that had been laid out to dry—the traditional cooking fuel of village India. A tangle of wires led to a charging station in a small room downstairs, where rows of lanterns, some yellow and some green, were hooked up to chargers. Each lantern will hold a charge for six to eight hours.

What made the idea so attractive was the financing model, Dubey said. With solar power, whether it's a single lantern or a 5,000-megawatt array, the biggest obstacle is the initial cost. Part of the production costs of TERI's lanterns is underwritten by a variety of often surprising corporate sponsors, including GE and Coca-Cola. As the program expands, the unit price will come down, but for most villagers the lanterns will still be out of reach. So TERI's solution is fee for service: rentals, not sales. Each village would have a charging station run by a local entrepreneur, who would rent out the lanterns for a couple of rupees a day—pegged to the amount an average family would otherwise spend on kerosene.

Mahavir Singh led me along a narrow path through the fields to a cluster of mud-walled houses and a small white temple. Green parakeets raised a racket in a large neem tree. There were several small holes around the base of the tree. Singh said they were cobra dens. About ten villagers are bitten every month, but most are cured by a holy man who lives nearby, with a poultice of neem leaves—highly prized in traditional Ayurvedic medicine—and hair from a cow's tail.

We came to a hut where three or four scrawny goats had their faces stuck in a feed bowl and a bad-tempered water buffalo strained at its heavy metal chain. The woman of the house invited us inside. Some children's T-shirts hung on a clothesline. The wall was covered with pictures of Hindu gods and gurus. Next to them was an Iberia poster that showed an airplane taking off over a chalet hotel in the Alps and a row of motorboats at anchor on a sapphire lake. The inspirational motto on the poster said: DON'T WAIT FOR YOUR SHIP TO COME IN, SWIM OUT FOR IT.

"The solar lanterns allow people to do many things," Dubey said. Women gathered in the evenings to discuss health and family issues; the embroiderers and carpet makers in a nearby hamlet were working longer hours and making more money. This woman's two sons were sitting in the glow of a lantern on the mud floor of another room. The thirteen-year-old, Ajay, was using the lamp to do his English homework.

It was dark now, and Singh had switched on his own lantern to guide us back through the snaky field. As we stood outside his house, making our goodbyes, I noticed something I hadn't seen earlier. He had water, he had energy, and now I saw that he also had the third element of prosperity: mobility, a car. An entry-level Tata Indica hatchback, to be precise. He said it was the only one in the village. I couldn't imagine how he had brought it here, along that potholed track.

I asked if he had heard of the Tata Nano, which was scheduled to be unveiled in Mumbai at the end of March and was already making headlines around the globe. At one lakh—$2,000—it would be the cheapest car in the world, and it would get more than fifty miles to the gallon. Some people said it would usher in a car-owning revolution. I asked Mahavir Singh if he might be interested in a Nano himself. He thought about it for a while but seemed skeptical. "Perhaps," he said. "But I'd need to see it first." A sentiment that many Indian environmentalists have echoed, albeit for different reasons.

The Man from Siam

Back in Delhi, I stopped at a newsstand and picked up a copy of *Auto India.* There was nothing in there about the Nano: Tata Motors was keeping its new baby under wraps for another couple of weeks.

Otherwise, *Auto India* looked much like any other car mag: reviews of the new E-Class Mercedes; glossy gatefold ads that said things like *Smooth. Suave. Sure. The Über-Cool Is Here.* So when I went to the Society of Indian Automobile Manufacturers (SIAM) to meet its director, Dilip Chenoy, I pretty much knew what to expect: lots of guy talk about sound systems and leather upholstery and zero to 60 in 5.9 seconds.

Instead, Chenoy started talking about milk.

"My favorite image in India is the milkman," he said. "He used to walk to your door with four bottles. Then he developed a carrier, so he brought eight. Then he came on a bicycle with two big drums. Then he progressed to a motorbike, with four drums, and then a small pickup truck with even more. From there to a larger truck, and then the mother dairy bought a big refrigerated truck. In the process India went from being the fifteenth-largest producer of milk in the world to being the largest. And without that shift in transportation you would not have been able to realize the dream."

He intended this as a parable, obviously, and it served as the prelude to an impassioned speech about India's development goals and the social benefits of greater mobility, ticking off the environmental pros and cons of various forms of private and public transportation.

"We have more than a billion people but fewer than a hundred million motor vehicles on the road," he began. "So our challenge is to figure out the most economically viable way of providing mobility, so that people can get to school, find employment in rural areas, become entrepreneurs. And it has to be sustainable in terms of emissions. That's what we're trying to do here, and the private car is only part of it."

Even though car ownership might increase, he wanted to put the numbers in context. He said, "There are fewer cars in India than Detroit produces—or used to produce—in a year. So the scale is totally different. And the primary use of a car here is for the service economy. As you will have seen, the cars here are loaded with stuff."

I asked why there was such a wide variation among Indian cities in "mode share"—the percentage of travelers using different kinds of transportation. I picked three cities, more or less at random, from a chart I'd been given by Partha Mukhopadhyay, a transportation expert at the Center for Policy Research in Delhi. All three were about the same size, close to 2.5 million people. In Kanpur, 16 percent of passengers travel by car. In Jaipur, the figure is 8 percent. In Nagpur, it's 3 percent. It depends to a large extent on the availability of public transportation, Chenoy explained, and that sector has historically been neglected in India. The term is also too narrowly defined, he added. "There's this mindset that public transportation equals a forty-two-seater bus. But it may also be a car, or a small van, or an SUV."

Presumably he'd seen me wince at the mention of SUVs. Too polite to sneer at my American preconceptions, he explained patiently that SUVs in India are generally not sold to highway hogs and soccer moms; three-quarters of them are sold in rural areas, where they may be used to haul goods, to take village kids to school, or as a "para-transit" option to compensate for the absence of buses. And Indian SUVs, made by companies like Tata and Mahindra, are subject to increasingly stringent fuel efficiency and emissions standards. The SUV as instrument of social progress and friend of the environment: an arresting notion.

"A lot of well-meaning people talk about gas-guzzlers and also about big luxury cars," he continued, warming to his theme of cultural relativism. "But let's not miss the wood for the trees. We're only talking about three thousand luxury cars a year." And those high-end vehicles, while they may be emblematic of India's new culture of conspicuous consumption, are very important from an environmental perspective, he said. They're the test-bed for all the technological innovations—things like common-rail diesel engines, homogeneous gas compression, variable valve timing, lightweight alloys and composites—that will later find their way into the mass market to increase efficiency and lower emissions.

"And all of the major Indian manufacturers are thinking that way?" I asked.

"Yes, yes!" he exclaimed, as if there were any other way to think.

The Über-Cool

The following day it took more than an hour to get to the spanking-new Tata Motors showroom in Okhla Phase II, just a mile or two from the train station where I'd begun my journey. The first part of the ride was deceptive, since it sped us along the broad, grassy avenues and past the heroically scaled government buildings that the British architect Edwin Lutyens laid out in the early twentieth century, when Delhi took over from Calcutta as the capital of the raj. After that, though, it was a tedious stop-start crawl through the morning traffic, with constant diversions around WORK IN PROGRESS signs that denoted new highway overpasses or extensions to the Delhi Metro.

As we inched forward along the ring road, countless two-wheelers—motorbikes and scooters—slalomed in and out of the traffic. Some carried four people: dad in front, older kid in the middle, mom riding sidesaddle with an infant on her lap, and only dad wearing a helmet.

Getting around Delhi is a tricky business; in the language of urban planning it's a "polynodal" city. It isn't like Mumbai, for example, which is built on a narrow north-south axis and where people have a long tradition of traveling from A to B by train. Delhi planners must anticipate the need for complicated, zigzag journeys. If they don't, people will shun public transportation.

Which is not to say that the buses aren't overcrowded; there just aren't enough of them. All of them carry hand-painted slogans, which, given the choking smog, appear to have been written by someone with a sense of humor. Some read WORLD'S LARGEST ECO-FRIENDLY CNG BUS SERVICE. The message on others comes in a number of variant spellings: PROPELD BY CLEAN FUEL, or PREPALLED BY CLEAN FULL. In 1998, following a series of Supreme Court rulings on air pollution, the city of Delhi ordered the conversion of all commercial passenger vehicles—buses, taxis, and autorickshaws—to compressed natural gas. But that raised a couple of big questions, Mukhopadhyay had said. Was it wise to mandate a particular fuel rather than to increase overall fuel efficiency? And could the plan be replicated in other cities, since that would mean a massive increase in India's capacity to produce and distribute natural gas? At best, he said, given the overall increase in congestion, Delhi had bought itself a short reprieve from even more polluted air.

The Metro, which I'd ridden the previous evening, was supposed to offer another remedy. It is squeaky clean, and the trains are punctual. There are notices about the specially designed features for "differently abled passengers." The first phase of the system, which opened in 2000, was finished on time—a distinct rarity for India, and indeed for any megaproject anywhere. Now the lines are being extended to the airport and to the satellite city of Gurgaon, a center of India's outsourcing boom. Yet the Metro, too, has its share of critics: much of the cost was borne by massive public subsidies; ridership predictions were overoptimistic; and the lines didn't go where they were needed, since they had been planned in the 1970s, when Delhi had just a fraction of its present population of 14 million.

So the number of private vehicles just keeps on rising. At the last count, in March 2007, there were almost 1.6 million cars in the city, Mukhopadhyay had told me, more than in India's other two megacities—Mumbai and Kolkata—combined. And there are twice that number of two-wheelers. More than 1,000 new vehicles hit the roads every day.

We arrived at the Tata showroom at last, and while I waited for the regional manager to get there I made small talk with a couple of the salesmen. They spoke about Ratan Tata, the head of the company, with something approaching reverence, as if he were not only a corporate titan but a personal guru, even a saint. It was those overloaded two-wheelers, one of the salesmen said, that had inspired Mr. Tata to come up with the idea of the Nano. He wanted something for families that was safe, affordable, and environmentally friendly.

One side of the showroom was given over to Tata, the other to Fiat; the two companies have a joint marketing arrangement. But I couldn't see any car that looked small enough to be a Nano. "No," the salesman said, "they haven't let anyone see it yet, not even us." He giggled with anticipation, like a kid waiting for Santa Claus.

His boss, Vishwas Kapoor, arrived at last, full of apologies for being late. "Stuck in traffic," he said, unnecessarily.

He walked me around. Many of the models on display were variants of the Indica, the car I'd seen at Mahavir Singh's house in Rajasthan. Nearby were several larger Tata cars and SUVs.

"What kind of mileage do all these vehicles get?" I asked, wondering how they would compare to the aptly named Nano, which looks a bit like a four-door version of the tiny, slope-fronted Smart car that is so popular in Europe.

The answer amazed me. The various models of Indica get anything from 35 to 50 miles per gallon. Even the ten-seat diesel Sumo Victa, the biggest of the SUVs, boasts more than 32 mpg on the highway. Impressive numbers by U.S. standards.

Around eleven o'clock the first walk-in customer of the day arrived, a smartly dressed Sikh in a red turban. He introduced himself as Manjeet Singh, a travel agent in the suburban neighborhood of Vasant Vihar. He already owned a small fleet of eight Indicas, which the company used for commercial purposes. He was here today to buy number nine.

I asked him the same question I'd asked the other Mr. Singh, back in the village of Badgujran: "Would you consider a Nano?"

"Not for myself," he said, "but a lot of my friends are thinking about it. It could be as a second car, or maybe a gift for their wife or their eighteen-year-old who has just learned to drive. And the two-wheelers are also interested, of course."

That was really the crux of the matter, Mukhopadhay had said. It all depended on how the Nano was marketed. The idea of millions of two-wheelers being replaced by four-wheelers appalled him, no matter how high the mpg. "But in the unlikely event the Nano becomes a lifestyle statement, switching people away from larger cars, it could be a great success," he said. "It will be a test of our social imagination about what car ownership means."

The Nano will appeal to both market segments, said the man from Tata, although no one really knows what to expect. The numbers will be small at first, just a few thousand cars a month; the main production facility, in the state of Gujarat, isn't even on line yet. As if to assuage my anxiety at the prospect of all those two-wheeler owners upgrading, he pointed at the Fiat poster on the wall. It showed a car swerving around a scenic bend, as cars always do in the ads. It was the latest incarnation of the humble Fiat 500, introduced in the 1950s as the Italian equivalent of Germany's VW Beetle or France's tin-can Citroën 2CV.

"How much does one of those cost in India?" I asked.

"With 100 percent import duty, about fourteen or fifteen lakhs," Kapoor said.

"Thirty thousand dollars? Who's going to pay that?"

"Oh, you'd be surprised. It's very popular. We get famous movie actors, models, fashion photographers in here all the time to buy it. In Europe it may be basic, but here it's a lifestyle product."

In India, he was suggesting, small can be über-cool.

What Would Gandhi Drive?

Prime Minister Manmohan Singh wound up his presentation of the National Action Plan on Climate Change last year with the homage to Gandhi that is obligatory for any Indian politician. In this instance he paraphrased one of the Mahatma's most famous sayings: "Earth provides enough to satisfy every man's need but not for every man's greed."

But virtually in the same breath, Singh set the bottom line of India's climate policy, and it seemed to sit oddly with Gandhi's philosophy: per capita greenhouse-gas emissions would never exceed those of the industrialized world. To put this in perspective, the current per capita level is only about one-twentieth that of the United States. But a population of 1.15 billion is a powerful multiplier. India is already the world's fourth-largest emitter of greenhouse gases, and its recent economic growth rates, if sustained, will mean a doubling of energy demand by 2020.

By the time you read this, India will have a new coalition government. Perhaps the Hindu nationalist Bharatiya Janata Party will come out on top this time; perhaps the Congress Party. Either way, there will be the same onward rush of economic growth, the same commitment to bring millions out of poverty. And given its agricultural and service-based economy, its lack of dependence on exports, India may be shielded from the worst of the global meltdown. Indian diplomats will take part in the climate negotiations that will lead, by the end of this year, to a replacement for the Kyoto Protocol, and they will insist, with good reason, that the United States take the lead in finding a solution. You created the problem, they will say; you solve it. Meanwhile, the monsoons will grow more erratic. There will be worse floods and more severe droughts. The glaciers will go on melting. There will be more coal-fired power plants and more hydro dams. More Indians will buy cars.

What would Gandhi make of it all? I wondered. I had a pretty good idea what he'd think if he stood on that footbridge at the Okhla train station and peered out into the murk at the cranes and the malls of NOIDA. There was no mystery about what he'd think of the ads for the E-Class Mercedes. But in other respects I was less sure.

What would Gandhi make of the Nano? And, come to that, what would he think of village kids going to school in an SUV instead of sitting with their teacher in the shade of a neem tree or, more likely, not going to school at all? What would he think of a water pyramid, or the chance to power the spinning wheel in his ashram with solar panels? These were not easy questions to answer, and in trying to do so it seemed wise to leave many of my Western preconceptions behind.

The rutted back roads of Rajasthan and the sleek flyways of the Delhi suburbs: at first they seemed worlds apart. Yet there was a common logic in the changes that were under way in both places, and it was summed up in that word *entrepreneur* that people kept using. There are the kinds of entrepreneurs, of course, who have created entire new Silicon Valleys in Bangalore and Hyderabad. There are those who will design the next generation of diesel engines and variable crankshafts. But there are also the jal sabhas with their account books and Mr. Singh with his lantern, Mr. Ram with his pyramid and Dilip Chenoy's parable of the milkman—all of them hints, however small, of how India might yet realize its dreams of development without tearing itself, and the rest of the planet, apart.

Contributors' Notes

Other Notable Science and
Nature Writing of 2009

Contributors' Notes

Gustave Axelson is the managing editor of *Minnesota Conservation Volunteer* (www.mndnr.gov/magazine), a nonprofit outdoors-advocacy magazine published by the Minnesota Department of Natural Resources. Axelson also freelances for the *New York Times, Backpacker, Men's Journal,* and *Midwest Living.* A graduate of the University of Illinois Urbana–Champaign journalism school and a Knight Digital Media Center fellow at the University of California Berkeley Graduate School of Journalism, he lives in St. Paul, Minnesota, with his wife, Amy, and two sons, Anders, seven, and Henrik, three, with whom he makes frequent paddling trips into the Boundary Waters Canoe Area Wilderness every year.

Burkhard Bilger has been a staff writer at *The New Yorker* since 2000. His articles have appeared in *The Atlantic, Harper's Magazine,* the *New York Times,* and other publications, and his book *Noodling for Flatheads* was a finalist for a PEN-Faulkner Award. A former senior editor of *Discover* and deputy editor of *The Sciences,* Bilger lives in Brooklyn with his wife, Jennifer Nelson, and his children, Hans, Ruby, and Evangeline.

George Black lives in New York with his wife, the writer Anne Nelson. In the course of his extensive travels on five continents, he has written about the civil wars in Central America in the early 1980s, the Velvet Revolution in Czechoslovakia, the Chinese democracy movement, the impact of climate change in Bangladesh, and many other topics. His next book, his sixth, is about the 1870 military-civilian expedition to Yellowstone—the first systematic exploration of the future national park—and its relationship to the Indian Wars in the West.

Brian Boyd is the University Distinguished Professor of English at the University of Auckland, where, inter alia, he teaches a course on literature and science. The world's leading authority on Vladimir Nabokov, he coedited *Nabokov's Butterflies: Uncollected and Unpublished Writings* (2000). Since then he has been working on evolutionary approaches to literature, especially in *On the Origin of Stories: Evolution, Cognition, and Fiction* (2009), in which "Purpose-Driven Life" is the Afterword; in the coedited *Evolution, Literature, and Film: A Reader* (2010); and in the forthcoming *On the Ends of Stories: Literature and Evolution.* He is also writing a biography of the philosopher of science Karl Popper.

Kenneth Brower, the son of the pioneering environmentalist David Brower, is a freelance writer and the author of many books and magazine articles on the environment and natural history. His first memories are of the Sierra Nevada and the wild country of the American West. His work has taken him to all the continents. He lives in Berkeley, California.

Jim Carrier is an award-winning journalist, civil rights activist, and filmmaker. In a forty-year career, Jim has written nine books, been published in *National Geographic* and the *New York Times,* written *Denver Post* series on the legacy of the atomic bomb and on the Marlboro Man, and produced multimedia projects for the Southern Poverty Law Center. He has roamed by Jeep through the American West and by sailboat across the Atlantic and Mediterranean. Now based in Madison, Wisconsin, he is at work on a film about the racial history of the banjo.

John Colapinto is a staff writer at *The New Yorker* and the author of the nonfiction book *As Nature Made Him: The Boy Who Was Raised As a Girl* and the novel *About the Author.* He is married and has an eleven-year-old son.

In the fifteen years that **Andrew Corsello** has been writing for *GQ,* his work has been nominated five times for the National Magazine Award, winning once, and anthologized three times in *The Best American Magazine Writing.* He lives in San Francisco with his wife, an Episcopal priest, and their two young sons.

Timothy Ferris has written a dozen books—among them *The Science of Liberty, The Whole Shebang,* and *Coming of Age in the Milky Way*—and made three nonfiction films: *The Creation of the Universe* (1986), *Life Beyond Earth* (1999), and *Seeing in the Dark* (2007). He produced the Voyager phonograph record, an artifact of human civilization containing music and sounds of Earth launched aboard the twin Voyager interstellar spacecraft;

now exiting the solar system, the Voyagers are the most distant probes ever created by humans. Called "the best popular science writer in the English language" by the *Christian Science Monitor* and "the best science writer of his generation" by the *Washington Post,* Ferris has received the American Institute of Physics prize and a Guggenheim Fellowship. His works have been nominated for the National Book Award and the Pulitzer Prize. A fellow of the American Association for the Advancement of Science, Ferris has taught in five disciplines—astronomy, English, history, journalism, and philosophy—at four universities. He is currently an emeritus professor at the University of California, Berkeley.

Tim Flannery has published more than 140 peer-reviewed scientific papers. His books include the landmark works *The Future Eaters* and *The Weather Makers,* which has been translated into twenty-five languages. In 2006 the book won the New South Wales Premier's Literary Prize, the O2 (a German environmental prize), and the Lannan Literary Award for Nonfiction. In 2007 he cofounded and was appointed the chair of the Copenhagen Climate Council, a coalition of community, business, and political leaders who have come together to confront climate change.

Jane Goodall, Ph.D., DBE, the founder of the Jane Goodall Institute and a UN Messenger of Peace, began her landmark study of chimpanzee behavior in what is now Tanzania in July 1960 under the mentorship of the famed anthropologist and paleontologist Louis Leakey. Her work at the Gombe Stream Chimpanzee Reserve became the foundation of primatological research and redefined the relationship between humans and animals. In 1977she established the Jane Goodall Institute (JGI), which is now a global leader in the effort to protect chimpanzees and their habitats. It also is widely recognized for establishing innovative community-centered conservation and development programs in Africa and Jane Goodall's Roots & Shoots, a global environmental and humanitarian youth program, which has groups in more than 120 countries. For more information, please visit www.janegoodall.org.

Philip Gourevitch, a longtime staff writer for *The New Yorker,* is the author of *The Ballad of Abu Ghraib* (2008), *A Cold Case* (2001), and *We Wish to Inform You That Tomorrow We Will Be Killed with Our Families: Stories from Rwanda* (1998), which won the National Book Critics Circle Award, the Los Angeles Times Book Prize, the George Polk Award for Foreign Reporting, and the Guardian First Book Award. His books, short fiction, essays, and reportage have been translated into a dozen languages. From

2005 to 2010 he was the editor of the *Paris Review.* He is writing a new book from Rwanda.

Elizabeth Kolbert is a staff writer for *The New Yorker* and the author of *Field Notes from a Catastrophe: Man, Nature, and Climate Change.* Her series on global warming, "The Climate of Man," from which the book was adapted, won a National Magazine Award and a National Academies Communications Award. She lives with her husband and three sons in Williamstown, Massachusetts.

Robert Kunzig is a senior environment editor at *National Geographic* and has been a science writer for nearly thirty years, including fourteen on the staff of *Discover.* He is the author of two books: *Mapping the Deep,* about oceanography, which won the science-writing prize of the Royal Society in 2000; and *Fixing Climate,* written with the climate scientist Wallace Broecker, which was published in 2008. Kunzig lives in Birmingham, Alabama, with his wife, Karen Fitzpatrick.

Jonah Lehrer is a contributing editor at *Wired* and the author of *How We Decide* and *Proust Was a Neuroscientist.* He has also written for *The New Yorker,* the *New York Times Magazine, Nature, McSweeney's,* and *Outside.*

Richard Manning is the author of eight books, including *One Round River,* which was named a significant book of the year by the *New York Times* in 1998. He has worked as a consultant on agriculture, poverty, and the environment to the McKnight Foundation, the Rockefeller Foundation, and the Food and Agriculture Organization of the United Nations. He has written for *Harper's Magazine, Proceedings of the American Philosophical Society, Wired, Men's Journal, OnEarth,* the *Los Angeles Times, The American Scholar, Frankfurter Allgemeine Zeitung,* and many other publications. Among his several awards are the University of Montana's Mansfield Center's Lud Browman Award for science writing and the C. B. Blethen Award for investigative journalism.

Kathleen McGowan is a contributing editor at *Discover.* She writes about neuroscience, genetics, and other subjects in science and medicine for publications such as *Psychology Today, Self,* and *Redbook.* She lives in New York City.

Evan Osnos joined *The New Yorker* as a staff writer in 2008. He is the magazine's correspondent in China, where he has lived since 2005. Previously he worked as the Beijing bureau chief of the *Chicago Tribune,* where he contributed to a series that won the 2008 Pulitzer Prize for investigative reporting. Before his appointment in China, he worked in the Middle East, reporting mostly from Iraq.

David Quammen is a freelance journalist and author whose eleven books include fiction, essay collections, and the nonfiction titles *The Song of the Dodo, Monster of God,* and *The Reluctant Mr. Darwin.* His short work has appeared in a range of journals, from *Rolling Stone* and *Outside* to the *New York Times Book Review* and *Harper's Magazine.* Presently he is a contributing writer for *National Geographic.* His current book project involves the ecology and evolution of scary viruses. Quammen lives in Montana with his wife, Betsy Gaines, a conservationist, and travels on assignment, by preference to jungles, deserts, and swamps.

Matt Ridley is the author of several books on genetics, evolution, and economics, including *The Red Queen, Genome,* and *The Rational Optimist.* He did research in zoology at Oxford University before becoming a journalist with *The Economist* and then as a freelance. He lives near Newcastle in northern England.

Felix Salmon is the finance blogger at Reuters.

Michael Specter, who has been on the staff of *The* New Yorker since 1998, writes frequently about science, public health, and the impact of new technologies on society. His book *Denialism* is out in paperback this fall.

Don Stap is the author of two works of natural history—*Birdsong* and *A Parrot Without a Name*—and a collection of poems, *Letter at the End of Winter.* He is a frequent contributor to *Audubon* and has written as well for such publications as *National Wildlife, Smithsonian, Orion, Living Bird,* the *North American Review,* and the *New York Times.* In addition, he is the recipient of a fellowship in creative writing from the National Endowment for the Arts. A native of Michigan, Stap has taught at the University of Central Florida since 1985, where he is a professor of English.

Dawn Stover is a freelance science and environmental writer based in the Pacific Northwest. She is an editor at large for *Popular Science,* where she was a staff editor for nineteen years. Previously she worked at *Harper's Magazine* and *Science Digest.* Her work has also appeared in the *New York Times, Scientific American, New Scientist, Science Illustrated, Conservation, Outside,* and *Backpacker.* She is a charter member of the Society of Environmental Journalists and a longtime member of the National Association of Science Writers.

Steven Weinberg is a professor of physics and astronomy at the University of Texas. His honors include the Nobel Prize in Physics and the National Medal of Science, election to numerous academies, and sixteen honorary doctoral degrees. He has written more than three hundred articles on elementary particle theory, cosmolology, and other scientific topics, and twelve books; the latest, *Lake Views: This World and the Universe,* is a collection of his essays from the *New York Review of Books* and other periodicals. Educated at Cornell, Copenhagen, and Princeton, he taught at Columbia, Berkeley, MIT, and Harvard, where he was the Higgins Professor of Physics, before coming to Texas in 1982.

Tom Wolfe has established himself as our prime fictional chronicler of America at its most outrageous and alive. He lives in New York City with his wife, Sheila; his daughter, Alexandra; and his son, Tommy.

Other Notable Science and Nature Writing of 2009

SELECTED BY TIM FOLGER

MARCIA ANGELL
Drug Companies & Doctors: A Story of Corruption. *New York Review of Books,*
January 15.
NATALIE ANGIER
The Art of Deception. *National Geographic,* August.
JOEL ACHENBACH
Will Yellowstone Blow Again? *National Geographic,* August.
ANTHONY AVENI
Apocalypse Soon? *Archaeology,* November/December.

CHARLES BOWDEN
Unseen Sahara. *National Geographic,* October.
Contested Ground. *Orion,* November/December.
OLIVER BROUDY
Dead Man Driving. *Men's Health,* December.
What If the Sun Could Kill You? *Men's Health,* July/August.
ALAN BURDICK
The New Web of Life. *OnEarth,* Fall.
LESTER R. BROWN
Could Food Shortages Bring Down Civilization? *Scientific American,* May.

ANDREW CURRY
Rituals of the Nasca Lines. *Archaeology,* May/June.

FRANS DE WAAL
The Empathy Instinct. *Discover,* October.
MARK DOWIE
Nuclear Caribou. *Orion,* January/February.

VIRGINIA MORRELL
Going to the Dogs. *Science,* August 28.
LIZA MUNDY
Deer Heaven. *Washington Post Magazine,* April 26.

DAVID NOLAND
NASA and Its Discontents. *Popular Mechanics,* February.

H. ALLEN ORR
Which Scientist Can You Trust? *The New York Review of Books,* March 26.

MICHAEL POLLAN
Sneaky Orchids. *National Geographic,* September.
HEATHER PRINGLE
Witness to Genocide. *Archaeology,* January/February.

HARRISON H. SCHMITT
From the Moon to Mars. *Scientific American,* July.
BILL SHERWONIT
Reflections on Thrush Songs, Newt Tracks, and Old-Growth Stands of
Trees. *Isle,* Fall.
NEIL H. SHUBIN
This Old Body. *Scientific American,* January.
MARK SLOUKA
Dehumanized. *Harper's Magazine,* September.
CHRISTOPHER SOLOMON
Foot. Loose. *Outside,* October.

GARY TAUBES
RNA Revolution. *Discover,* October.

WILLIAM T. VOLLMANN
Expectations. *San Francisco Panorama,* December 8.

ALEC WILKINSON
What Would Jesus Bet? *The New Yorker,* March 30.
DAVID WOLMAN
Turning the Tides. *Outside,* January.

CARL ZIMMER
The Entangled Bank. *Discover,* November.